PENGUIN BOOKS

LUCY

Dr Donald C. Johanson is perhaps one of the most influential and best known of American paleoanthropologists, following his dramatic discovery in 1974 of a 3-million-year-old hominid skeleton, now known as 'Lucy', in the Hadar Valley of Ethiopia.

He was born in Illinois in 1943 of Swedish immigrant parents and grew up in Connecticut. He took M.A. and Ph.D. degrees in human paleontology at the University of Chicago in 1970 and 1974. After his discoveries in 1974, Dr Johanson returned to the field in Ethiopia in 1975, when he found 'The First Family', a unique collection of thirteen individuals who died in a single geological moment. Also in 1975 Dr Johanson was appointed curator of physical anthropology at the Cleveland Museum of Natural History and, beginning in 1976, he developed a laboratory of physical anthropology that attracted scholars from all over the world. Since then he has carried out field research in Yemen, Saudi Arabia, Egypt, Jordan and Tanzania.

In 1981 he became the founding director of the Institute of Human Origins in Berkeley, California, and he was a professor of anthropology at Stanford University until 1990, as well as a frequent guest lecturer at universities across the United States and abroad. He is the author, with Maitland Edey, of *Blueprints: Solving the Mystery of Evolution* (1989) and, with James Shreeve, of *Lucy's Child* (1989; Penguin 1991). He has also written numerous scientific and popular articles.

Maitland A. Edey is a former editor of *Life* magazine and Time–Life Books. Of his eleven previously published books, two have been on paleoanthropology.

DONALD C. JOHANSON
AND
MAITLAND A. EDEY

LUCY

The Beginnings of Humankind

PENGUIN BOOKS

PENGUIN BOOKS

Published by the Penguin Group
Penguin Books Ltd, 27 Wrights Lane, London W8 5TZ, England
Penguin Books USA Inc., 375 Hudson Street, New York, New York 10014, USA
Penguin Books Australia Ltd, Ringwood, Victoria, Australia
Penguin Books Canada Ltd, 10 Alcorn Avenue, Toronto, Ontario, Canada M4V 3B2
Penguin Books (NZ) Ltd, 182–190 Wairau Road, Auckland 10, New Zealand

Penguin Books Ltd, Registered Offices: Harmondsworth, Middlesex, England

First published in Great Britain by Granada Publishing 1981
Published in Penguin Books 1990
5 7 9 10 8 6 4

Printed in England by Clays Ltd, St Ives plc

To
Sally Johanson
and
Helen Edey

Contents

JOHANSON: You know, this is a real mess. Two kinds in South Africa. Two kinds in East Africa. Now maybe two kinds in Ethiopia—
WHITE: Maybe one kind in Ethiopia.
JOHANSON: Maybe two kinds. That makes six different kinds of hominids. That doesn't make any sense at all.
WHITE: What do you want to do, reorganise it?
JOHANSON: Would you be willing?
WHITE: I would.
JOHANSON: It might require the naming of a new species.
WHITE: It probably would.
JOHANSON: That would raise one hell of a stink.
WHITE: Yup.

Cleveland, 1977

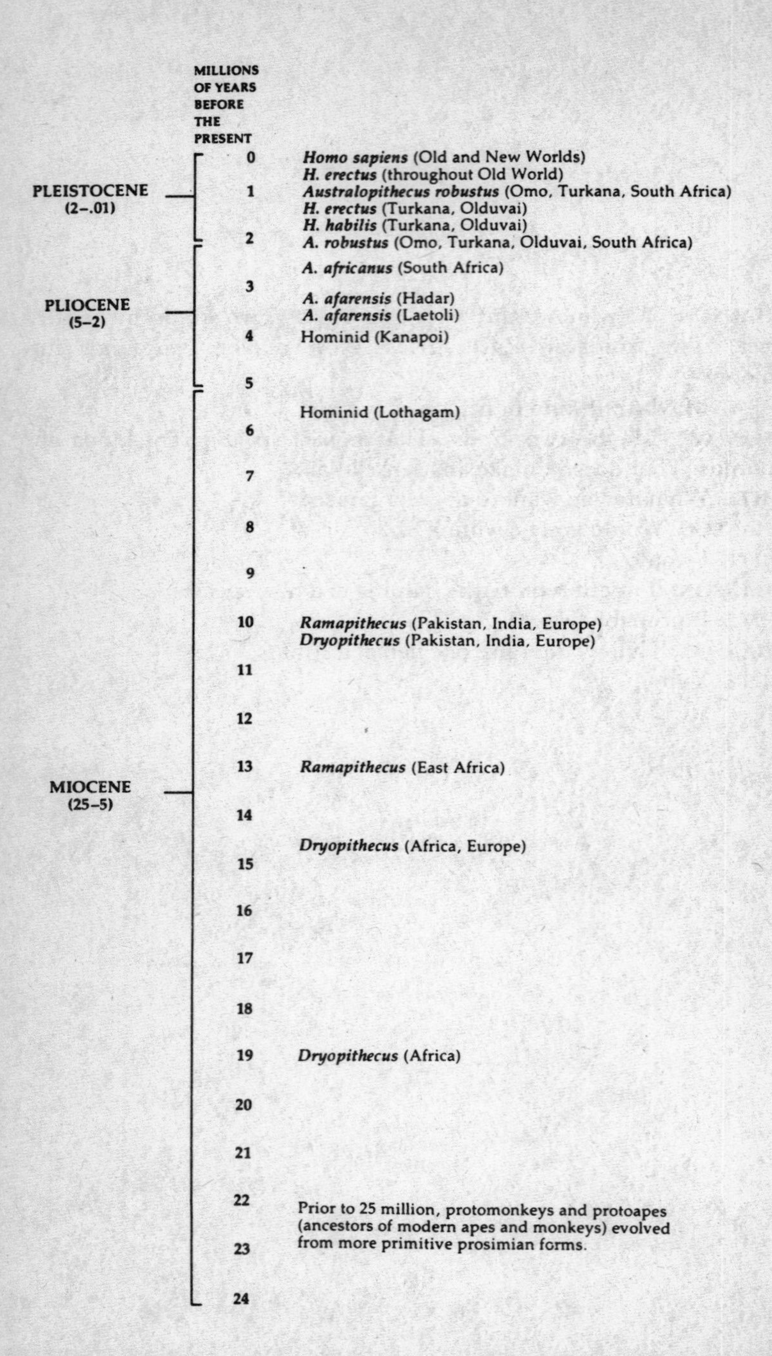

MILLIONS OF YEARS BEFORE THE PRESENT
PLEISTOCENE (2–.01)
PLIOCENE (5–2)
MIOCENE (25–5)
0 Homo sapiens (Old and New Worlds)
H. erectus (throughout Old World)
1 Australopithecus robustus (Omo, Turkana, South Africa)
H. erectus (Turkana, Olduvai)
H. habilis (Turkana, Olduvai)
2 A. robustus (Omo, Turkana, Olduvai, South Africa)
A. africanus (South Africa)
3
A. afarensis (Hadar)
A. afarensis (Laetoli)
4 Hominid (Kanapoi)
5
Hominid (Lothagam)
6
7
8
9
10 Ramapithecus (Pakistan, India, Europe)
Dryopithecus (Pakistan, India, Europe)
11
12
13 Ramapithecus (East Africa)
14
Dryopithecus (Africa, Europe)
15
16
17
18
19 Dryopithecus (Africa)
20
21
22
Prior to 25 million, protomonkeys and protoapes (ancestors of modern apes and monkeys) evolved from more primitive prosimian forms.
23
24

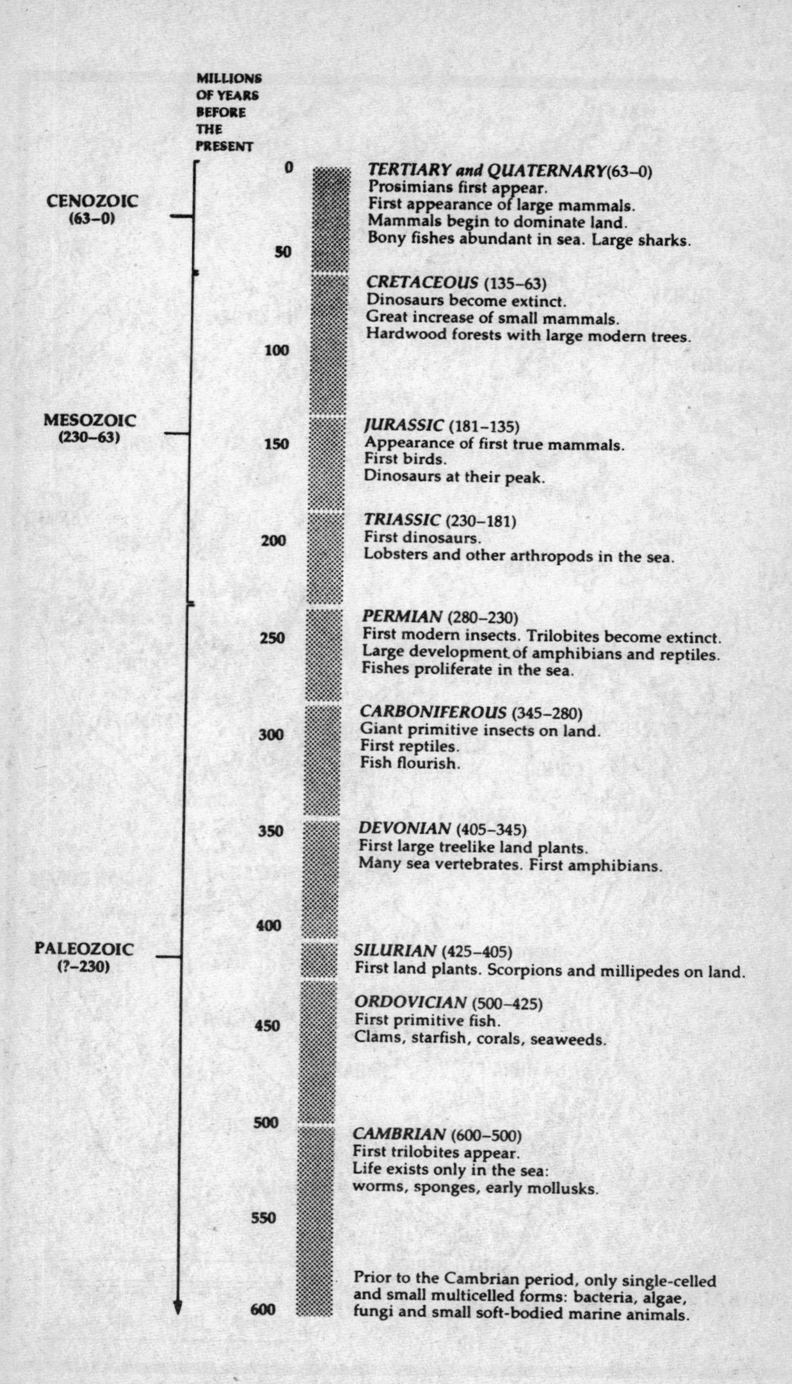
MILLIONS OF YEARS BEFORE THE PRESENT
CENOZOIC (63–0)
MESOZOIC (230–63)
PALEOZOIC (?–230)
0
50
100
150
200
250
300
350
400
450
500
550
600
TERTIARY and QUATERNARY (63–0)
Prosimians first appear.
First appearance of large mammals.
Mammals begin to dominate land.
Bony fishes abundant in sea. Large sharks.
CRETACEOUS (135–63)
Dinosaurs become extinct.
Great increase of small mammals.
Hardwood forests with large modern trees.
JURASSIC (181–135)
Appearance of first true mammals.
First birds.
Dinosaurs at their peak.
TRIASSIC (230–181)
First dinosaurs.
Lobsters and other arthropods in the sea.
PERMIAN (280–230)
First modern insects. Trilobites become extinct.
Large development of amphibians and reptiles.
Fishes proliferate in the sea.
CARBONIFEROUS (345–280)
Giant primitive insects on land.
First reptiles.
Fish flourish.
DEVONIAN (405–345)
First large treelike land plants.
Many sea vertebrates. First amphibians.
SILURIAN (425–405)
First land plants. Scorpions and millipedes on land.
ORDOVICIAN (500–425)
First primitive fish.
Clams, starfish, corals, seaweeds.
CAMBRIAN (600–500)
First trilobites appear.
Life exists only in the sea:
worms, sponges, early mollusks.
Prior to the Cambrian period, only single-celled and small multicelled forms: bacteria, algae, fungi and small soft-bodied marine animals.

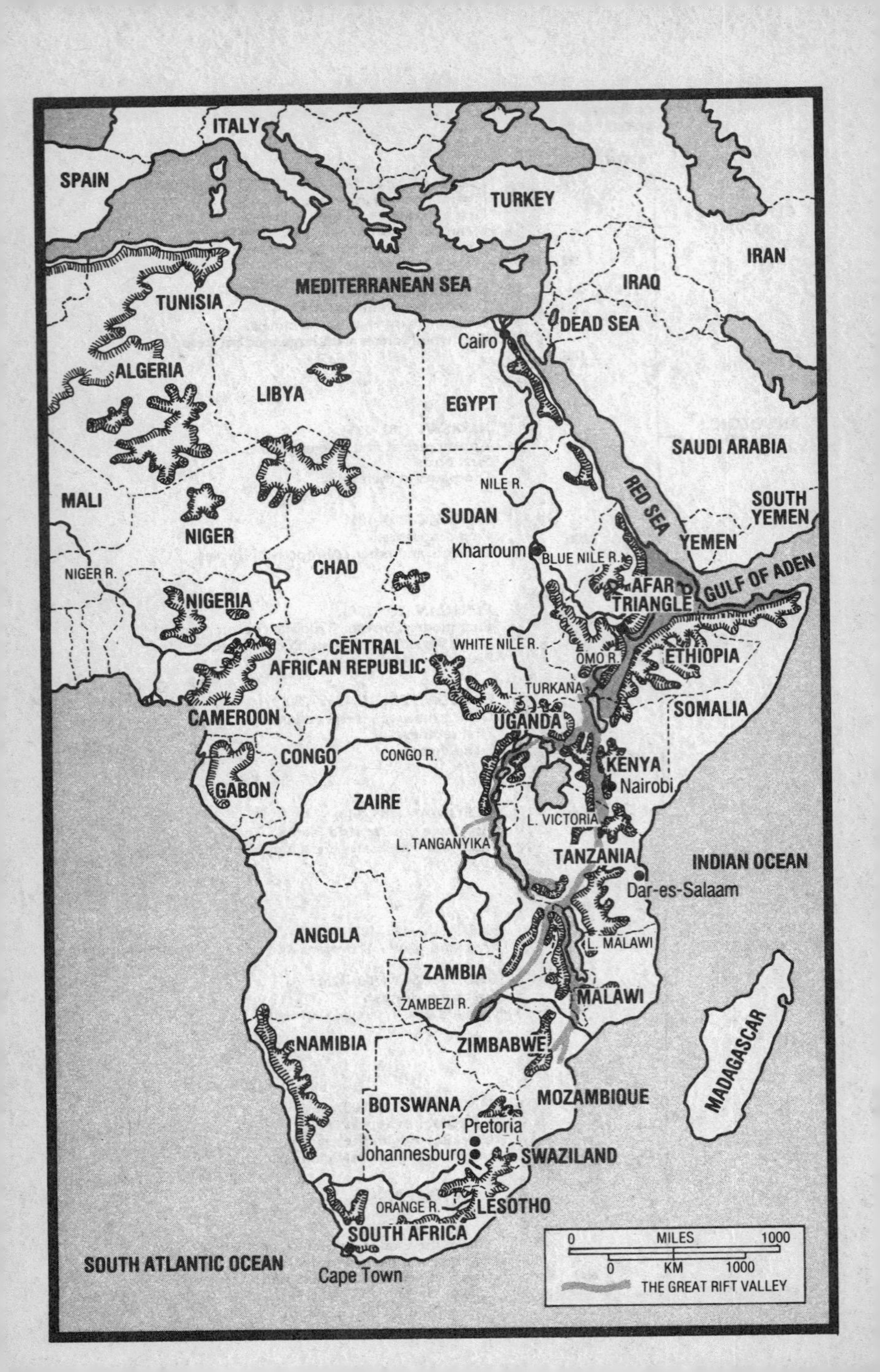
ITALY
SPAIN
TURKEY
IRAN
MEDITERRANEAN SEA
IRAQ
TUNISIA
DEAD SEA
Cairo
ALGERIA
LIBYA
EGYPT
SAUDI ARABIA
NILE R.
RED SEA
MALI
SUDAN
SOUTH YEMEN
NIGER
YEMEN
Khartoum
BLUE NILE R.
GULF OF ADEN
NIGER R.
CHAD
AFAR TRIANGLE
NIGERIA
CENTRAL AFRICAN REPUBLIC
WHITE NILE R.
OMO R.
ETHIOPIA
L. TURKANA
CAMEROON
UGANDA
SOMALIA
CONGO
CONGO R.
KENYA
Nairobi
GABON
ZAIRE
L. VICTORIA
L. TANGANYIKA
TANZANIA
INDIAN OCEAN
Dar-es-Salaam
ANGOLA
L. MALAWI
ZAMBIA
ZAMBEZI R.
MALAWI
MADAGASCAR
NAMIBIA
ZIMBABWE
MOZAMBIQUE
BOTSWANA
Pretoria
Johannesburg
SWAZILAND
ORANGE R.
LESOTHO
SOUTH AFRICA
MILES
0
1000
0
KM
1000
THE GREAT RIFT VALLEY
SOUTH ATLANTIC OCEAN
Cape Town

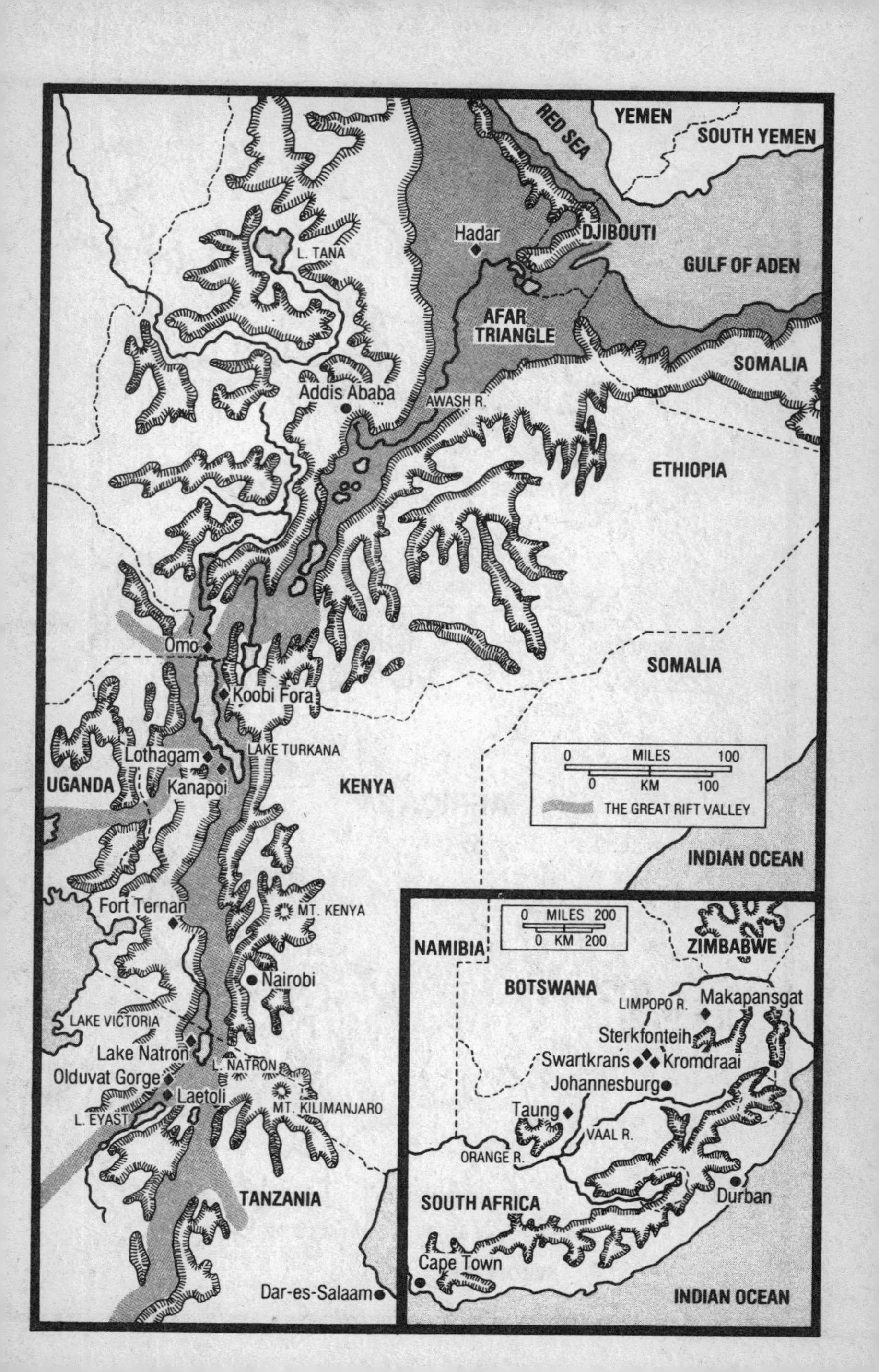
RED SEA
YEMEN
SOUTH YEMEN
Hadar
DJIBOUTI
L. TANA
GULF OF ADEN
AFAR TRIANGLE
SOMALIA
Addis Ababa
AWASH R.
ETHIOPIA
Omo
SOMALIA
Koobi Fora
LAKE TURKANA
Lothagam
UGANDA
Kanapoi
KENYA
0 MILES 100
0 KM 100
THE GREAT RIFT VALLEY
INDIAN OCEAN
Fort Ternan
MT. KENYA
Nairobi
LAKE VICTORIA
Lake Natron
L. NATRON
Olduvat Gorge
Laetoli
L. EYASI
MT. KILIMANJARO
TANZANIA
Dar-es-Salaam
0 MILES 200
0 KM 200
NAMIBIA
ZIMBABWE
BOTSWANA
LIMPOPO R.
Makapansgat
Sterkfonteih
Swartkrans
Kromdraai
Johannesburg
Taung
VAAL R.
ORANGE R.
SOUTH AFRICA
Durban
Cape Town
INDIAN OCEAN

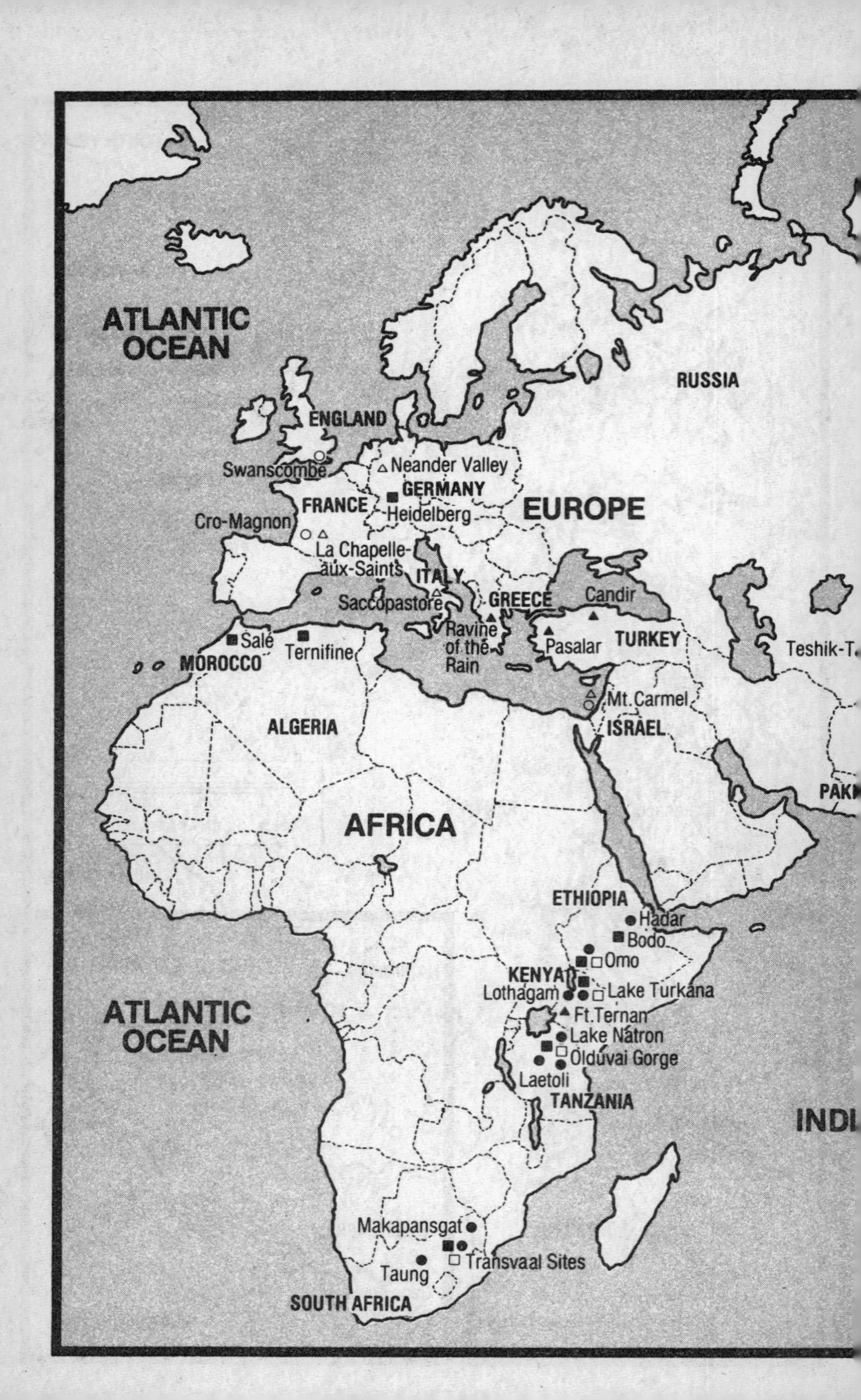
ATLANTIC OCEAN
RUSSIA
ENGLAND
Swanscombe
Neander Valley
GERMANY
FRANCE
Heidelberg
EUROPE
Cro-Magnon
La Chapelle-aux-Saints
ITALY
Saccopastore
GREECE
Candir
Ravine of the Rain
Pasalar
TURKEY
Teshik-T
Salé
Ternifine
MOROCCO
Mt. Carmel
ISRAEL
ALGERIA
PAK
AFRICA
ETHIOPIA
Hadar
Bodo
Omo
KENYA
Lothagam
Lake Turkana
Ft. Ternan
ATLANTIC OCEAN
Lake Natron
Olduvai Gorge
Laetoli
TANZANIA
INDI
Makapansgat
Taung
Transvaal Sites
SOUTH AFRICA

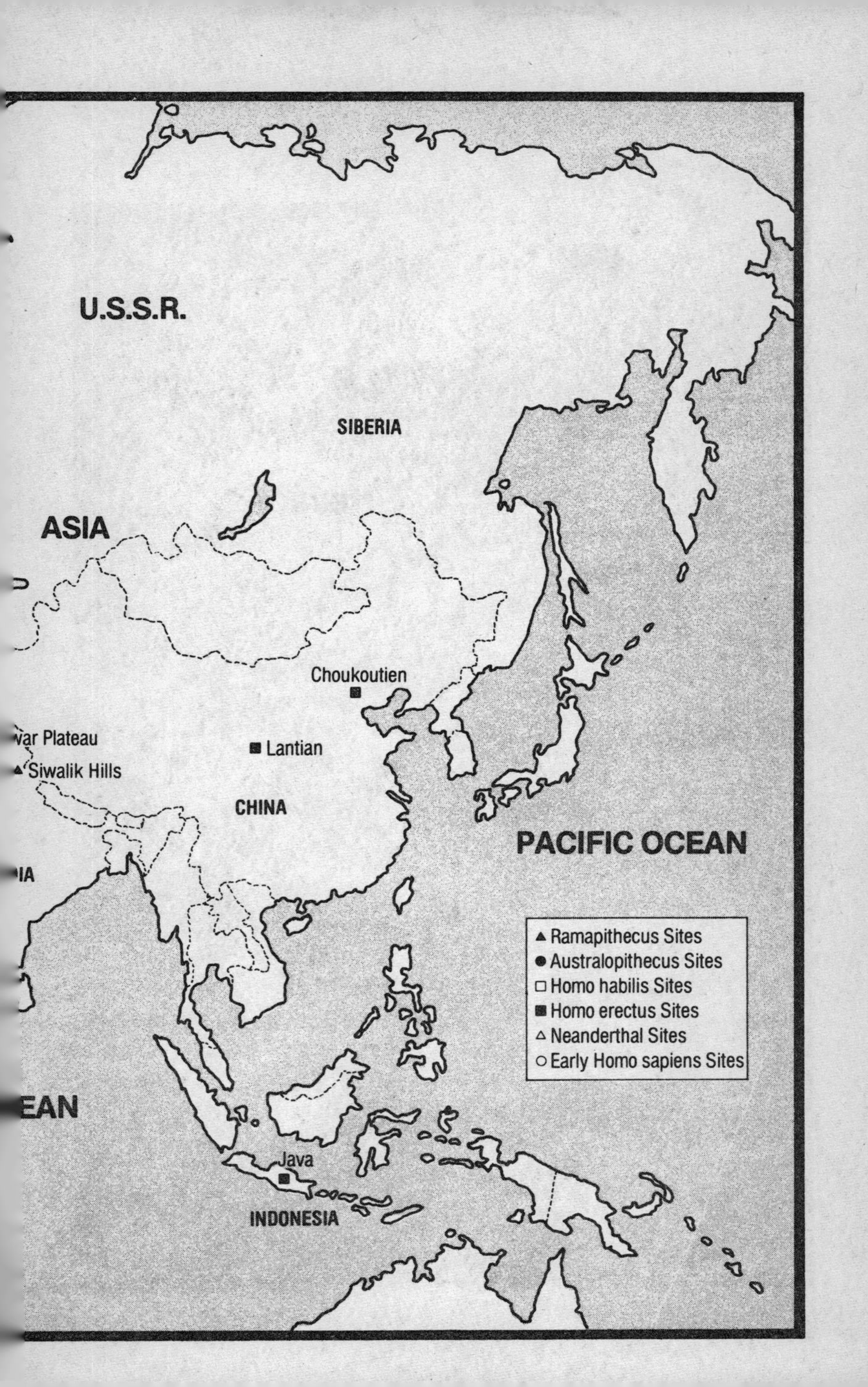

U.S.S.R.
SIBERIA
ASIA
Choukoutien
war Plateau
Lantian
Siwalik Hills
CHINA
PACIFIC OCEAN
IA
Ramapithecus Sites
Australopithecus Sites
Homo habilis Sites
Homo erectus Sites
Neanderthal Sites
Early Homo sapiens Sites
EAN
Java
INDONESIA

Prologue

In some older strata do the fossilized bones of an ape more anthropoid (manlike) or a man more pithcoid (apelike) than any yet known await the researches of some unborn paleontologist?

T. H. HUXLEY

On the morning of November 30, 1974, I woke, as I usually do on a field expedition, at daybreak. I was in Ethiopia, camped on the edge of a small muddy river, the Awash, at a place called Hadar, about a hundred miles northeast of Addis Ababa. I had been there for several weeks, acting as coleader of a group of scientists looking for fossils.

For a few minutes I lay in my tent, looking up at the canvas above me, black at first but quickly turning to green as the sun shot straight up beyond the rim of hills off to the east. Close to the Equator the sun does that; there is no long dawn as there is at home in the United States. It was still relatively cool, not more than 80 degrees. The air had the unmistakable crystalline smell of early morning on the desert, faintly touched with the smoke of cooking fires. Some of the Afar tribesmen who worked for the expedition had brought their families with them, and there was a small compound of dome-shaped huts made of sticks and grass mats about two hundred yards from the main camp. The Afar women had been up before daylight, tending their camels and goats, talking quietly.

For most of the Americans in camp this was the best part of the day. The rocks and boulders that littered the landscape had bled away most of their heat during the night and no longer felt like

Don Johanson searches for fossils in a gully at Hadar. The deposits there are what is left of about two million years of sediments laid down in the bed of an ancient lake which has long since evaporated. Subsequent rains have produced gullies in the deposits, bringing to the surface fossils like those in the foreground.

stoves when you stood next to one of them. I stepped out of the tent and took a look at the sky. Another cloudless day; another flawless morning on the desert that would turn to a crisper later on. I washed my face and got a cup of coffee from the camp cook, Kabete. Mornings are not my favorite time. I am a slow starter and much prefer evenings and nights. At Hadar I feel best just as the sun is going down. I like to walk up one of the exposed ridges near the camp, feel the first stirrings of evening air and watch the hills turn purple. There I can sit alone for a while, think about the work of the day just ended, plan the next, and ponder the larger questions that have brought me to Ethiopia. Dry silent places are intensifiers of thought, and have been known to be since early Christian anchorites went into the desert to face God and their own souls.

Tom Gray joined me for coffee. Tom was an American graduate student who had come out to Hadar to study the fossil animals and plants of the region, to reconstruct as accurately as possible the kinds and frequencies and relationships of what had lived there at various times in the remote past and what the climate had been like. My own target – the reason for our expedition – was hominid fossils: the bones of extinct human ancestors and their close relatives. I was interested in the evidence for human evolution. But to understand that, to interpret any hominid fossils we might find, we had to have the supporting work of other specialists like Tom.

'So, what's up for today?' I asked.

Tom said he was busy marking fossil sites on a map.

'When are you going to mark in Locality 162?'

'I'm not sure where 162 is,' he said.

'Then I guess I'll have to show you.' I wasn't eager to go out with Gray that morning. I had a tremendous amount of work to catch up on. We had had a number of visitors to the camp recently. Richard and Mary Leakey, two well-known experts on hominid fossils from Kenya, had left only the day before. During their stay I had not done any paperwork, any cataloguing. I had not written any letters or done detailed descriptions of any fossils. I *should* have stayed in camp that morning – but I didn't. I felt a strong subconscious urge to go with Tom, and I obeyed it. I wrote a note to myself in my daily diary: *Nov. 30, 1974. To Locality 162 with Gray in AM. Feel good.*

As a paleoanthropologist – one who studies the fossils of human ancestors – I am superstitious. Many of us are, because the work we do depends a great deal on luck. The fossils we study are extremely rare, and quite a few distinguished paleoanthropologists have gone a lifetime without finding a single one. I am one of the more fortunate. This was only my third year in the field at Hadar, and I had already found several. I know I am lucky, and I don't try to hide it. That is why I wrote 'feel good' in my diary. When I got up that morning I felt it was one of those days when you should press your luck. One of those days when something terrific might happen.

Throughout most of that morning, nothing did. Gray and I got into one of the expedition's four Land-Rovers and slowly jounced our way to Locality 162. This was one of several hundred sites that were in the process of being plotted on a master map of the Hadar area, with detailed information about geology and fossils being entered on it as fast as it was obtained. Although the spot we were headed for was only about four miles from camp, it took us half an hour to get there because of the rough terrain. When we arrived it was already beginning to get hot.

At Hadar, which is a wasteland of bare rock, gravel and sand, the fossils that one finds are almost all exposed on the surface of the ground. Hadar is in the center of the Afar desert, an ancient lake bed now dry and filled with sediments that record the history of past geological events. You can trace volcanic-ash falls there, deposits of mud and silt washed down from distant mountains, episodes of volcanic dust, more mud, and so on. Those events reveal themselves like layers in a slice of cake in the gullies of new young rivers that recently have cut through the lake bed here and there. It seldom rains at Hadar, but when it does it comes in an overpowering gush – six months' worth overnight. The soil, which is bare of vegetation, cannot hold all that water. It roars down the gullies, cutting back their sides and bringing more fossils into view.

Gray and I parked the Land-Rover on the slope of one of those gullies. We were careful to face it in such a way that the canvas water bag that was hanging from the side mirror was in the shade. Gray plotted the locality on the map. Then we got out and began doing what most members of the expedition spent a great deal of their time doing: we began surveying, walking slowly about, looking for exposed fossils.

Some people are good at finding fossils. Others are hopelessly bad at it. It's a matter of practice, of training your eye to see what you need to see. I will never be as good as some of the Afar people. They spend all their time wandering around in the rocks and sand. They have to be sharp-eyed; their lives depend on it. Anything the least bit unusual they notice. One quick educated look at all those stones and pebbles, and they'll spot a couple of things a person not acquainted with the desert would miss.

Tom and I surveyed for a couple of hours. It was now close to noon, and the temperature was approaching 110. We hadn't found much: a few teeth of the small extinct horse *Hipparion*; part of the skull of an extinct pig; some antelope molars; a bit of a monkey jaw. We had large collections of all these things already, but Tom insisted on taking these also as added pieces in the overall jigsaw puzzle of what went where.

'I've had it,' said Tom. 'When do we head back to camp?'

'Right now. But let's go back this way and survey the bottom of that little gully over there.'

The gully in question was just over the crest of the rise where we had been working all morning. It had been thoroughly checked out at least twice before by other workers, who had found nothing interesting. Nevertheless, conscious of the 'lucky' feeling that had been with me since I woke, I decided to make that small final detour. There was virtually no bone in the gully. But as we turned to leave, I noticed something lying on the ground partway up the slope.

'That's a bit of a hominid arm,' I said.

'Can't be. It's too small. Has to be a monkey of some kind.'

We knelt to examine it.

'Much too small,' said Gray again.

I shook my head. 'Hominid.'

'What makes you so sure?' he said.

'That piece right next to your hand. That's hominid too.'

'Jesus Christ,' said Gray. He picked it up. It was the back of a small skull. A few feet away was part of a femur: a thighbone. 'Jesus Christ,' he said again. We stood up, and began to see other bits of bone on the slope: a couple of vertebrae, part of a pelvis – all of them hominid. An unbelievable, impermissible thought flickered through my mind. Suppose all these fitted together? Could they be parts of a single, extremely primitive skeleton? No such skeleton had ever been found – anywhere.

'Look at that,' said Gray. 'Ribs.'

A single individual

'I can't believe it,' I said. 'I just can't believe it.'

'By God, you'd better believe it!' shouted Gray. 'Here it is. Right

Each time a fossil is found – and collected – in a new locality at Hadar, that locality must be numbered, and all pertinent information entered on a master map of the whole area. Only in this way can an increasingly complex torrent of geological and paleontological information be made useful to the scientists.

A single individual?

here!' His voice went up into a howl. I joined him. In that 110-degree heat we began jumping up and down. With nobody to share our feelings, we hugged each other, sweaty and smelly, howling and hugging in the heat-shimmering gravel, the small brown remains of what now seemed almost certain to be parts of a single hominid skeleton lying all around us.

'We've got to stop jumping around,' I finally said.'We may step on something. Also, we've got to make sure.'

'Aren't you sure, for Christ's sake?'

'I mean, suppose we find two left legs. There may be several individuals here, all mixed up. Let's play it cool until we can come back and make absolutely sure that it all fits together.'

We collected a couple of pieces of jaw, marked the spot exactly and got into the blistering Land-Rover for the run back to camp. On the way we picked up two expedition geologists who were loaded down with rock samples they had been gathering.

'Something big,' Gray kept saying to them. 'Something big. Something *big*.'

'Cool it,' I said.

But about a quarter of a mile from camp, Gray could not cool it. He pressed his thumb on the Land-Rover's horn, and the long blast brought a scurry of scientists who had been bathing in the river. 'We've got it,' he yelled. 'Oh, Jesus, we've got it. We've got The Whole Thing!'

That afternoon everyone in camp was at the gully, sectioning off the site and preparing for a massive collecting job that ultimately took three weeks. When it was done, we had recovered several hundred pieces of bone (many of them fragments) representing about forty percent of the skeleton of a single individual. Tom's and my original hunch had been right. There was no bone duplication.

But a single individual of what? On preliminary examination it was very hard to say, for nothing quite like it had ever been discovered. The camp was rocking with excitement. That first night we never went to bed at all. We talked and talked. We drank beer after beer. There was a tape recorder in the camp, and a tape of the Beatles song 'Lucy in the Sky with Diamonds' went belting out into the night sky, and was played at full volume over and over again out of sheer exuberance. At some point during that unforgettable evening – I no longer remember exactly when – the new fossil picked up the name of Lucy, and has been so known ever since, although its proper name – its acquisition number in the Hadar collection – is AL 288-1.

'Lucy?'

That is the question I always get from somebody who sees the fossil for the first time. I have to explain: 'Yes, she was a female. And that Beatles song. We were sky-high, you must remember, from finding her.'

Then comes the next question: 'How did you know she was a female?'

'From her pelvis. We had one complete pelvic bone and her sacrum. Since the pelvic opening in hominids has to be proportionately larger in females than in males to allow for the birth of large-brained infants, you can tell a female.'

And the next: 'She was a hominid?'

'Oh, yes. She walked erect. She walked as well as you do.'

'Hominids all walked erect?'

'Yes.'

'Just exactly what is a hominid?'

That usually ends the questions, because that one has no simple answer. Science has had to leave the definition rather flexible because we do not yet know exactly when hominids first appeared. However, it is safe to say that a hominid is an erect-walking primate.

* In this book the general term 'man' is used to include both males and females of the genus *Homo*.

All the surface gravel from the Lucy site was brought down to the Awash River, where it was spread out on long cloths and picked over bit by bit to ensure that no smallest part of Lucy would be overlooked. The person working here is Claude Guillemot of the French team.

That is, it is either an extinct ancestor to man,* a collateral relative to man, or a true man. All human beings are hominids, but not all hominids are human beings.

We can picture human evolution as starting with a primitive

apelike type that gradually, over a long period of time, began to be less and less apelike and more manlike. There was no abrupt crossover from ape to human, but probably a rather fuzzy time of in-between types that would be difficult to classify either way. We have no fossils yet that tell us what went on during that in-between time. Therefore, the handiest way of separating the newer types from their ape ancestors is to lump together all those that stood up on their hind legs. That group of men and near-men is called hominids.

I am a hominid. I am a human being. I belong to the genus *Homo* and to the species *sapiens*: thinking man. Perhaps I should say wise or knowing man – a man who is smart enough to recognize that he is a man. There have been other species of *Homo* who were not so smart, ancestors now extinct. *Homo sapiens* began to emerge a hundred thousand – perhaps two or three hundred thousand – years ago, depending on how one regards Neanderthal Man. He was another *Homo*. Some think he was the same species as ourselves. Others think he was an ancestor. There are a few who consider him a kind of cousin. That matter is unsettled because many of the best Neanderthal fossils were collected in Europe before anybody knew how to excavate sites properly or get good dates. Consequently, we do not have exact ages for most of the Neanderthal fossils in collections.

I consider Neanderthal conspecific with *sapiens*, with myself. One hears talk about putting him in a business suit and turning him loose in the subway. It is true; one could do it and he would never be noticed. He was just a little heavier-boned than people of today, more primitive in a few facial features. But he was a man. His brain was as big as a modern man's, but shaped in a slightly different way. Could he make change at the subway booth and recognise a token? He certainly could. He could do many things more complicated than that. He was doing them over much of Europe, Africa and Asia as long as sixty or a hundred thousand years ago.

Neanderthal Man had ancestors, human ones. Before him in time was a less advanced type: *Homo erectus*. Put him on the subway and people would probably take a suspicious look at him. Before *Homo erectus* was a really primitive type, *Homo habilis*; put him on the subway and people would probably move to the other end of the car. Before *Homo habilis* the human line may run out entirely. The next stop in the past, back of *Homo habilis*, might be something like Lucy.

All of the above are hominids. They are all erect walkers. Some were human, even though they were of exceedingly primitive types.

Others were not human. Lucy was not. No matter what kind of clothes were put on Lucy, she would not look like a human being. She was too far back, out of the human range entirely. That is what happens going back along an evolutionary line. If one goes back far enough, one finds oneself dealing with a different kind of creature. On the hominid line the earliest ones are too primitive to be called humans. They must be given another name. Lucy is in that category.

For five years I kept Lucy in a safe in my office in the Cleveland Museum of Natural History. I had filled a wide shallow box with yellow foam padding, and had cut depressions in the foam so that each of her bones fitted into its own tailor-made nest. *Everybody* who came to the Museum – it seemed to me – wanted to see Lucy. What surprised people most was her small size.

Her head, on the evidence of the bits of her skull that had been recovered, was not much larger than a softball. Lucy herself stood only three and one-half feet tall, although she was fully grown. That could be deduced from her wisdom teeth, which were fully erupted and had been exposed to several years of wear. My best guess was that she was between twenty-five and thirty years old when she died. She had already begun to show the onset of arthritis or some other bone ailment, on the evidence of deformation of her vertebrae. If she had lived much longer, it probably would have begun to bother her.

Her surprisingly good condition – her completeness – came from the fact that she had died quietly. There were no tooth marks on her bones. They had not been crunched and splintered, as they would have been if she had been killed by a lion or a saber-toothed cat. Her head had not been carried off in one direction and her legs in another, as hyenas might have done with her. She had simply settled down in one piece right where she was, in the sand of a long-vanished lake edge or stream – and died. Whether from illness or accidental drowning, it was impossible to say. The important thing was that she had not been found by a predator just after death and eaten. Her carcass had remained inviolate, slowly covered by sand or mud, buried deeper and deeper, the sand hardening into rock under the weight of subsequent depositions. She had lain silently in her adamantine grave for millennium after millennium until the rains at Hadar had brought her to light again.

That was where I was unbelievably lucky. If I had not followed a hunch that morning with Tom Gray, Lucy might never have been found. Why the other people who looked there did not see her, I do not know. Perhaps they were looking in another direction. Perhaps the light was different. Sometimes one person sees things that another misses, even though he may be looking directly at them. If I had not gone to Locality 162 that morning, nobody might have bothered to go back for a year, maybe five years. Hadar is a big place, and there is a tremendous amount to do. If I had waited another few years, the next rains might have washed many of her bones down the gully. They would have been lost, or at least badly scattered; it would not have been possible to establish that they belonged together. What was utterly fantastic was that she had come to the surface so recently, probably in the last year or two. Five years earlier, she still would have been buried. Five years later, she would have been gone. As it was, the front of her skull was already gone, washed away somewhere. We never did find it. Consequently, the one thing we really cannot measure accurately is the size of her brain.

Lucy always managed to look interesting in her little yellow nest – but to a nonprofessional, not overly impressive. There were other bones all around her in the Cleveland Museum. She was dwarfed by them, by drawer after drawer of fossils, hundreds of them from Hadar alone. There were casts of hominid specimens from East Africa, from South Africa and Asia. There were antelope and pig skulls, extinct rodents, rabbits and monkeys, as well as apes. There was one of the largest collections of gorilla skulls in the world. In that stupefying array of bones, I kept being asked, What was so special about Lucy? Why had she, as another member of the expedition put it, 'blown us out of our little anthropological minds for months'?

'Three things,' I always answered. 'First: what she is – or isn't. She is different from anything that has been discovered and named before. She doesn't fit anywhere. She is just a very old, very primitive, very small hominid. Somehow we are going to have to fit her in, find a name for her.

'Second,' I would say, 'is her completeness. Until Lucy was found, there just weren't any very old skeletons. The oldest was one of those Neanderthalers I spoke of a little while ago. It is about seventy-five thousand years old. Yes, there *are* older hominid fossils, but they are all fragments. Everything that has been reconstructed from them has had to be done by matching up those little pieces – a tooth here, a bit of jaw there, maybe a complete skull from somewhere else, plus a leg

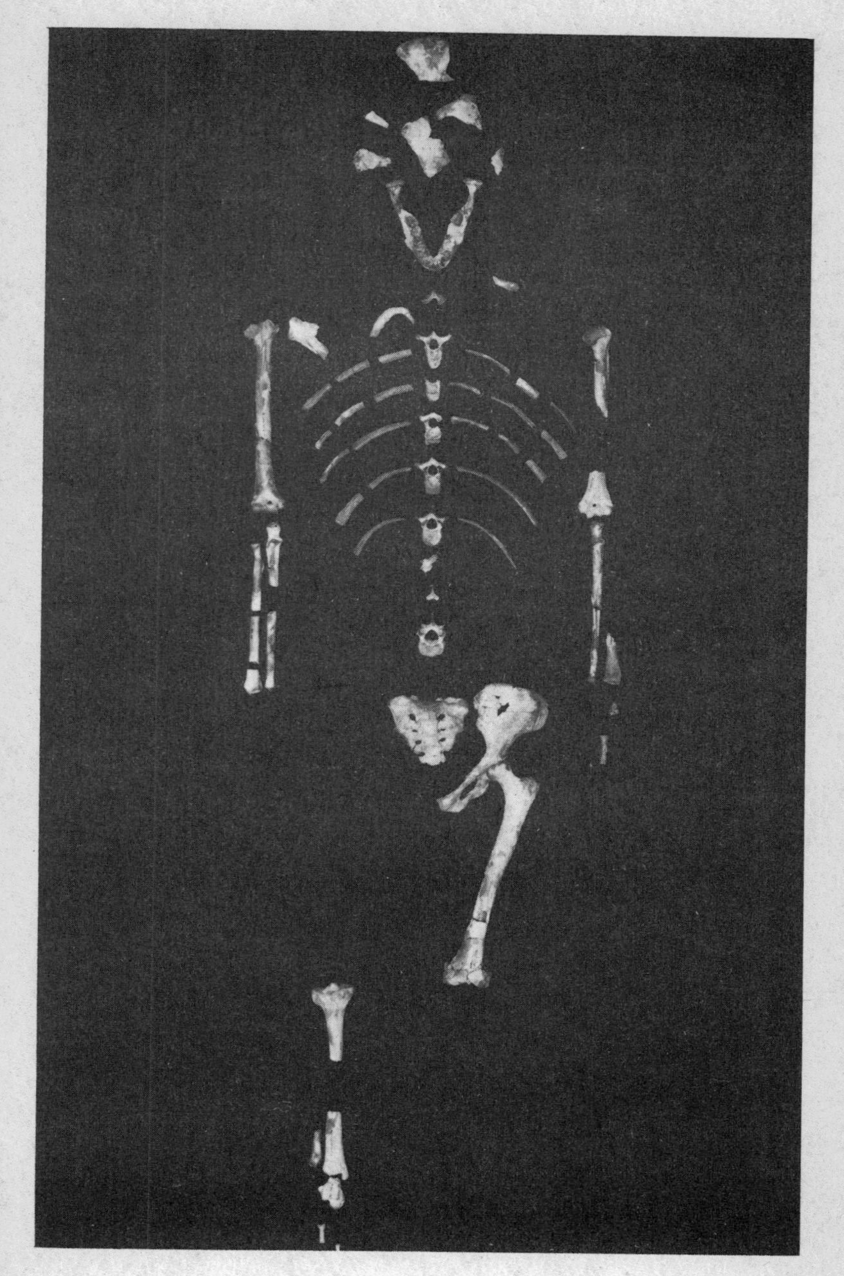

bone from some other place. The fitting together has been done by scientists who know those bones as well as I know my own hand. And yet, when you consider that such a reconstruction may consist of pieces from a couple of dozen individuals who may have lived hundreds of miles apart and may have been separated from each other by a hundred thousand years in time – well, when you look at the complete individual you've just put together you have to say to yourself, "Just how real is he?" With Lucy you know. It's all there. You don't have to guess. You don't have to imagine an arm bone you haven't got. You *see* it. You see it for the first time from something older than a Neanderthaler.'

'How much older?'

'That's point number three. The Neanderthaler is seventy-five thousand years old. Lucy is approximately 3.5 million years old. She is the oldest, most complete, best-preserved skeleton of any erect-walking human ancestor that has ever been found.'

That is the significance of Lucy: her completeness and her great age. They make her unique in the history of hominid fossil collecting. She is easy to describe, and – as will be seen – she makes a number of anthropological problems easier to work out. But exactly what is she?

The rest of this book will be devoted to answering that question. Unique Lucy may be, but she is incomprehensible outside the context of other fossils. She becomes meaningless unless she is fitted into a scheme of hominid evolution and scientific logic that has been laboriously pieced together over more than a century by hundreds of specialists from four continents. Their fossil finds, their insights – sometimes inspired, sometimes silly – their application of techniques from such faraway disciplines as botany, nuclear physics and microbiology have combined to produce an increasingly clear and rich picture of man's emergence from the apes – a story that is finally, in the ninth decade of this century, beginning to make some sense. That story could not even begin to be told, of course, until Charles Darwin suggested in 1857 that we *were* descended from apes and not divinely created in 4004 B.C., as the Church insisted. But not even Darwin could have suspected some of the odd turns the hominid story would take. Nor could he have guessed which apes we are descended from. Indeed, we are not entirely sure about that even today.

PART ONE
Background

1 The Early Fossil Finds

Unless we wilfully close our eyes we may with our present knowledge approximately recognize our parentage; nor need we feel ashamed of it.

CHARLES DARWIN

Apologists emphasize that man cannot be a descendant of any living ape – a statement that is obvious to the verge of imbecility – and go on to state or imply that man is not really descended from any ape or monkey at all, but from an earlier common ancestor. In fact, the common ancestor would certainly be called an ape or monkey in popular speech by anybody who saw it. Since the terms ape and monkey are defined by popular usage, man's ancestors were apes or monkeys (or successfully both). It is pusillanimous if not dishonest for an informed investigator to say otherwise.

GEORGE GAYLORD SIMPSON

When Darwin dropped his blockbuster, he did it, as it were, by remote control. He was an extremely shy man, plagued by bad health. He had no stomach for getting into the kind of all-out fight that instinct told him would flare when his theory of evolution ran headlong into the political and religious orthodoxy of the day. The Church of England had immense political and social leverage above and beyond its magisterial control over the souls of Britain's people. Indeed, Darwin was so reluctant to confront that juggernaut that for several years he did not publish his theory at all. He kept on amassing greater and greater evidence for what his flash of insight, gained years before in the Galápagos Islands, had told him was true: all species had evolved. They had been evolving since the beginning of life. They were related by descent, and those lines of descent could be traced in fossils. The fossils with which Darwin chose to back up his theory were as uncontroversial as he could make them: obscure marine organisms, small barnacles, long-extinct clams, and so on. Only once in *The Origin of Species* did he even hint that evolution had anything to do with people. At the very end he inserted one sentence: 'Light will be thrown on the origin of man and his history.'

That was enough, and the fight was on. Luckily for Darwin, he had an able companion: the aggressive and imaginative scientist Thomas Henry Huxley. While Darwin lurked in his home like a timid and anxious turtle, Huxley was out front, arguing. He took on a distinguished Anglican bishop in public debate, and demolished him. He even succeeded in making England's Prime Minister, Benjamin Disraeli, look foolish. It was Huxley who publicly associated humans with apes. He pointed out the great range of similarity between man and what he recognised to be man's closest living relatives, the gorilla and chimpanzee. From this he reasoned that the three had a common – and not terribly remote – ancestor. Since those apes were found living only in Africa, Huxley suggested that fossils of the joint ancestor might be found there also.

Unfortunately, there were no known fossil ancestors from Africa. In fact, ancestors of any sort were in short supply. There was only one: part of a skull and some limb bones that had been dug out of a cave in the Neander Valley in Germany shortly before Darwin's classic was published. Although this specimen, known thereafter as Neanderthal Man, excited wide interest at the time of its discovery, it achieved no respectability whatsoever. For nineteenth-century minds unaccustomed to thinking about the possible plasticity of the human skeleton, this one seemed too brutish to qualify as a bona fide evolutionary example. It had an abnormally thick skull that was longer and narrower than that of contemporary man, and heavy eyebrow ridges. German anatomists – and they were the most learned in the world – went after it vigorously. It was the skull of an elderly Dutchman, said Dr Wagner of Göttingen. No, said Dr Mayer of Bonn; he was a Cossack soldier chasing Napoleon's retreating army; he got lost, wandered into the cave and died there. The French expert Pruner-Bey had a different idea: this was the skull of a powerfully built Celt somewhat resembling a modern Irishman, and with a low mental organization. The coup de grâce was delivered by the renowned Rudolf Virchow, who declared that Neanderthal Man's oddities were individual pathological deformities that had nothing to do with primitiveness. Rather, said Virchow in a detailed review, they were the result of rickets in childhood and arthritis in old age, with some severe blows on the head suffered sometime in between!

There was also the question of the fossil's age. For all that anybody could tell, it might have been alive and well during Napoleon's time. Neanderthal Man was stowed away and ignored.

But not quite forgotten. The idea that Neanderthal Man might be an actual human persisted. Stubborn students of the past kept digging in caves and riverbeds. They found Cro-Magnon Man, so named for the locality in southern France where his bones were first unearthed. Many specimens were recovered, among them complete skeletons that were so virtually indistinguishable from those of today that even the most skeptical had to concede that they were humans. The argument then shifted to how old they were. But as the science of geology advanced, along with ways of estimating time by observing the evolution of different kinds of mammals in different rock strata or in the built-up debris in cave floors, that argument too subsided. Although true dates were impossible to calculate, comparative – or relative – ones were becoming increasingly easy to arrive at, and the first estimates of an evolutionary time scale began to be possible.

It is obvious the fossil lying in the uppermost levels of a cave floor must be younger than something buried deeper. The same is true of bones or stone tools found in sand or gravel layers deposited by rivers; their position with respect to one another is critical to any understanding of their relative age. In the latter case, if geologists can come up with reliable estimates of how long it takes for a given layer of gravel to be deposited, then a clue to true age will exist.

A great deal of this basic spadework was done in the latter half of the nineteenth century. By the end of it, a crude but useful calendar of events was worked out for Western Europe. It revealed that the Cro-Magnon 'caveman' had been living there for forty or fifty thousand years and may have disappeared as recently as ten thousand years ago. The Cro-Magnons were not the lurching subhumans of the comic strips. They were men like us, men who could paint beautiful pictures and presumably could think complex thoughts, men with the beginnings of religious beliefs, men who routinely made a variety of useful stone tools and clearly had an elaborate culture, in many respects more sophisticated than some of the so-called 'savage' cultures found in isolated parts of the earth today.

Older, more primitive humans were also found. Neanderthal Man, it turned out, was real after all. He *was* somewhat different from us, and he should have been. If the theory of evolution had any validity whatsoever, then human fossils would have to reveal an increasing retreat toward primitiveness as one tracked them deeper into time. Neanderthal Man did go deeper, fifty or a hundred thousand years deeper, maybe two hundred thousand – those figures became more uncertain the deeper one went, and increasingly alarming. Neanderthalers were innately disturbing to minds that had long been attuned to thinking in terms of thousands of years and had only recently been stretched to think in terms of tens of thousands of years. Thinking in terms of hundreds of thousands of years was still very difficult.

That may explain why the announcement in 1893 by a Dutch scientist, Eugène Dubois, that he had found a half-million-year-old apeman in Java was so dismaying.

Anyone who thinks I am lucky at finding fossils should consider Dubois. His luck was unbelievable. What are the odds that a young Dutch anatomy professor who scarcely knew anything about fossils, who had never actually seen a hominid fossil, who had never been outside Holland, could go halfway around the world to a place where no fossils had ever been collected, just on a logical hunch, and find something?

It is as if somebody were to announce, 'I'm a rare-gem prospector. I don't know anything about gems. I haven't seen any. But that's what I want to be. I have never been prospecting in the field, but I notice that rubies come from certain mountain formations in certain latitudes in Burma and Thailand. Therefore I plan to explore similar mountains in similar latitudes in Mexico in the hope of finding emeralds.'

The odds against success in a chain of logic as tenuous as that must be up in the millions. But Dubois did find his emerald: the Java ape-man, *Pithecanthropus erectus*.

His logic was simplicity itself. As a boy he had heard about the Neanderthal fossil that had been taken from a limestone cave near Düsseldorf two years before he was born. He read everything he could about Neanderthal Man, believed strongly in the theory of evolution, and gradually became convinced that Neanderthal was an early type of human, despite the doubts expressed by scientists. If that was so, he reasoned, then older, more apelike forms should exist somewhere. The place to look for them would not be Europe, where the weather would have been too cold to support such creatures, and where ice sheets would have destroyed all trace of them. Therefore he would look in the tropics. He would pick a place like Sumatra – known to contain a species of large manlike ape, the orangutan. He would do his prospecting in caves that he had been told existed there.

What Dubois hoped to find was a 'missing link'. Like many of his contemporaries with a certain amount of scientific curiosity, he had read Darwin but had gotten some of Darwin's ideas wrong. If men were descended from apes, as Darwin and Huxley had suggested, then the way to prove it was to find a creature that stood halfway between the two, something that presumably was a blend of man and orangutan or man and chimpanzee. Darwin's idea, of course, was quite different. He was not thinking of parallel linkages but of vertical ones, of chains of relationships connected through time. To Darwin, the close affinities of man and ape did not suggest the existence of an in-between type. Rather, they meant that man and ape had a common ancestor to which each was connected by separate links of its own. What that ancestor might have looked like, neither Darwin nor Huxley was prepared to say. Still, the missing-link concept became widely popular, and it was on the strength of it that Dubois, against the urgings of his family and his professional colleagues, gave up his teaching career and went to the Dutch East Indies.

Unable to raise the money to finance an expedition, he joined the Dutch Army as a military doctor and had himself posted to Sumatra. His medical duties were light, and over two years he managed to explore a number of caves there, but with little success. Then he contracted malaria, was transferred to Java and was placed on inactive duty. Now, with plenty of time on his hands, he was free to devote all his energies to some fossil-bearing strata he had found on a bend of the Solo River, a small, slow stream at a place called Trinil. The Dutch Government became interested in what he was doing and gave him convict labor from the prisons to help him dig. The trouble with that work force was that its members stole fossils almost as fast as they found them and sold them to Chinese traders, who ground those 'dragon bones' into powder for shipment to China, where they brought high prices as medicines and aphrodisiacs. This practice was stopped only when Dubois's foreman persuaded the Dutch authorities to ban the sale of fossils.

The dig at Trinil now began to produce a flow of interesting material, including extinct mammals hitherto unknown to science.

Like every zealot, Dubois had a remarkable ability to soak up information about what he was interested in – and in the matter of fossils he was obsessed. He became knowledgeable in identifying his finds, and was soon shipping off crates of them to Holland. But he worked the Solo River bank for an entire year before he found what he was really looking for: a primate fossil. This was one very large molar tooth. He could not decide whether it belonged to an extinct giant chimpanzee or to an orangutan. While he was puzzling over that, he made another find within a yard of the first. This was the top of a primate skull. It was extremely thick; too low and heavy for the skullcap of a man, he decided, too large and rounded for the skull of an orangutan. 'A manlike ape' was his conclusion.

How manlike that ape might be came with the jolting discovery the next year of an upper leg bone, a femur. It was virtually identical with that of a modern man, and indicated that its owner had walked erect. Although this all-important leg bone and another tooth were found about fifty feet away from where the skullcap had been found, Dubois concluded that all had once belonged to the same individual. He sent off a triumphant cable to Europe that he had found the missing link, and followed shortly himself with his fossils in a box.

Instead of triumph, he experienced bitter controversy. For years his fossils were the subject of furious debate. Some thought he had mistakenly joined the skull of some kind of extinct ape with the leg bone of a man who had died and been fossilized much later. They argued and argued about the skullcap, feeling that it was far too primitive to be that of a man. Dubois believed otherwise. He clung to his missing-link hypothesis, insisted that all his fossil parts came from the same individual, and took them to England, where they were subjected to an exhaustive review by Sir Arthur Keith, already emerging as the successor to the great German, Rudolf Virchow, as the premier paleontologist of the day. Keith's mind was elastic enough to accept the possibility of a man more primitive and smaller-brained than any hitherto known. The longer he studied Dubois's fossils, the more certain he became that they represented a human being and not a missing link. That infuriated Dubois, leaving Keith with the impression that Dubois was 'impatient of his critics;

One and one-half million years of evolution in humans has produced a higher and rounder skull, a much larger brain, and a distinct chin, which is missing in *Homo erectus*. There has also been a gradual disappearance of *erectus* heavy eyebrow ridges.

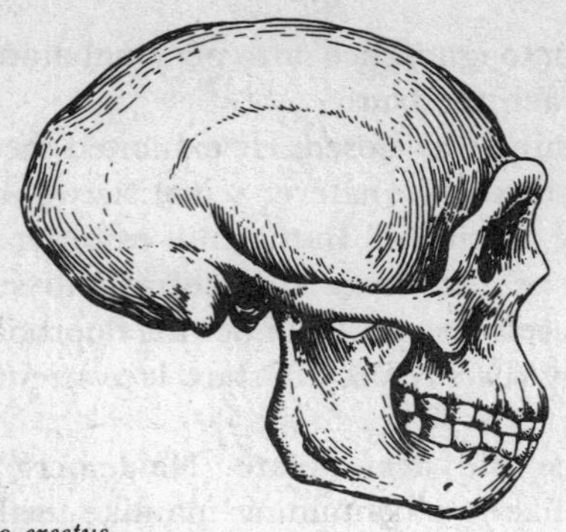
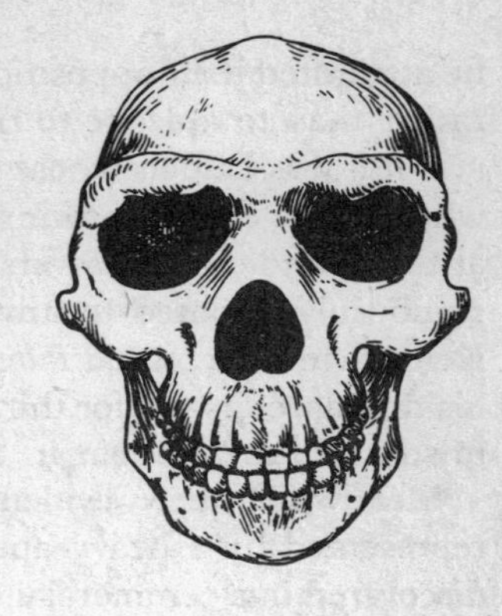

Homo erectus
1,500,000–200,000

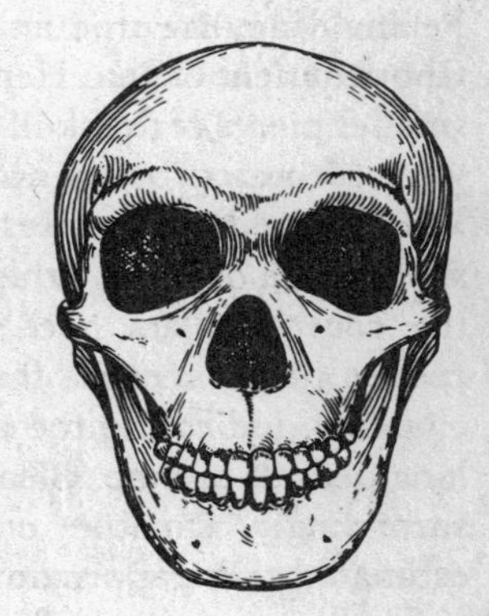

Homo sapiens neanderthalensis
200,000–50,000

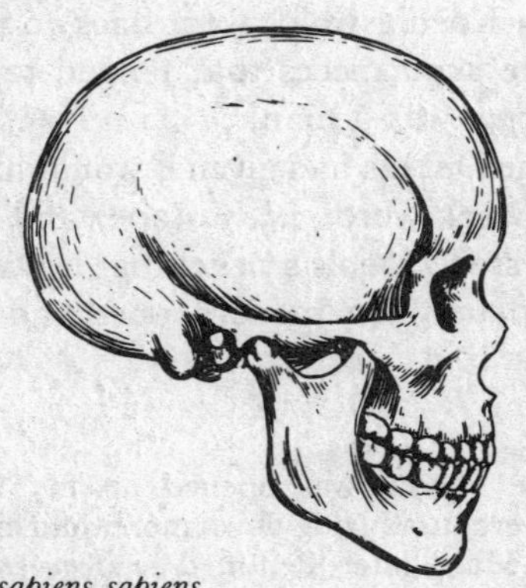
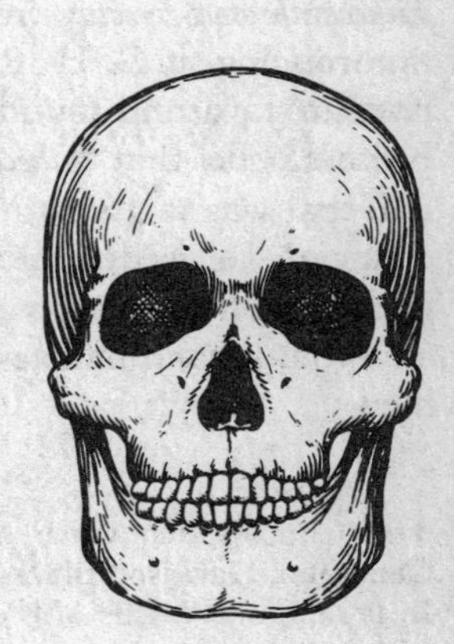

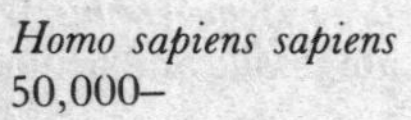

Homo sapiens sapiens
50,000–

he attributed their opposition to ignorance or to personal animosity, rather than to a desire to reach the truth'.

But it was Dubois whose mind was closed. He exhibited the fossils wherever he could, showing them to whatever scientists would look at them, and the controversy continued. In the end, embittered and stubbornly opposed to any new evidence that seemed to support Keith's view, he buried *Pithecanthropus erectus* under the floorboards of his dining room and for thirty years refused to show it to scientists or to speak to them about it.

Meanwhile, new evidence did accumulate. Heidelberg Man, represented by a heavy apelike jaw containing manlike teeth, was discovered in a commercial sandpit in Germany. He was followed by Peking Man, the product of a decade of digging in a series of caves in Choukoutien, China. Here the sample was much larger: 5 skulls, 15 smaller pieces of the skull or face, 14 lower jaws and 152 teeth. These were recovered before work there was stopped by the outbreak of World War II. The caves also contained a great many implements, some made of stone, others of the bones and antlers of animals that the cave dwellers had slaughtered. They had also presumably cooked them, because the cave floors contained layer after layer of charcoal, indicating the presence of fires that had been kept lit for long periods of time. Clearly, those cave dwellers were humans, and surprisingly capable ones considering their age, which was estimated to be approximately half a million years.

Finally, and ironically, forty years after Dubois had extracted *Pithecanthropus erectus* from the bank of the Solo River, another anthropologist, G. H. R. von Koenigswald, went back to the same river in Java and found more skull pieces that tended to confirm Keith's belief that *Pithecanthropus* was a man.

If that was so, then the name Dubois had given it would have to be changed. Derived from the Greek word *pithecos* (ape) and *anthropos* (man), it all too clearly expressed Dubois's belief that it was indeed an evolutionary halfway step between modern apes and modern men.

The Choukoutien Cave outside Peking was opened up in 1927 by a Canadian, Davidson Black, who eventually dug down more than eighty feet. In it he found fossils of 'Peking Man', later identified as *Homo erectus*. The white squares marked on the cave wall are to help locate fossils and artifacts in the various levels of debris dug out of the cave.

The entire naming process had, in fact, gotten into a muddle. Heidelberg Man and Peking Man had earned Latin names of their own, as had one or two others from the Mediterranean and Africa, each scientist having decided that his own fossil was something special, deserving at least a distinct species name, if not a genus name. Heidelberg Man, for example, was named *Homo heidelbergensis*. His finder recognised that he was a man and thus belonged in the genus *Homo*, but decided to put him in a species of his own. Peking Man surfaced with the name *Sinanthropus pekinensis*: Chinese Man from Peking. The implication here was that the Peking fossils were manlike but not quite manlike enough to be called *Homo*. And yet they certainly were not apes. The solution: a new genus.

That is not as foolish as it may sound. The only way a scientist can start to understand anything is to describe it, to measure it and name it. That is what those people were trying to do. We must remember how ignorant everybody was about hominid evolution between 1900 and 1925. There was only a handful of fossils, not enough to make good sense of. Nobody knew exactly what they were – just some kind of early humans. Nobody knew how they were related. Nobody knew exactly how old they were. Furthermore, they were all different. Not *very* different, but different. That raises an important question: how much difference is needed for it to be a significant difference?

If I were to go out into the street today and collect the skulls of the first ten people I saw, I would come back with a sample showing wide variations in brain size and facial measurements. And if, half a million years from now, somebody were to dig up those same ten skulls, and if they came from different parts of the world, and if some of them were so crushed that one could not really measure them properly, and if some had their teeth missing, and if some others were represented only by a few teeth, and if – on top of that – there could be no certainty that all of them really were half a million years old, what would he do?

He would concentrate on the differences, and measure everything carefully. He would study this one and compare it with that one, and what would impress him would be those differences. It took a long time for anthropologists to get it through their own skulls that populations are extremely variable. Therefore one has to have a big sample – men, women and children – before one can begin to recognise the features that are common to them all. With such a sample, one begins to get not only a sense of the consistent features of the population but also a sense of its variability. One gets what is known as a range of variation. Anything within that range, whether it is the size of the brain, the nature of the teeth, the shape of the pelvis – anything within those acceptable limits is a member of that species. Anything outside them – then we are talking about a different species.

Peking Man made such an approach possible. There were fragments from enough individuals in that Chinese cave to provide the beginnings of a range of variation, whereas a single find like Heidelberg Man or Dubois's Java ape-man could not. Later discoveries, notably some excellent fossils from East Africa, have now clarified the picture to a point at which all the above-mentioned fossils are almost universally recognised as being highly variable members of a single wide-ranging species that was directly ancestral to modern man. *Pithecanthropus*, *Sinanthropus* and *Homo heidelbergensis* are no more. All of them have been given the name *Homo erectus* (erect-walking man). They were *Homo*, albeit somewhat different from ourselves. Their brains were smaller, their skulls were thicker, their eyebrow ridges more prominent, their jaws heavier. In fact, all their bones were heavier. They were exceptionally powerful people, both the men and the women, with muscles to match their thick frames. Although a *Homo erectus* male would have been too small to excel as a professional football player, a properly motivated one probably would have been devastating at lacrosse or hockey, two of the most physical sports now played by medium-sized men. It was this toughly built, medium-sized, medium-brained direct ancestor whose evolutionary escalation from *erectus* to *sapiens* is now believed to have taken place between four hundred thousand and one hundred thousand years ago.

The foregoing conclusions have taken only a few pages to describe, but to arrive at them took forty or fifty years of study. By the end of that time, an accumulation of fossils and cultural evidence had made abundantly clear that modern man had an ancestor who was already widely spread over the world at least half a million years ago, perhaps much longer.

Learning something like that does not come overnight. It takes time to sink in. First you have to have the raw materials, the fossils. Then you have to sort them out. Unfortunately, they do not come to hand in any neat way. You have to take what is given to you. Each time you get something new, you do some re-sorting. You do some hoping and some guessing. You do the best you can.

Then you must make the kind of mental adjustment that is necessary before you can believe what you see. That is sometimes the hardest part of all. You can get dug into a particular mind-set and not want to change. A scientist may have spent his life arguing that a certain combination of tooth and jaw characteristics or a certain skull size can mean only one thing. When some new evidence comes along to challenge his idea, it is hard for him to accept it. I think that one of the most difficult things for people to accept about the Java ape-man was how old he was. He was at least five hundred thousand years old. That makes him five times the age of the Neanderthalers everyone was having such a hard time digesting. Now consider Lucy. She is six times as old as Dubois's ape-man. *Six times!*

People ask what filled that enormous gap.

Several things did. An earlier type of man, for one. And two or three kinds of australopithecines – a new name, and an extremely important one.

Australopithecines were early hominids that were *not* men. I have already said that one could be a hominid and not be a man. A couple of million years ago, there were types walking about in Africa that were so primitive and had such queer teeth and such small brains that they could not qualify as humans. The big question was: were they ancestors or cousins? Everyone was arguing about that. To make matters worse, there were two or three kinds of them, large and small versions of an erect-walking near-human that was given the genus name *Australopithecus*.

Having found in Lucy something that was at least as old as the australopithecines and in some respects even more primitive, I had to ask myself: is Lucy an australopithecine? For a long time I did not know. When her bones had been cleaned off, sorted out and arranged on a table, my first thought was that she might be. But I could not be sure. She was so small. Her teeth were not right. She was like a two-legged ape, except that she was no ape. She was a tremendously perplexing, terribly intriguing little hominid. But whether she was an australopithecine or not would have to await a prolonged study of her bones. At that time the whole australopithecine question was in a mess.

There again, I have to note how lucky I have been. The time to get involved in a science is when it is in a mess, the way the Plio-Pleistocene was back in the early 1970s. That is when young people, coming in, have the best chance of doing something interesting – of helping to unscramble the mess.

2 South Africa: The First Man-Apes

Man's predecessors differed from living apes in being confirmed killers; carnivorous creatures, that seized living quarries by violence, battered them to death, tore apart their broken bodies, dismembered them limb from limb, slaking their ravenous thirst with the hot blood of victims and greedily devouring livid writhing flesh.

RAYMOND DART

We are Cain's children . . . Man is a predator whose natural instinct is to kill with a weapon. The sudden addition of the enlarged brain to the equipment of an armed already-successful predatory animal created . . . the human being.

ROBERT ARDREY

One cannot help but feel that Dart tends to minimise the social nature which even early man must have had to survive and to care for young who needed even more time for the adaptation to a group life. He paints a picture too starkly overshadowed by struggle to be quite believable.

LOREN EISELEY

The Australopithecine Mess actually started in 1924. And it began in a totally unexpected place. As we have seen, most of the energies of those seeking the roots of man had been directed at Europe. They had succeeded in identifying Cro-Magnon and Neanderthal Man, and had pushed humanity's roots back to a point a hundred thousand years ago. Later the search extended to Java and China, and the new horizon became half a million years. Nobody had looked for hominid evidence in Africa, even though South Africa was known to contain large numbers of extremely old mammal and reptile fossils. In fact, an eccentric Scot named Robert Broom had gained an international reputation from his studies of those fossils; he was able to show how the first mammals had evolved from earlier reptilian forms. But Broom had never found any hominid fossils, and nobody else had ever gone looking for them. It had not occurred to anyone to do so, on the reasonable assumption that none would be found. Hominids are descended from apes, apes live in tropical forests, and for millions of years there have been no tropical forests in South Africa.

Indeed, this assumption was so strong that when a young South African woman who was interested in fossils saw what she took to be the skull of an extinct baboon on the mantelpiece of a friend, she commented on that.

'Oh, no,' said the friend, 'that could not be a baboon. There are no monkey or ape fossils in South Africa.' He went on to explain that the skull had come from a lime works that he owned at Taung, then a part of the Bechuanaland Protectorate. When chunks of limestone were blasted out, fossils were occasionally found in the rock. This was one of them; he didn't know what it was.

Her curiosity piqued, the young woman passed the story along to her anatomy professor, Dr Raymond Dart, then teaching at the University of Witwatersrand in Johannesburg. Dart told her that her friend was quite right about apes – they did not occur in South Africa and never had – but that he was wrong about baboons. They were large monkeys adapted to ground-dwelling in dry habitats. They were common then in South Africa and had been for hundreds of thousands of years.

Dart, as it happened, was extremely interested in fossils himself. He asked a favor of the quarry owner: if any more fossiliferous rock turned up, might some be shipped to him? In due course, two large boxes of broken limestone arrived at his house. He found nothing of interest in the first, but when he opened the second his eye was struck by a roundish piece among the jagged hunks. He recognised this as an endocranial cast – limestone that had slowly solidified inside a skull, filling it completely and producing an exact replica of the size and shape of a long-vanished brain. In his own words: 'The convolutions and furrows of the brain and the blood vessels of the skull were plainly visible.' Visible to a trained eye, that is. Dart had made an intensive study of endocranial casts during his medical studies in London and was particularly well equipped to recognise one when he saw it, even though the skull itself was missing. It had been blasted loose during the quarrying, and was either lost or lying deeper in the box.

Dart's first thought was that this had to be the endocranical cast of a baboon. But in looking at it more closely, he decided that it was too large to fit inside the skull of any baboon he had ever seen. Furthermore, its shape was unusual; the forebrain was bigger in proportion than would have been the case in a baboon.

With his knowledge that apes – chimpanzees and gorillas – were more intelligent than baboons and had larger brains, it crossed his mind that, against all odds, some kind of previously unknown, extinct, open-country ape might have lived in South Africa in the distant past. He began rummaging around in the box of rocks for a piece that fitted the brain cast. If he could find that, he would have the skull itself. Unfortunately, this was the day he had agreed to be the best man at a friend's wedding. Indeed, the ceremony was to take place at his own house. He was still poking around in the box when the wedding guests began to arrive and the groom started banging on the door for him to finish dressing. He tore himself loose, dusted off his trousers, put on his coat and tie, went through the ceremony – and the instant it was over, rushed back to the box of rocks. Within a minute he had found the piece into which the endocranial cast fitted. He stood staring at that second piece of rock, realising that he was looking from the broken-open back into the inside of a small head. Turning it around to look at the face, he found it covered by a thick encrustation of limestone mixed with sand and gravel, a cementlike substance known as breccia. Because of this covering of breccia it was impossible to make out the external features of the skull at all, although he knew a face was there; it was dimly visible – inside out, so to speak – from the rear.

Dart was not a paleontologist. He had little experience in the techniques of fossil preparation and no access to books that could tell him how to proceed. But proceed he did, very intelligently. Not knowing how fragile the skull might be, and fearful that hard chisel blows might smash it, he embedded the skull in a box of sand to steady it and cushion the effect of hammering. Then he proceeded very gingerly to chip away the breccia. First he used a small chisel, later one of his wife's knitting needles which he had filed to a sharp triangular point. After seventy-three days of chipping and picking,

The reason Raymond Dart's identification of the Taung Baby as a hominid was unacceptable to many British scientists was that with its low cranium, prognathous face and no chin, it does look superficially more like a young chimpanzee than a human child. However, its back teeth are large in proportion to its front teeth (a human characteristic); it lacks the pointed canines and diastema (tooth gap) that are marks of apes.

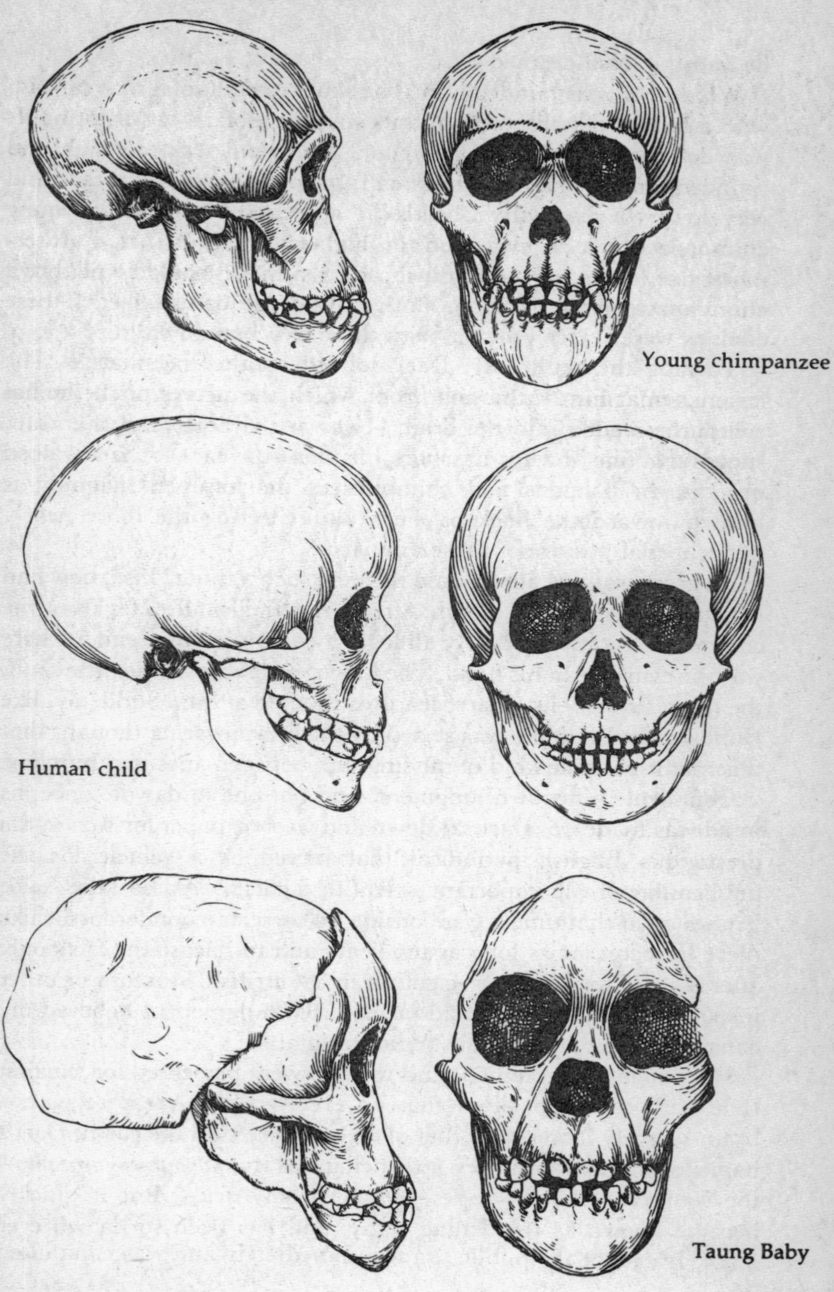
Young chimpanzee
Human child
Taung Baby

he had the fossil cleaned off.

What he saw astounded him. The skull was that of a six-year-old, with a full set of baby teeth and its six-year molars just erupting. It was definitely not a young baboon. The skull was too high and rounded, and the face was too small; baboons have long snouts and very low crowns to their heads. It appeared more like a young chimpanzee, but seemed even too high-crowned for that. Furthermore, one look at its teeth convinced Dart that it could be neither a chimpanzee nor a baboon. Both have large canine teeth; these canines were scarcely larger than those of a human child.

Turning the skull over, Dart noticed another peculiarity. The foramen magnum – the hole from which the nerves of the spinal column ascended into the brain – was at the bottom of the skull, suggesting that during its short life this six-year-old had walked upright. In baboons and chimpanzees the foramen magnum is located toward the back of the skull, reflecting the more nearly quadrupedal posture of those animals.

An erect-walking ape? Could there be such a thing? If so, how had it gotten way down in South Africa, two thousand miles from the common habitat of apes? By all logic it could not have, and yet here was the evidence in his hand. The longer he looked at the little skull, the more strongly its bizarre features shouted at him. Suddenly, like Dubois before him, he was seized by the overpowering thought that this might be some kind of missing link between apes and humans.

Confident in his own judgment, and not one to dawdle once his mind was made up, Dart sat down and wrote a paper for *Nature*, the prestigious English periodical that served as a vehicle for the publication of all important scientific articles. As he later said: 'Although at that time it was considered correct to ponder fossils like these in secret for as long as ten years and to release the facts only after a conclave of senior scientists in the British Museum or other international institutions had pronounced judgment, I believed my conclusions about the brain to be irrefutable.'

His paper was accepted, and readers were presented for the first time with the proposition that an erect-walking creature with a brain scarcely larger than that of an ape once trod the earth. Dart's name for this extraordinary little being was *Australopithecus africanus* – the southern *(Australo)* ape *(pithecus)* from Africa. But it quickly became known as the Taung Baby, and has been so known ever since. It caught the public's fancy immediately and was catapulted into the headlines as 'the missing link', found at last. It even became the subject of music-hall jokes: 'Who was that girl I saw you with last night; is she from Taung?'

Dart's publication *had* been rather hasty. He had checked his paper with nobody. The reaction of the scientific community was one of predictable caution. The center of the paleoanthropological world was now England, and at its center stood Sir Arthur Keith, a sixty-year-old tower of influence, Keeper of the Hunterian Collection at the Royal College of Surgeons and past president of the Anthropological Institute. Keith's first pronouncement was 'Professor Dart is not likely to be led astray. If he has thoroughly examined the skull we are prepared to accept his decision'.

Keith soon had second thoughts: 'Of his [Dart's] knowledge, his power of intellect and of imagination there could be no question; what rather frightened me was his flightiness, his scorn of accepted opinion, the unorthodoxy of his outlook.'

Still later, Keith dismissed the Taung skull out of hand: 'At most it represents a genus in the chimpanzee or gorilla group . . . The Taung ape is much too late in the scale of time to have any place in man's ancestry.' Just what Keith meant by 'late' must have puzzled Dart. His estimate of the age of the Taung skull was a whopping million years. That seemed pretty old to him.

Others in the Big Four of British anthropology – Sir Grafton Elliot Smith, Sir Arthur Smith Woodward and Dr W. L. H. Duckworth – were less positive in their rejection of Dart's claim, but still disappointingly evasive. Elliot Smith's failure to come out more strongly in his behalf was particularly painful to Dart because it was under him that Dart had studied endocranial casts. Elliot Smith was well aware that Dart knew a prodigious amount about the size and shape of the heads of primates. Still, Dart's astonishing assertion was too much for him to swallow.

The only outright support Dart got from any scientist was from a colleague, the redoubtable Robert Broom, legendary classifier of South African mammal and reptile fossils. Broom wrote a letter to Dart congratulating him on his *Nature* article. Then, two weeks later, Broom '. . . burst into my laboratory unannounced. Ignoring me and my staff, he strode over to the bench where the skull reposed and dropped on his knees "in adoration of our ancestor".' A controversial character and a born fighter himself, Broom was impatient with Dart for his failure to defend his fossil find more vigorously. He himself did. He wrote to *Nature* strongly supporting Dart's view, and was able to flush out some others who felt as he did.

One in particular was W. J. Sollas, professor of geology and paleontology at Oxford University. His name should be noted carefully, for it will crop up again as the story of fossil man unfolds. Sollas was delighted to hear that Broom took issue with Woodward and Keith. The reason: Sollas hated them both. He was a thin-skinned and contentious men who did not take kindly to Woodward's dismissal of some of his work. He had spent five years perfecting a way of making copies of fossils embedded in rock. In a demonstration before a scientific audience he had been publicly humiliated by Woodward, who waved off the procedure as a 'toy' and consigned it to oblivion. He was equally enraged by Keith's often-reiterated statements that certain geologists should be more helpful to paleoanthropologists in dating their specimens. Sollas was well aware that 'certain geologists' could refer only to him. Therefore, when he detected Keith's hand in what appeared to him to be an attempt to prevent the publication of Broom's letter in *Nature*, he immediately fired off a scathing note to Broom: 'Keith has great influence [in suppressing your letter. He is] indeed the most arrant humbug and artful climber in the anthropological world. He makes the rashest statements in the face of evidence. Never quotes an author but to misrepresent him, generalises on single observations, and indeed there is scarcely a single crime in which he is not adept. Journalism, my dear boy, journalism pure and simple . . . He has gone up like a rocket and will come down like the stick.'

But Keith did not come down. He stood fixed in the anthropological firmament. Sollas had nothing more to say, and Dart was left with only the support of Broom, a somewhat questionable asset. Broom was an odd man, one of the most elusive, exasperating, arrogant and unconventional figures in a science that has produced more than its share of them – and one of the most brilliant. Born into a Scottish family that was on intimate terms with poverty and tuberculosis, he suffered from severe chest troubles as a small boy and was sent to spend much of his time with his grandmother, who took in lodgers at the seaside town of Millport. One lodger, an old man in his eighties, was interested in natural history. When he died he left his microscope to little Robert, who thereafter developed a feverish interest in all things having to do with natural history, and particularly the ancient past, which he had discovered in the form of fossil shells in a nearby limestone quarry. With only four years of grammar-school education, he found work as an assistant in a laboratory at Glasgow University. He eventually enrolled there as a medical student, graduating in 1889 with honors in midwifery.

Broom was an excellent doctor – observant, sensible and particularly adept at all things having to do with the birth of babies and the care of mothers. With a medical skill to support him wherever he went, he travelled the world – first the United States, then the wilder parts of Australia. He advertised as a *locum tenens* (temporary replacement) when he needed work, served as a doctor in remote cattle districts and in gold-mining towns, preferring an austere life next to nature to a more rewarding one in a large city. His wife, a servant girl whom he had married on a quick trip to Scotland, trailed along after him. She remains a dim figure, sometimes suffering from illness, never complaining, saying nothing about his countless infidelities – for Broom was a notorius womaniser.

He was a true loner, a wanderer. Because of his impoverished boyhood he had developed a deep and lasting contempt for the 'gentry', who represented the enormous inequality of wealth and the social callousness that had made his parents' life so grim in Scotland. He was scornful of stupid men in positions of authority, and he had met many of them in medical schools and museums. His dealings with them were sly and devious. Later, when he settled in South Africa and continued to move about restlessly while he homed in on his true life's work of organising the reptile and mammal fossils of the Karroo desert area, his abrasiveness and deviousness sometimes caused him difficulty in getting access to collections he wanted to study. When he succeeded, it was occasionally rumored that the collections would have small gaps here and there after he left. Consequently, during one period he was forbidden access to the collection of the South African Museum.

Broom's knowledge of fossils was profound, and a combination of logic, infallible memory, passionate interest and instinct; he possessed an innate sense of what fitted – of what went with what, what had descended from what, what had to be filled in to make a workable animal out of a fragment. He supported himself on the side by selling fossils to the British Museum. When his relations with became tangled he turned to the American Museum of Natural History in New York, and was accused of skullduggery for selling specimens that were not rightfully his to sell.

It was this thorny and unpredictable man, himself suspect in the remote halls of British paleontology, who was Dart's only vocal supporter through the 1920s. Dart himself did little more than continue to fiddle with the Taung skull. After four years of careful scratching and prying, he succeeded in separating the lower jaw from the upper. For the first time he was able to get a look at the chewing surfaces of the teeth. More than ever he was convinced of the rightness of his original diagnosis – that his fossil was a hominid. But the distant and haughty Keith gave not an inch. He continued to insist that Dart's claim was totally without merit.

To do Keith justice, he was caught on the horns of a terrible dilemma that had nothing to do with Dart or the Taung Baby at all. Not only he but all of British paleoanthropology was engaged in a desperate struggle to interpret fossils that they had studied exhaustively but could make no sense of whatsoever. Several of these were casts and fragments of the various Heidelberg and Java types that had been accumulating slowly since the turn of the century. All of them pointed in one direction: ancestral man had had a much smaller brain than modern man, and it apparently got smaller – and more apelike – the deeper one travelled in time. But his teeth and jaws were not like those of apes, even as much as half a million years ago. They had a different configuration entirely.

The fossil human jaw resembled the modern one in having its teeth arranged in the mouth in a bow-shaped curve, with the widest part of the curve at the very back where the third molars were. In apes, whose jaws were longer, the teeth on one side of the mouth tended to be parallel with those on the other, forming two sides of a rectangle, with the front teeth forming the third side. Furthermore, among adult apes and particularly among males, the upper canines were so large that they extended downward like fangs into the row of lower teeth, and there had to be gaps in the lower row to accommodate them. In

The problem presented by Piltdown Man becomes clear when it is compared with *Homo erectus* and modern *Homo sapiens*. The Piltdown cranium is like that of modern man in shape and size, and not like *erectus* at all. But its jaw is more primitive than either, having the boxlike configuration and huge pointed canine of a large ape. Piltdown appears to be made up of a human skull and an ape jaw arbitrarily joined together – which is what it ultimately turned out to be.

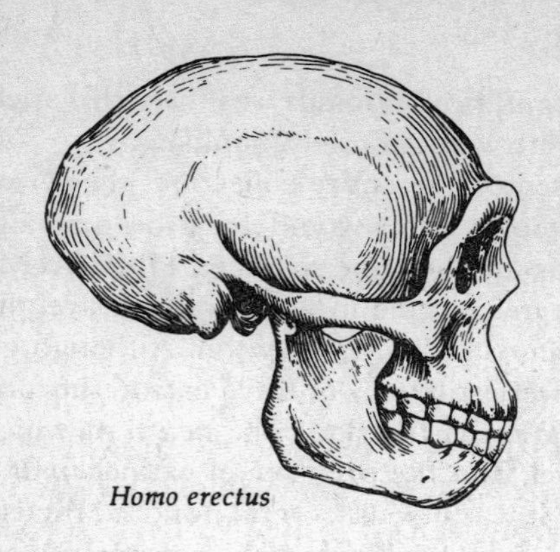
Homo erectus

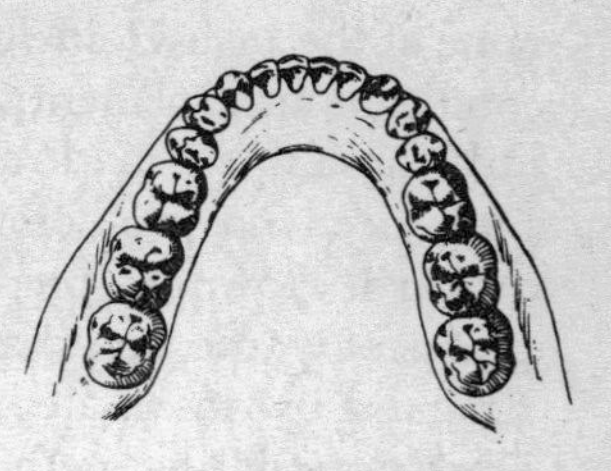
Homo erectus jaw

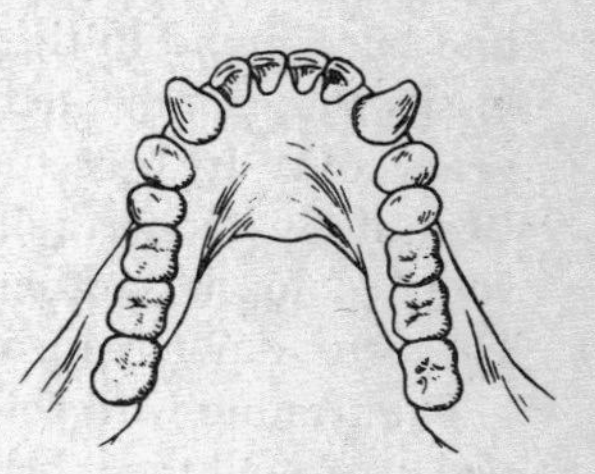
Piltdown jaw

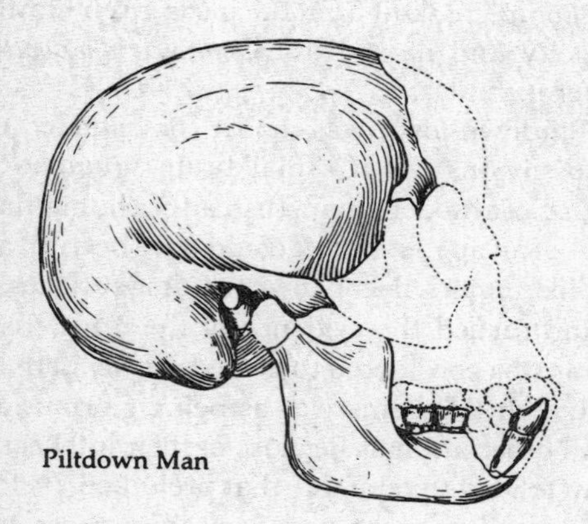
Piltdown Man

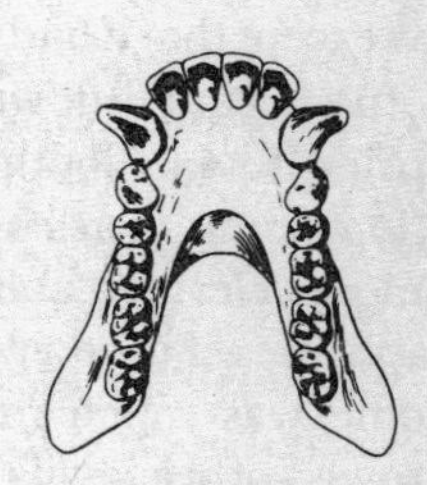
Ape jaw

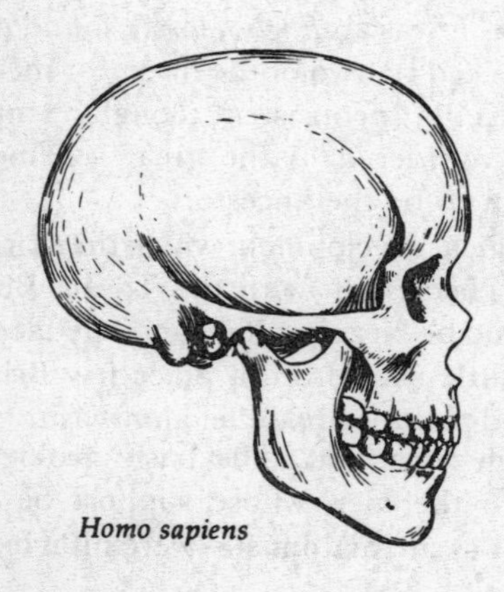
Homo sapiens

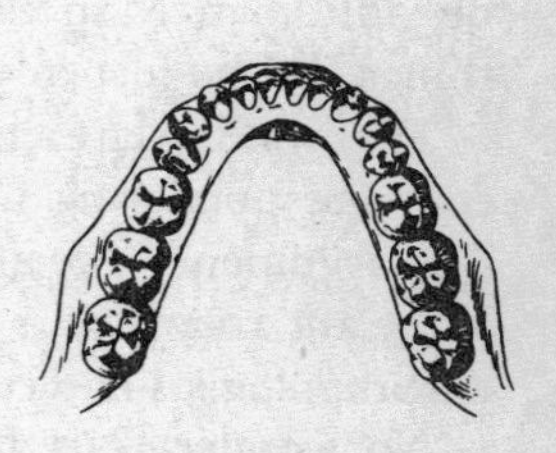
Homo sapiens jaw

none of the Java-Heidelberg and similar fossils was there this apelike condition. In all of them the teeth were definitely manlike.

If these fossils had represented the entire collection, there would have been no problem. Interpretation would have been straightforward. However, there was another: the so-called Piltdown Man, pieces of whose skull had been collected in 1912 from a gravel pit in England by an amateur scientist, Charles Dawson. Although good laboratory dating methods were still forty or fifty years in the future, nevertheless the dark coloration of the skull, the depth at which it had been found in the gravel, and the presence of extinct mammal fossils with it indicated that it was at least several hundred thousand years old. It was turned over to Keith, who reconstructed a beautifully and virtually complete skull from the parts given him. He vouched for its authenticity and named it *Eoanthropus dawsoni*, or Dawson's dawn man, after its finder.

The trouble with the Piltdown fossil was that it contradicted the evidence of all the other fossils. Instead of a small brain, it had a very large one, comparable to that of a modern man. Instead of a human jaw, it had one shaped like that of an ape; in fact, it could have been an ape jaw except that it had manlike molars, flat on their grinding surfaces.

In short, Piltdown Man turned the evidence of the other fossils upside down. Not that this was considered to be necessarily bad. For a long time men have thought of themselves as being glorious and unique, elevated above all other animals because of their intellectual capacities. Piltdown Man tended to reinforce that prejudice. Its lofty dome was much more satisfying to human egos than those low-browed other fossils. Better by far to have a large brain and an ape's face than vice versa. Furthermore, it was an *English* fossil, found only a few miles from London and backed by two of the foremost men in British science, Woodward and Keith. Inconsistent though he might be, Piltdown Man had to be considered in the great scheme of things, for he just might turn out to be the ancestor.

That was the latest anthropological development which the Britishers were pondering when Dart checked in with the Taung Baby. The argument – the ancestor could be large-brained and ape-faced – left scant room for a shadowy little skull from a place few British anthropologists had ever bothered to think about, let alone visit.

No wonder Dart had a rough time of it. The basic geology of South Africa was unfamiliar to the men whose support he was seeking. His credentials – except as an anatomist – were nil. He was backed by a wayward and suspect man. His claim for the age of the Taung skull was insane. A million years? How did he know that? He didn't. It was a guess based on a hunch derived from speculation about the age of some extinct animal fossils from the same limestone cave. Finally – and this was the hardest to swallow – he claimed erect manlike posture for his ape-brained little creature, all on the basis of the position of a hole in the bottom of the skull. If there had been a pelvis or a leg bone to support this last and wildest claim, perhaps he would have been listened to with more respect, but all Dart had was that one skull. It was not enough.

Nevertheless, in 1931 Dart was invited to London to an anthropological congress and given an opportunity to describe and defend his fossil. The star attraction at the congress was a presentation of the first finds of Peking Man beginning to come out of the Choukoutien cave in China. They were a sensation, and they were presented in a masterly way with many photographs and with sectional maps of the cave dig. Dart came on after that dazzlement. All he had to say was what he had said six years before. He said it timidly, and by his own admission he failed miserably. To cheer him up some colleagues took him off to dinner afterward, and he sent his wife, Dora, back to their hotel in a cab. She took the Taung skull with her and left it in the back seat.

It was recovered the next day after having been driven about London most of the night with a series of fares, any one of whom could have carried it off. Finally the taxi driver noticed the little package and turned it over to the police. They opened it, found a skull and began wondering if they had a murder on their hands. But by that time Dart had notified the police himself and was able to reclaim the Taung Baby. He returned to South Africa, the credentials of his fossil no better than they had been when he left.

The popular story is that Dart was driven out of paleoanthropology by frustration and disappointment. The truth is that his career in teaching and neurological research had always come first with him, and that it had been only by chance that the Taung Baby had been flung into his lap. He did abandon fossil hunting for nearly twenty years, but only to pursue what to him were more important goals, one of them being Marjorie Frew, head librarian of the Witwatersrand Medical Library. In 1936, five years after his disastrous trip to London, he divorced Dora and married Marjorie. Those had been five disorderly years for British anthropology. More and more evidence of Peking Man was pouring out of China. As a supporter of Piltdown Man, Sir Arthur Keith must have been unnerved by the Peking fossils. So must Sir Arthur Smith Woodward, Keeper of Geology at the British Museum. Woodward was even more deeply involved with Piltdown than Keith was. He had actually been on hand when one of the skull fragments was recovered by the amateur digger Charles Dawson from a muddy pit in Sussex. Also present was a bright young French priest/anthropologist, Pierre Teilhard de Chardin, who came up with part of an elephant tooth that same afternoon in the same trench. According to Teilhard, Woodward 'jumped on the piece with the eagerness of a boy, and I could see all the fire which his coldness conceals.'

Teilhard was referring to Woodward's usual demeanour, one of intense preoccupation and utterly without humour. He could afford nothing so frivolous. His origins were plain. He was not a true poverty case like Broom, but he was still at a distant remove from the upper class. Paleontology provided for him, as it had done for Huxley and would do for other obscure but brainy and ambitious young men, a rare door to distinction in a society structured to keep them in their place. Here, in a new science, they could swim upward in a froth of fresh ideas blown by Darwin and kept alive by public interest and controversy. It is remarkable about Woodward, Keith and Elliot Smith that all three of them were knighted.

That kind of recognition did not come easily. The infighting among the bright young men was sometimes bitter. Careers had to be forged in a climate where only supreme ability and tenacity could prevail. Like Huxley, Woodward won prizes, wrote blizzards of scientific papers, and by 1901, at the age of thirty-seven, had ascended to the post of keeper at the British Museum. His devotion to his work was obsessive, and made him so oblivious to the world around him that he scarcely noticed where he was going. One day, head down, he charged full tilt into an exhibit case, slipped and broke his leg. He insisted on setting it himself, and for the rest of his life he walked with a limp. Not long afterward he had a similar accident, this time breaking his arm.

For all that, Woodward was a superb paleontologist. At the height of his career he was the world's greatest authority on fossil fishes. This did not, however, make him an authority on hominid skulls. There Keith had the edge on him.

When it came time to reconstruct the Piltdown skull, Keith and

Woodward differed. Using the pieces that Dawson had found and which made up an almost complete left half, Keith simply followed the same dimensions in rounding out the right half of his reconstruction. Woodward, arguing that the left brain in advanced man is larger than the right, made his accordingly, and the resultant brain capacity was considerably less than Keith's. This put a direct challenge to Keith's competence at fitting skull fragments together. He met it with a grandstand flourish. A modern skull whose brain capacity had previously been measured was smashed to bits; Keith reassembled it to within three or four cubic centimeters of its original capacity. He won that skirmish, but it was a minor one in the general rejoicing that British anthropology at last had a skull of its own that could more than compete with all those from France and Germany. Older, and yet more manlike – the combination was irresistible. Both Keith and Woodward put their professional reputations on the line in endorsing Piltdown Man. Woodward, in fact, devoted the last thirty years of his life to a major work on the fossil. Its title: *The First Englishman*.

Despite all that, there had been rumblings from the beginning about the credentials of Piltdown Man. Paleontologists from the United States, with no national axe to grind, were able to view him more dispassionately. They claimed that the jaw and skull did not – could not – fit together. Now, with the evidence beginning to build from Peking, that possibility grew: the early brain might well have been small, not large, and the dentition manlike. What went through the minds of Keith and Woodward through the 1920s and up into the 1940s can only be guessed at. If either doubted anything about Piltdown Man, they kept their suspicions to themselves. The fossil was put in safekeeping. Those who wished to look at it could do so, but they were not allowed to touch it or subject it to any tests.

Meanwhile the Taung Baby remained in scientific limbo. One matter that kept it there was that it was the skull of a child. Human infants and ape infants tend to look more alike than the corresponding adults do. As an ape grows, its face and jaw grow much faster than its brain – far faster than the jaw development of a human. It was this that enabled many skeptics to continue to doubt the manlike hints in the Taung skull.

The point was hammered home with painful force by Sir Grafton Elliot Smith, who stated, 'It is unfortunate that Dart has had no access to infant chimpanzees, gorillas or orang-utans of an age corresponding to that of the Taung skull for had such material been available he would have realised that the posture of the head, the shape of the jaws, and many other details of the nose, face and cranium on which he relies for proof of his contention that *Australopithecus* was nearly akin to man, were essentially identical with the conditions met in the infant gorilla and chimpanzee'.

Dart had studied the dentition of the Taung Baby with extraordinary care and knew it was not like that of a chimpanzee or gorilla. That the three titans should stubbornly fail to see this must have been hard to bear. That Smith, his old mentor in the study of endocranial casts, should hammer down the lid of the coffin was surely excruciating. He had only Broom to commiserate with him.

During those same five years, Broom's own career had suffered. He was getting old. His great work on reptile and mammal fossils was done. The museums and universities were hostile to him. He drifted from one small medical practice to another, and began slowly to sink into poverty. In 1933 he could not afford the railroad fare from Maquassi, where he was then living, to Johannesburg to make a speech before the South African Association for the Advancement of Science – even though he was president of the Association. Funds had to be provided for him.

When Raymond Dart learned of this, he was outraged that the most distinguished scientist in South Africa – however controversial his past might have been – should be permitted to fall so low. He wrote letters to top members of the government, among them General Jan Smuts, who had just been elected Prime Minister. Smuts, long an admirer of Broom because he felt Broom's accomplishments brought credit to South Africa, applied some pressure. A few months later Broom received a grudging letter from the head of the Transvaal Museum at Pretoria, offering him a post as Assistant in Paleontology. Because of his age, the post was not to be considered a permanent one, and during his occupancy of it Broom was forbidden to collect fossils for himself or anybody else. His old enemies in the museums had long memories.

The offer was humiliatingly small, but Broom accepted it immediately. With a sinecure at last, even though it was a tiny one, he was free to retire from the practice of medicine and devote himself full-time to fossils. He struck up an acquantance with a remarkable Transvaal farmer, Sidney Rubidge, who had been collecting fossils for years and had formed a small private museum on his farm. With

Broom's help its collection soon outstripped that of the Transvaal Museum, to which Broom was supposed to be devoting all his time. He further confounded his superiors in 1936 when he got wind of the existence of fossils in a nearby limestone cave at Sterkfontein. A man named Barlow, foreman at the lime works, was selling them to tourists as souvenirs.

Broom hastily got in touch with Barlow, confirmed that fossils were indeed coming out of the quarry in profusion – and learned that some of them were being burned in the lime kilns and destroyed as fast as the rock was fed into the kilns. He explained to Barlow what a dreadful scientific opportunity was being wasted. Barlow apologised and promised to save anything else he came across. A couple of days later, he handed Broom an endocranial cast. Broom went through the recently blasted quarry rubble with great care and was able to recover fragments of an almost complete cranium. He cleaned them off, assembled them and realised instantly that they were australopithecine.

A second fossil like the Taung Baby had at last been found!

This news, when it reached Dart, both irritated and delighted him. Some of his own students had been interested in the Sterkfontein cave. Without saying anything to him, they had gone to Broom. Broom had calmly walked in and hornswoggled the authorities once again by lifting out the prize from under their noses.

On the other hand, the skull *was* australopithecine. In a stroke it validated the long and lonely claim that Dart had been arguing so fruitlessly for more than ten years. Moreover, he realised that if Broom had not acted promptly, the fossil would very probably have followed unnumbered others into oblivion in the kiln. Finally, Broom's fossil was that of an adult. No longer could there be that stain on the legitimacy of the Taung Baby – that its youth made it indistinguishable from a young ape.

On balance, Dart was overjoyed. A distinguished paleontologist had vindicated him. But he would have to wait another decade for the sweetest accolade of all. In 1947, Sir Arthur Keith, then approaching eighty and near the end of a fifty-year career as one of the famous anatomists in the world, spoke out: 'When Professor Dart . . . claimed a human kinship for the juvenile australopithecine, I was one who took the view that when the adult form was discovered it would prove to be nearer akin to the living African anthropoids, the gorilla and chimpanzee . . . I am now convinced that Professor Dart was right and I was wrong.'

That was a notable concession for one who had been so icily condescending for so long. Still, it did not go the whole way. Keith could not quite bring himself to call the australopithecines hominids. To him they were 'ground-dwelling anthropoids, human, in posture, gait and dentition, but still anthropoid in facial physiognomy and size of brain'. That, of course, is a good description of what they were, but the last step, use of that awful word 'hominid', stuck in his throat. There was – and still is – a peculiar resistance to considering such primitive creatures as related to humans. Keith's word for

THE FINDERS OF FOUR NOTABLE FOSSILS

J. C. Fuhlrott
Neanderthal Man, 1856
Fuhlrott was a German science teacher who recognised fossils dug from a quarry near his home as being the bones of a man, but not those of a modern man. This was the first nonmodern human fossil ever found; it underwent decades of controversy.

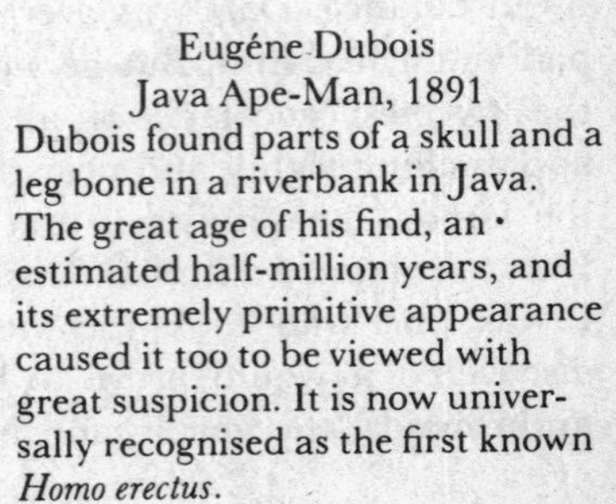

Eugéne Dubois
Java Ape-Man, 1891
Dubois found parts of a skull and a leg bone in a riverbank in Java. The great age of his find, an estimated half-million years, and its extremely primitive appearance caused it too to be viewed with great suspicion. It is now universally recognised as the first known *Homo erectus*.

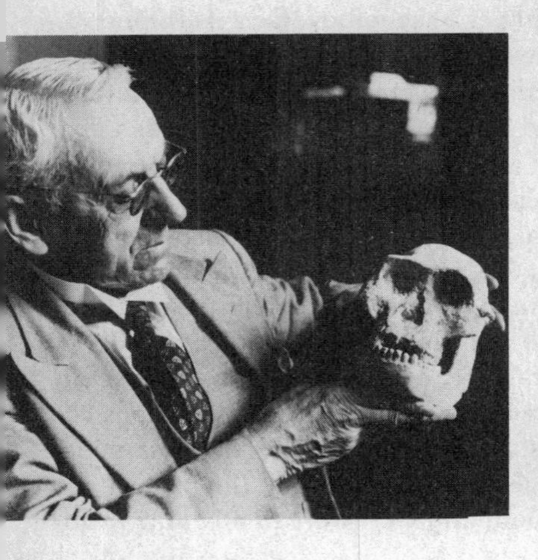

Robert Broom
Paranthropus, 1938
A second type of australopithecine was found by Broom at Kromdraai in South Africa. To him it seemed different enough to justify the creation of a new genus for it. It is now widely recognized as a robust australopithecine, descended from *africanus*.

Raymond Dart
Taung Baby, 1942
Estimated by its finder to be as much as a million years old, this fossil was so extremely primitive that many scientists were convinced for years that it was an ape. It is, in fact, the first *Australopithecus africanus*, an authentic hominid, many of whose fossils have been found since.

them: 'Dartians'.

What took that Olympian head so long to give its half-nod? Several things. First, the information coming out of South Africa was not as clear or convincing as it might have been. Broom himself caused some confusion by declaring that his new fossil and the Taung Baby were not exactly alike. The confusion would deepen; a third australopithecine skull would be found – and would be different from the other two.

As usual, Broom was at the center of the action. He was now over seventy. By rights he could have been tending roses quietly in a garden. Instead, wearing the formal black suit and starched wing collar that had been the trademark of the country doctor for many years, he stumped about at Sterkfontein in the hot sun, trudging up gullies and poking into rock formations at a clip that wore out some of his younger colleagues. Hearing of a schoolboy, Gert Terblanche, who reportedly had found some fossil teeth at Kromdraai, a farm near Sterkfontein, Broom immediately went there, only to find that the boy was at school. He talked to Gert's sister and persuaded her to walk him out to the hillside where the fossils had been found. There he picked up a tooth and insisted on proceeding to the school, another mile walk in the hot sun. The principal summoned Gert, in whose pocket were four fossil teeth which Broom bought for a shilling each. Next, Broom wanted to go immediately to the place where the teeth had come from, but classes had another couple of hours to go. He was persuaded instead to give all the students a lecture on fossils and how to recognise them in limestone deposits. This unexpected visit to an obscure country school by the famous scientist, and his subsequent lecture, were enthralling. Further classes were suspended, and Broom and Gert were finally able to go to the fossil site.

Gert pointed out the place where he had hammered the teeth from a skull embedded in a limestone matrix. The skull had been smashed in the process, but Broom was able to remove what was left and recover some fragments. All this material, plus one more tooth that Gert was persuaded to part with for five chocolate bars, gave Broom enough to reconstruct his second australopithecine skull.

To his great surprise, it was markedly different from the one he had found at Sterkfontein. It was somewhat larger, its bone thicker. It had a much thicker jaw, very large, heavily enameled molar teeth, and marks on the bone that indicated the presence of extremely powerful chewing muscles.

Broom had already fancied he could detect slight differences between the Sterkfontein fossil and the Taung Baby. With the sense of proprietorship that seems to be born in the finders of fossils, he decided, after some initial indecision, that the differences were great enough to justify a new name. He chose *Plesianthropus transvaalensis* (near-man from the Transvaal). When he turned to the Kromdraai fossil, there was no need for indecision. The differences here were plain. The Kromdraai ape-man seemed more primitive than either of the others, not so much 'near' as dimly 'foreshadowing' the human condition. He settled on the name *Paranthropus* (toward man) *robustus*. The *robustus* he tacked on to underscore the specimen's

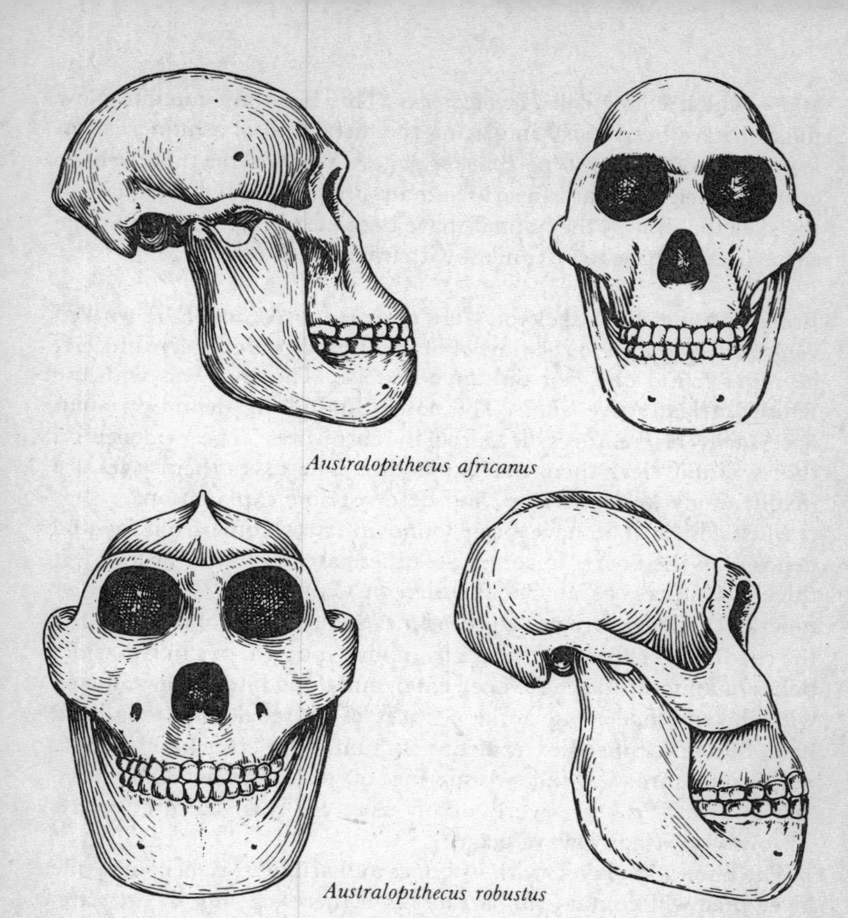

Recognition that there were two distinct types of australopithecine in Africa came with the discovery of more fossils. Shown here are Mrs Ples, a gracile type (*Australopithecus africanus*) found by Robert Broom in South Africa and a robust type (*Australopithecus robustus*). The robust type has a more massive jaw and larger molars than the gracile; it also has a prominent bony ridge on top of its skull.

larger size and generally more massive bone structure.

Like Dart, Broom had no difficulty in recognising all these South African fossils as erect-walking hominids, although he desperately needed some postcranial bones (other parts of the skeleton) to persuade others. He was rewarded in August 1938, and was able to write to Keith in London: 'Two weeks ago I was delighted to get the distal end of a femur [lower end of a thighbone] of the Sterkfontein ape – which I now call *Plesianthropus*. This is almost human. Now today I have been lucky in getting the distal end of a humerus [an arm bone] of the new type *Paranthropus* . . . You will see that we have gotten something that is nearly human. I hope soon to have much of the skeleton, but as the animals have been clearly devoured by large carnivores we have to be content with fragments.'

Broom's hopes for a skeleton were premature. Although he worked vigorously right up to the time of his death at age eighty-five in 1951, he never found one, nor did anybody else. The fault was with the South African cave sites. The fossils found in them were not necessarily of creatures devoured by carnivores. The evidence on that was not clear then, nor is it today. The caves themselves are maddeningly uninformative, and deserve some explanation.

Most fossils that have been found in caves come from layered deposits in the floors. In some cases these are only three or four feet thick. In others, as at Choukoutien in China, they are eighty or ninety. The layers are not all the same. Some consist of rubbish that the cave dwellers have left there, trampling it down, pushing it aside: bones of animals that have been eaten and flung into corners, mixed with flaked stone tools, antlers, traces of charcoal from long-dead fires, bits of fossilised excrement containing the crunched bones of mice and shrews – all whispering of past human habitation. 'Throwaway' is the word man uses for his culture today. Throwaway it has always been.

The habitation layer, with its bones and artifacts, is, of course, the layer that will contain human fossils if there are any in the cave. Many layers contain no bones at all. They represent periods when the cave was uninhabited, times when nothing but blown-in dust settled and solidified on the cave floor. It may have piled many feet deep, indicating a change in climate that may have forced people out of the region entirely for tens of thousands of years. That ebb and flow of humanity may be traced by the presence of a second habitation floor higher up, followed by another exodus. The layer-cake analogy returns: a fossil horizon, a dust horizon, a mud horizon from the rising of a nearby river, another fossil horizon – one on top of another. The archeologist has only to do his homework conscientiously, noting the position of everything as he digs deeper, to reconstruct the history of that cave.

In South Africa the situation is not so neat. There is no straightforward story of layering. The caves were formed underground by the action of subsurface water on certain minerals, leaching them away until large caverns inside the hillside were created. Analogously, if a lump of sugar were encased in a block of porous nonsoluble material like polystyrene, and water allowed to seep in, gradually the sugar would dissolve and leak out, leaving a space in the center.

Some such process occurred in certain dolomite-limestone hillsides in South Africa. Caverns began to form inside the hill. Cavern growth would have been accelerated by rain falling on the hillside and percolating downward through cracks in the limestone. Gradually those cracks would have been enlarged by the trickling water until there were actual openings to the surface. Now there would be a true cave with an entrance. At first the entrance would be too small to admit anything but dust and such small animals as mice and bats. But by the water action it would grow bigger and bigger, finally admitting large prey animals and possibly hominids.

It is reasonable to ask why a normal sectioning of the cave deposits cannot be made. The answer is that the deposits cannot be read clearly. The caves are filled solid – from the bottom to the top – with un unreadable accumulation of dirt, pebbles and other debris, mixed with blocks of dolomite that have fallen from the roofs and walls. This jumble of material has been cemented together by a solution of limestone that had dried and hardened until the whole thing is a solid

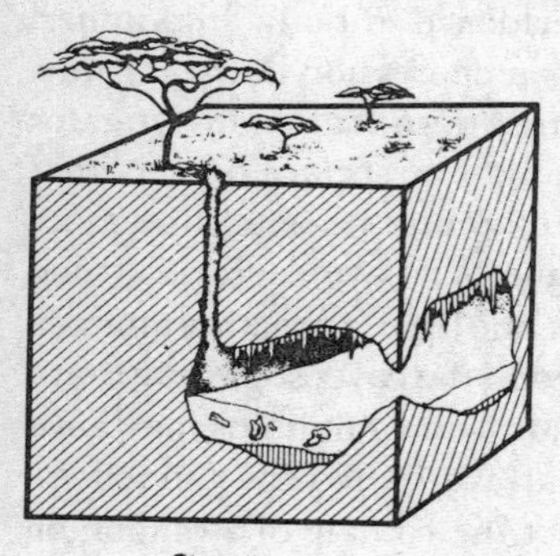

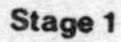

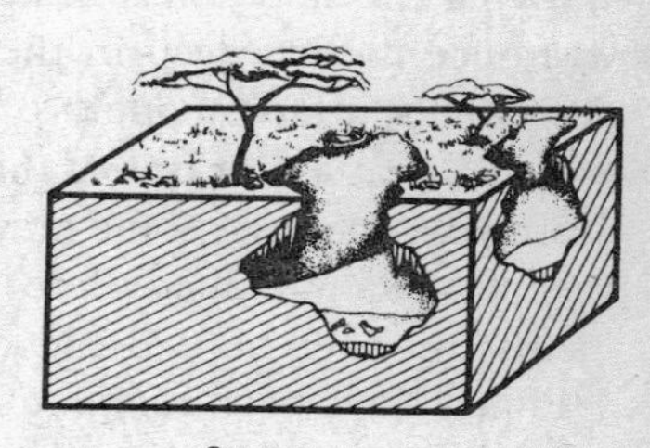

The limestone caves of South Africa were created by the leaching away of soluble rock under the influence of rain and ground water. Openings to the surface eventually appeared. In time these became large enough to admit animals. It is hard to date the fossils in the caves because it is not clear how or when the deposits inside were created.

block of concrete, with fossils scattered through it like raisins in a pudding. Scientists trying to analyze the contents of such a cave cannot tell whether a fossil walked in, fell in, was dragged in by a predator, or was dropped in from above through one of the enlarged cracks.

A great deal of effort and ingenuity has been devoted to cracking the enigma of the South African caves. The associated mammal fossils have been analyzed and compared with those from other places. This was at first extremely frustrating, because many of the mammals uncovered were extinct species not known from anywhere else. There was nothing to compare them with. More recently, the fossil fauna of Eastern Africa have become better known – and promise to provide a key.

The cave breccia – the concrete itself – has been analyzed exhaustively. It proves only that the various caves are not alike and probably have had different histories. In some of them the breccia is pink, in others brown or orange. Does this have to do with past differences in the external climate? Nobody can be sure.

Analysis has revealed that there have been intrusions of a chocolate-colored breccia into the pink in one cave. It has been established that it is much younger. How much younger, nobody is prepared to say.

There is suspicion that the bottom has fallen out of at least one cave into a deeper cave below it, making a single larger one and hopelessly scrambling the geological evidence in both. In another, the ceiling has collapsed, to form an open depression in the hillside. This, in turn, has been eroded and scattered, so that an investigator studying the slope finds himself tramping over surface breccia that was once hidden deep in the hill. Determining its age through its position is utterly hopeless. In still another cave there is strong evidence that a river flowed through it at one time, attracting animals during dry seasons. Some may have died in the cave; some may have been hunted and killed there by hominids; some hominids may have themselves been hunted and killed by carnivores. No one can tell.

Efforts have been made to reconstruct the climate of a million or two million years ago by a microscopic examination of the degree of rounding or weathering of individual grains of sand in the breccia. The results so far: inconclusive.

All that can be said definitely about the caves is that some are older than others. Paleontologists have determined this by comparing the number of extinct animal species and the number of surviving species that occur in fossil form in each. In three caves the breccia contain no surviving species at all. They are Taung, Sterkfontein and Makapansgat. That would make them older than Kromdraai and a fifth cave, Swartkrans, where some surviving species are found.

Analysis of the hominids in the five caves, however, produces some information of stunning impact. The large, robust primitive *Paranthropus* is found only in the two youngest caves. The slender, more manlike little *Australopithecus/Plesianthropus* is found only in the three older ones. How can that be?

Failure to resolve this question, which had all of anthropology confounded even as it was reluctantly coming to terms with the idea that there were erect ape-men in South Africa, was another of the reasons that made Sir Arthur Keith so slow to acknowledge the legitimacy of the first australopithecine. A final reason was the onset of World War II. For six years South Africa was effectively cut off from the rest of the scientific world, and the flow of information dried up. So did the work itself. Only Broom continued with some desultory scratching at Sterkfontein.

With the war over, and in an effort to lift this blanket of confusion, Raymond Dart allowed himself to be persuaded to re-enter the field. He picked a dramatic spot. Two hundred miles north of Johannesburg, in a remote and rugged part of the country, is a place called Makapansgat. It gets its name from a large dolomite cave that was the scene of a notable nineteenth-century massacre: the last stand of the African chief Makapan, driven there by revengeful Boers in reprisal for atrocities committed on white farmers by the native Africans.

With about three thousand of his tribesmen Makapan holed up in this cave. Defended by natural escarpments and narrow approaches which the Africans further strengthened with piles of rock, they held out against the besieging whites for several weeks, but were eventually driven out by thirst. About two thousand were killed in skirmishes outside the cave, another thousand inside. Makapan and his people were obliterated. For years afterward their bones lay all about, disturbed only by the ubiquitous lime workers, who came and cut away part of the cave face.

In 1936 the South African government declared Makapansgat a historic monument. A visitor to the site at that time examined the spot where the lime workers had been blasting and noticed that they had laid bare a layer of ash and soot underneath a solid covering of stalagmite-like dripstone that had slowly accumulated on the cave floor, effectively concealing the ash layer from view until then. When the ash was examined, it proved to contain the skull of an extinct baboon. Clearly the cave had been inhabited long before Makapan died there.

Dart had an able young student, Phillip Tobias, the finder of the baboon skull. He showed it to Dart and said, 'Doesn't that make Makapansgat much older than anyone thought?'

'Yes, it does,' said Dart.

'Isn't it worth investigating?' persisted Tobias. 'You should go back into the field.'

Dart eventually did. He was unable to do anything during the war. But by 1947 he had raised enough money to begin working at Makapansgat on a large scale. He immediately began finding animal fossils. Twelve years later, he had sifted through ninety-five tons of breccia and had come up with 150,000 pieces of fossil bone. The vast majority of this material belonged to various other mammal species, but a very small percentage of it was hominid. Of particular interest to him were forty-two smashed baboon skulls, twenty-seven of which had been broken in on the left side. This odd coincidence puzzled him. Statistically, it was not to be accounted for by random accident. For example, could twenty-seven of those forty-two baboons have been hit on the left side of the head by falling from the roof of the cave?

It did not take Dart long to come up with an answer. The mind that Keith had earlier charged with unorthodoxy now hit on an idea almost as riveting as his original claim that the Taung Baby was not an ape. He decided that those baboon skulls had been smashed by australopithecine hunters, and that the killers were right-handed.

Dart had already begun to speculate about the life-style of his little hominids. That they were erect walkers was now pretty well established. This meant that they spent all or most of their time on the ground, presumably wandering about in the dry bushy veldt environment that was scarcely different from the South African veldt of today. How did they manage to survive? he asked himself, noting that they lacked the large canine teeth of that other savanna dweller, the baboon, and that they surely were not as good as baboons at climbing trees or rock faces. And yet they would have been exposed to the same large carnivores as baboons. How did they protect themselves?

That peculiar sample of smashed baboon skulls suggested an

answer: weapons. According to Dart, the Makapansgat hominid was a brutal little fellow who survived in a hard world only because of his murderous instincts developed by his dependence on weapons. He wrote: '*Australopithecus* lived a grim life. He ruthlessly killed fellow australopithecines and fed upon them as he would any other beast, young or old. He was a flesh eater and as such had to seize his food when he could and protect it night and day from other carnivorous marauders.'

To justify this picture of a ferocious weapon-wielding little killer, weapons had to be found. Dart sought stone ones in vain for years, never finding a single one. He then began to analyze the animal bones found at Makapansgat and decided that there was an unnaturally large concentration of knobby leg bones, antelope horns and large-toothed jaws. These, he announced, were the weapons of *Australopithecus*, who had selected them and brought them to the cave, and from them had developed an entire arsenal of clubs, daggers, scrapers, picks and saws. For this non-stone culture Dart found a name: osteodontokeratic – which means bone-tooth-horn.

In the twenty or thirty years that have passed since the osteodontokeratic culture was announced by Dart, so much more has been learned about the probable life-styles and capabilities of australopithecines that the whole thing has become something of an embarrassment to anthropologists, who honor Dart for his dazzling recognition of the first australopithecine, but shake their heads over this later aberration.

It is doubly embarrassing for anthropology that Dart, while he was at the height of his osteodontokeratic concoctions, should have met a traveling American journalist, the late Robert Ardrey. Ardrey had a vivid imagination and was immediately taken by Dart's vision of ancestral beings whose frightful propensity for killing their own kind separated them from all other living creatures. He had a somber view of modern man, and it suited him perfectly to trace our worst instincts – expressed in murder and warfare – to Dart's idea that they got their start with an australopithecine who began systematically to club baboons to death.

This, unfortunately, is anthropology by analogy. Ardrey readily acknowledged that he had no expertise in fossils, that he could not tell a humerus from a tibia. But he did claim a knowledge of statistics. When he was shown by Dart how many baboons in his sample had had their skulls crushed in a particular way – more than chance seemed to explain – he became an instant believer. He published an eloquent but badly flawed best seller, *African Genesis*, which planted the killer-ape idea so firmly in many minds that it still lingers there. Dart's evidence, he wrote, would 'establish the carnivorous predatory australopithecines as the unquestioned antecedents of man and as the probable authors of man's constant companion, the lethal weapon'.

Given the nature of hunting and the behaviour of baboons, that seems unlikely. Baboons are so alert and swift that it is almost impossible for humans to catch them by chasing them, let alone kill them with blows on a particular part of the head. That right-handed australopithecines were able to approach baboons – which obligingly waited for them – and then tap them on the left side of the skull does not have the degree of probability that Ardrey thought it did.

Ardrey then went on to cite another case of 'intentional armed assault [on a small skull piece of a juvenile australopithecine] that showed two small round perforations about an inch apart. The holes could not have been of animal origin, since no carnivore has canines set so closely together . . . The living australopithecine, three quarters of a million years ago, had been struck by something. Not only had he been struck once. He had been struck twice'.

That skull piece with two holes in it also caught the eye of C. K. Brain, an experienced geologist and director of the Transvaal Museum, who thought that predators, possibly leopards, were more likely to have been the killers of australopithecines than their fellow australopithecines were. To him the holes did look suspiciously like tooth marks. Instead of merely speculating about that, he made a study. He measured the distance between the lower canines (the stabbing teeth that would have made the puncture) of a number of predators then common in South Africa: lions, leopards and some smaller cats. The punctures fitted the leopard tooth-gap exactly.

Brain thought some more about leopards. He knew that those animals drag their kills into trees, where they spend the next few

One theory of how australopithecine fossils got into South African caves is that leopards carried their hominid kills into trees growing beside sinkholes. As a leopard devoured its victim, the bones would fall into the hole along with the much more numerous bones of baboons and other animals also killed by leopards. This theory seems more plausible than the one which holds that australopithecines were the killers of the other animals – as well as each other – and dragged their prey into the caves themselves.

days consuming them. As they do so, the bones fall to the ground. In a dry country like the South African highveldt, trees are often found growing around the dolomitic-limestone fissures that lead to caves or are themselves in the process of becoming caves. Many of them hold rainwater and become natural sinks or small reservoirs. That is what encourages the growth of trees around their edges. With their roots extending down to the water, they fare better than neighboring trees. They are often the largest ones for miles around and become ideal storage places for leopard kills. Over some thousands of years, as trees grew and died and grew again about a cave entrance, and as generation after generation of leopards hauled kills into them, there would have been a large assortment of bones falling into the cave beneath. Among them could have been hominid bones.

Brain considered the leopard possibility and concluded that it could better account for the presence in a limestone cave of a large number of animal bones and a small number of hominid bones than Dart's idea that the animal bones were weapons and that the broken skulls were those of the victims of predatory australopithecines. He did not stop there. He made a study of several thousand goat remains outside Hottentot villages, and found that after the goats had been eaten by people and the leftovers thrown away to be scavenged by dogs, what remained did not differ significantly from what appeared to have been left in Makapansgat, where the original eaters may have been large carnivores and the scavengers may have been hyenas. Were the original carnivores hominids? Brain does not know. All that his work can do is suggest that Dart's conclusions about clubs and saws and crushed skulls are probably incorrect.

Others have suggested that hyenas themselves were the collectors. Hyenas do carry bones about to gnaw on. They even eat them. They have immensely powerful jaws and formidable digestive juices. They crack and grind up small bones with gusto; these pass through the alimentary tract and emerge as whitish feces. But the hyena's predilection does not identify it as the agent that stored bones in the caves. Large caves are not the natural habitat of hyenas. Overall, the hyena argument seems slim. A better one can be made for the leopard.

This problem inched its way up through the 1940s and into the 1950s without a good answer. Brain and others began getting better and better geological data on the caves, but it was clear to all that a good explanation for the bones in them would depend on a better understanding of the hominids themselves.

By this time, as in the earlier case of all the Heidelberg-Peking-Java types, the naming of the South African hominids had gotten out of hand. There now were:

Australopithecus africanus: The Taung Baby, the original australopithecine, named by Raymond Dart in 1925.

Plesianthropus transvaalensis: From the Sterkfontein cave, the second South African australopithecine ever found. Although it closely resembled the Taung Baby, its finder, Robert Broom, gave it its own name. Broom and a gifted assistant, J. T. Robinson, subsequently found a number of other fossils of this same type by blasting at Sterkfontein.

Australopithecus prometheus: This is the fossil found by Dart at Makapansgat. It was very much like the two preceding ones. But Dart, driving hard now on his osteodontokeratic culture, and excited by traces of fire in the cave floor (those traces subsequently turned out to be much younger), came to the conclusion that the 'killer apes' from Makapansgat knew the use of fire as well as the use of weapons. So, another name, from Prometheus, the fire bringer of Greek myth.

Paranthropus robustus: The name given by Broom to the larger and seemingly much more primitive type represented by the skull and teeth provided him by the schoolboy Gert Terblanche at Kromdraai.

Paranthropus crassidens: Also named by Broom. He and Robinson found numerous examples of this creature at the fifth and last of the South African cave sites, Swartkrans, less than a mile from Sterkfontein. It too was primitive and robust, but Broom, following a personal determination to name more extinct species before he died than any other scientist, ignored its resemblance to the Kromdraai fossils. He was willing to put them both in the same genus, *Paranthropus*, but insisted on a new species name for this one.

So the anthropological music box had turned once again, but it had ground out the same melody as before; only the grace notes were different. Once more there was a hatful of names crying for simplification.

The difference was that the Heidelberg-Peking-Java men had been scattered over immense distances throughout the Old World. The five South African fossil sites were all concentrated in one small area of one continent. More than that, three of them – Sterkfontein,

Kromdraai and Swartkrans – were within three miles of one another. On sober second thought, did anybody really believe that five different species of ape-man from three separate genera could all have erupted in this one small area when they were unknown in any other place on the globe? Did each site have a different species of ape-man? That seemed overwhelmingly unlikely.

Broom died. Dart grew old and retired. New men, like their students Tobias and Robinson, began to emerge as strong voices in South African anthropology. They became increasingly familiar with all the fossils from all the sites. Slowly the picture began to clarify itself. As usual when the sample was small, the differences between fossils was what had stood out. Now, with the sample grown large, certain similarities began to assert themselves. Eventually it became almost universally recognised that there were only two types of early hominid in South Africa, not five.

What to name them?

If an anthropologist today were given this charge, and a clean slate to go with it, he might well have gone along with Broom and his *Plesianthropus* (near-man). For that is what all those fossils were: near-man. Dart's name, *Australopithecus* (southern ape), was misleading; it was now almost universally accepted that they were not apes. Nevertheless, by the rules of nomenclature – the scientific rules for naming things – Dart had got there first. By right, therefore, his name had to be accepted. From the 1950s on, *all* the South African ape-men would be known as australopithecines.

They were divided into two types – a slender 'gracile' type and a burlier, more primitive-appearing 'robust' type. The former, of which the Taung Baby was the original, the 'type specimen', retained its original name of *Australopithecus africanus*. With it went all the fossils from Makapansgat and Sterkfontein; it had become clear that they were gracile too.

The big, burly specimens from Kromdraai and Swartkrans, although measurably different from the gracile species, were obviously the same general kind of creature. They too were erect-walking ape-men. The solution: put them in the same genus but into a different species. They came out with the name *Australopithecus robustus*, and that is what they are still called.

Today, anthropologists talk easily about the robust and gracile australopithecines. But those names and those simplifications emerged only after a generation of debate and controversy.

3 East Africa: A Date at Last

I am determined to wear out, not to rust out.

ROBERT BROOM

I don't think you fully appreciate what I did in order to do the work I am now doing. There is no single course of study which could fit a person to do what I am doing. I did a tripos in modern languages, then a tripos in archeology and anthropology, and thereafter a post-graduate course in vertebrate zoology and geology, and then a year in anatomy at the Royal College of Surgeons which I combined with a course in meteorology.

LOUIS LEAKEY

Robert Broom died in 1951. He had been suffering from bulbar paresis (degeneration of the spinal nerves) and asthma and had great trouble in breathing. Nevertheless, he continued to work until the last few weeks of his life with a vigor that astonished everybody. In the previous year he had written a book and published fifteen scientific papers. He published another in 1951, but that was his last. He went downhill rapidly thereafter and died in April – literally worn out, not rusted out.

By the end of his life he had become very famous. He lectured at the University of Witwatersrand, corresponded with distinguished anthropologists around the world, but remained the irreverent, iconoclastic, thorny Scot that he had always been. His sympathies still ran with the common man. 'The average workman, quarry man or miner,' he said, 'has usually quite as much mental ability as the average university professor, and he often has more enthusiasm in the search for new scientific facts.' He went on to point out that the best fossil hunter in the United States, S. W. Williston, was a railway labourer. On women he also ran true to form. At seventy-four he said, 'I am only reminded of age when I climb a hill or catch a bus. My heart still goes pit-a-pat when I see a pretty girl and we still have quite a lot at Pretoria. They say sex-attraction still continues in Paradise – what a time I am looking forward to.'

I was eight years old when Broom died. I had never heard of him. I had never heard of australopithecines either, but I was already becoming interested in anthropology, although the word was unknown to me. I am the son of Swedish immigrants who settled in Chicago, where my father worked as a barber. He died when I was only two, and my mother took me to live in Hartford, Connecticut. There I had the luck to find a surrogate father in the person of Paul Leser, a neighbor who taught anthropology at the Hartford Seminary Foundation. Leser had no children of his own, and encouraged me to visit him and browse through his library, which contained numerous works on cultural anthropology. I can still remember my astonishment at learning that the world was full of people who lived close to nature – Sudanese, Pygmies, Asian islanders – getting their living from a direct interchange with their environment and not from the supermarket on the corner. Most children with inquiring minds become aware gradually of the diversity of human cultures, but this fact seems to have hit me all at once. It made a strong impression on me. I also became fascinated with the distant past. In high school I decided to become an anthropologist.

I was good at chemistry, and Leser tried to persuade me to concentrate on that. 'You'll starve as an anthropologist,' he said. 'There's no future in it except teaching, and even that's very shaky.' He pointed out that chemists were in great demand and that good ones with a bent for business could rise high in corporations. So I obediently continued with chemistry at the University of Illinois – but it soon bored me. More and more I found myself hanging around the anthropology department. Eventually I switched my major to anthropology, and worked for several summers as an archeologist on digs in the Midwest.

At about that time I heard about a travelling-scholar program that permitted students at any of the Big Ten universities (plus the University of Chicago) to apply for special work at any of the others. I had become dissatisfied with the anthropology courses offered at Illinois, so I decided to apply to Chicago, where F. Clark Howell headed the anthropology program.

Howell was then only thirty-nine years old, but already had wide experience as a teacher and field worker. Unassuming and uncontroversial – notably so in a field that has long been dominated by flamboyant personalities – he avoided disputes and concentrated on rigorous site development.

In the past a man going into the field would raise a little money, go out and find new fossils and decide he needed a little geology done. So he would go home, raise a little more money and get a geologist. The next year he would realise he needed an archeologist, and so on. It was like putting patches on a pair of pants. Clark's approach was totally different. He understood that proper fieldwork needed a team of qualified experts from the start. People do this routinely today, but I have to give Clark the credit for instituting the multidisciplinary approach to site development.

A good example of this would be his work at Torralba and Ambrona, two sites in Spain that had been excavated on and off by a Spanish nobleman, an amateur who amused himself by digging up elephant bones. Clark brought a team there in 1961. He needed good dating, so he got a palynologist – an expert in recognising fossil pollen grains – to do some sampling. He also got a geologist and an archeologist. They dug up the sites inch by inch, plotting everything they found on a series of maps that they made as they went down, a new map for every foot. They ended up with a three-dimensional picture of the whole thing, and were able to reconstruct what had happened there four hundred thousand years ago.

What a spot. Bands of *Homo erectus* would wait in the valleys between the hills for the big game herds that migrated south for the winter. They drove the game into swamps by setting grass fires. They even trapped elephants that way. The elephants got mired and the hunters were able to kill them. It is all there to be seen: the stone tools the hunters left lying around, the traces of burned grass, the animal fossils. There were a great many elephants, far too many to be accounted for by their getting stuck there by accident. Furthermore, their bones were all mixed up – proof that people had been chopping those elephants up and moving them around, smashing their bones to get at the marrow. If one analyzes a pile of bones intelligently, one may learn some interesting things. For instance, one elephant Clark found was only half an elephant. Half its legs and ribs were lying by themselves in one place. The other half was gone. Since one does not find half an elephant dying in a swamp, it was obvious that somebody had cut up that other half and carried the pieces away.

Howell's work at Torralba and Ambrona is regarded by professionals as a model of site development. Not only did it prove that bands of *Homo erectus* lived in Spain about four hundred thousand years ago – something that had not been demonstrated previously for Western Europe – but it showed in considerable detail what they had been doing; all this without his finding a single hominid fossil.

In the late 1960s, Howell put his multidisciplinary methods to work again, this time at the Omo River in southern Ethiopia, and again made a signal contribution to anthropological knowledge by the precise dating of a whole series of geological strata and by the matching of the evolutionary development of several kinds of mammals to those strata. He found a few hominid fossils, but they were far from spectacular. Howell's failure to come up with arresting fossils of human ancestors that land on the front pages of newspapers has left him virtually unknown outside the profession. Inside it, he is internationally respected. I shared that respect and phoned Clark for an appointment to discuss a possible transfer as a traveling scholar.

'What exactly do you want to do?' Howell asked when we met.

'I want to go to Africa and look for hominid fossils.'

'That's not easily done,' he said.

'I'll do whatever I have to do to get started. I'd like to start by studying under you.'

After some further questioning, Howell accepted me.

I had been rather nervous about making such a direct approach to the renowned Howell. The ease with which I was taken into his program astonished me.

'He was a bright kid,' said Howell later to one of my friends, 'so I took him.'

During the 1950s, three vitally important developments occured in paleoanthropology.

The first was the publication of a paper by Wilfrid Le Gros Clark in the *Journal of the Royal Anthropological Institute* in 1950.

During the previous decade Le Gros Clark had replaced the mighty Keith as the dean of the English anatomists, Keith having grown old and retired. That did not prevent Robert Broom from writing to Keith whenever he felt like it – caustic, as always, at Keith's reluctance to grant full hominid status to the South African australopithecines. Broom and Keith were exactly the same age. Whenever Broom got a letter back from Keith, he would examine the increasingly wavery handwriting and say smugly to a fellow South African scientist, George Findley: 'Poor old Keith, he seems to be getting rather shaky.' Broom's own handwriting, right up to his death, was as firm as a young man's.

Now that Keith was retired, Broom began to bombard Le Gros Clark with letters about the South African fossils, and they found a sympathetic eye. Nevertheless, Le Gros Clark was cautious, and said he would have to suspend judgment until he had more evidence in the form of actual casts, careful drawings and photographs of the originals.

By return mail Broom shot back: 'You say anatomists in England will have to suspend judgment until casts etc. etc. .. English judgment may be of a high order, but when *Australopithecus* was discovered in 1924 England did not suspend judgment. Four English scientists at once expressed their opinion that it was a chimpanzee.'

Le Gros Clark returned a soothing letter, again pressing for more details on the fossils, which Broom eventually sent him. They apparently convinced Le Gros Clark of the right of the australopithecine claim, because he immediately set out to convince others. Following a visit to South Africa, his first act was to make a detailed analysis of all the teeth in the ape jaw, and follow that with a similar analysis of the teeth in the human jaw, noting the differences between them. He came up with a list of about a dozen significant differences. Not only were these consistent for all apes and all humans, but they were also conspicuous and unmistakable. In effect, Le Gros Clark was saying, 'Take any three or four characteristics on my list and I will tell you instantly whether their possessor is a man or an ape.' When australopithecine traits were introduced into that comparison, their resemblance to those of men or apes should be plain.

Le Gros Clark's list of characteristics depended mostly on morphological observations (examinations of the shape and function of each tooth). He supplemented this work with biometrical studies (precise measurements: height, width and so on). Broom learned about this work and waited impatiently for the results. Meanwhile he got wind that the world's leading authority on apes and ape behaviour, Sir Solly Zuckerman, was making his own biometric measurements to prove that australopithecines were apes.

Broom, who knew instinctively *how* an animal used its teeth and how that use was reflected in their shape and evolutionary development, was respectful of morphological studies; he had been doing them all his life. Knowing also of the remarkable size differences that existed not only among his fossil individuals but also in modern man, he was scornful of biometry. 'I regard all biometricians in the field of morphology as fools,' he said flatly. Hearing of an American scientist whom he respected despite his having strayed into the biometrical minefields, he said of him, 'He has a small brain, but it is remarkably good for its size.'

With Zuckerman kicking up more and more biometric dust, Le Gros Clark thought it appropriate to challenge him to produce a full set of chimpanzee teeth that bore any resemblance to a set of australopithecine teeth. Because of the differences between the two, Zuckerman found it difficult to do this. He continued to concentrate on his statistical studies although professional statisticians began to question whether his figures had been calculated properly. Broom, of course, could not have cared less whether they were right or wrong. To him the important thing was that without an understanding of morphology, such figures were meaningless. You could find a cow, he said, whose teeth were the same size as a donkey's. Would that make them the same?

To give Zuckerman his due, there *were* resemblances between ape skulls and australopithecine skulls. The brains were approximately the same size, both had prognathous (long, jutting) jaws, and so on. What Zuckerman missed was the importance of some traits that australopithecines had in common with men. Charles A. Reed of the University of Illinois had summarised Zuckerman's misunderstandings neatly in a review of the australopithecine controversy: 'No matter that Zuckerman wrote of such characters as being "often inconspicuous"; the important point was the presence of several such incipient characters in functional combinations. This latter point of view was one which, in my opinion, Zuckerman and his co-workers failed to grasp, even while they stated they did. Their approach was extremely static in that they essentially demanded that a fossil, to be considered by them to show any evidence of evolving toward living humans, must have essentially arrived at the latter status before they would regard it as having begun the evolutionary journey.' In other words: if it wasn't already substantially human, it could not be considered to be on the way to becoming human.

In reply to Zuckerman and in order to demonstrate that the

latter's statistical approach was invalid, Le Gros Clark then published a comparison study in which he checked the australopithecine data against his list of ape-human differences and found that in virtually every respect they resembled the human model and not the ape model. *

Australopithecus was a hominid, a potential ancestor! A distinguished English scientist had at last examined all the evidence in a methodical way and found it to be so. After twenty-five years, Raymond Dart was vindicated.

The second great event of the 1950s had to do with Piltdown Man. Some years earlier, a dentist named Alvan T. Marston had dug up a fossil skull at Swanscombe in Kent. He became convinced that his skull was older than the Piltdown one, and began a strenuous campaign to have Swanscombe Man established as the 'First Englishman'.

Marston had some good arguments. As a dentist he had a working knowledge of how teeth and jaws functioned, and he was highly suspicious about the association of Piltdown's manlike cranium with an apelike jaw. To him it seemed more likely that a man and an ape had come together by accident in the Piltdown dig. That set him to speculating whether the Piltdown skull was really as old as it had been made out to be. His own *was* old, he knew that. Twenty-six kinds of Middle Pleistocene mammal fossils had been found with it. All he had to do was prove that the Piltdown skull was younger, and pride of place would be his. He argued his case noisily for a number of years without success.

In the late 1940s, science came to his rescue. Somebody had discovered that buried bones gradually soaked up fluorine from the surrounding earth; the longer they lay buried, the more fluorine they would contain. This was not a straightforward measurement; it depended on how rich in fluorine the surrounding earth was – and that was known to vary widely from place to place. But if a number of fossils from a certain locality tested the same and one didn't, it would be clear that there was something exceptional about that one. Either it belonged to a different time or it had come from a different place.

When Marston heard about the fluorine method, he asked Dr Kenneth Oakley of the British Museum to test the Swanscombe fossils. Oakley did so. Man and animals all proved to contain the

*For an analysis of the Le Gros Clark study, see page 266.

same amount of fluorine, indicating that they belonged together and were the same age. Mid-Pleistocene was the verdict.

Since an Early Pleistocene date or even one back in the Pliocene had been claimed for Piltdown Man, Marston demanded that it too be tested, along with its associated animal fossils. Again Oakley agreed. The group of bones he was given to test was a mixed one. A couple of them, actually found in the same dig with Piltdown Man, were extinct mammoth and mastodon teeth from the Pliocene. Fossils of more recent species – horses, red deer and beaver from nearby deposits with a comparable fluorine level – were added to enlarge the sample. To Oakley's astonishment, the younger mammals contained more fluorine than the older ones. It did not occur to him that the older ones might have been planted surreptitiously in the pit to make Piltdown Man appear old. That suspicion would come later. He could only conclude that either the fossil history in the Piltdown area was hopelessly mixed up or the specimens had been very carelessly collected. In either case, the animal fossils were useless in dating the human fossil – whose own fluorine date was even younger, certainly younger than Swanscombe Man. Because fluorine testing was still in its infancy and subject to some error, Oakley did not come out and say, as Marston obviously hoped he would, that Swanscombe Man was the First Englishman. That was left to others to infer, which frustrated Marston mightily. He wanted a bugle call, not an inference.

Oakley's report, published in 1950, merely said, 'Piltdown Man, far from being a primitive type, may have been a late specialised hominid which evolved in comparative isolation.' However, he had noticed something else about the fossil. In drilling into one of the teeth to get a sample for his fluorine test, he had noticed that its dark brown color – one of the characteristics that had been cited originally as a mark of great age – seemed to be only on the surface. Inside, the dentine was as white as that of a man freshly dead.

Now, a forgotten act by the original discoverer, Dawson – dead himself by this time – was recalled. In order to harden the fossil for better preservation, Dawson had dipped the various pieces into a solution of potassium bichromate. He had done this soon after finding them, he told Sir Arthur Smith Woodward at the time. This did not harden them, Woodward noted; it merely darkened them somewhat. Woodward the expert excused Dawson the amateur as having done the dipping in good faith but out of ignorance.

Marston, still thrashing about for evidence to shorten Piltdown Man's age, seized on this straw. He immediately performed a similar dip with a bit of fossil animal bone that he had collected on the Piltdown Common. It promptly turned from grayish to the dark iron-brown of the Piltdown skull. He returned to the attack, shouting that general estimates of Piltdown's age were patently wrong. The fluorine test proved it. The geology of the Piltdown dig itself was a total jumble and could not be relied on. The mastodon teeth had something fishy about them. Finally, all one had to do to make any piece of fossil bone look old was dip it in the right chemical solution.

He got nowhere. In fact, his insistence that the cranium and jaw could not fit together worked against him when Oakley's fluorine test showed that although they were not old, they were the same age. There was not much more that Marston could do. He wrote a long and detailed paper, drawing on his dental knowledge to show that whatever the age of the jaw and skull, they *could not* be joined to make up one individual. The cranium, he wrote, was that of a modern man at least forty years old; the jaw was that of a young ape, with the immature molars and premolars of an adolescent. That was his last blast. Thereafter he subsided into a baffled silence.

But the good that Marston did was not interred with his paper. It lived after him, igniting a spark in the head of J. S. Weiner of Oxford, who in 1950 began to dig into the whole history of Piltdown Man in a methodical way. He traced a long trail of sloppy records, superficial examinations and sheer ineptitude; a trail that had blundered into the thicket of prejudice that early man *should* look like Piltdown and not like a low-browed ape, and never blundered out again; a trail that eventually became as deeply stained as the fossil itself. Although dozens of men had fretted, each in his own way, about the Piltdown fossil, Weiner was the first to even whisper that that banner on the flagpole of British paleoanthropology, that star in its diadem, might be a fake.

Once the awful word was spoken, Piltdown Man came down with a crash. Weiner, Oakley and Le Gros Clark examined him in detail, and discovered an appalling fact. To make his teeth in his jaw look like human teeth, someone had filed his molars flat. Actually they were ape molars that originally had high pointed cusps. Those were now gone, and when the teeth were examined under a microscope the method of their removal was revealed. On their biting surfaces were the telltale scratch marks of a file. The dark surface brown of the teeth and jaw was revealed to be a paint of some kind, apparently hastily applied to make them resemble the color of the cranium.

New and better tests were run on the bone itself, showing that the skull could not have been from the Upper Pleistocene. It is now known to be only about five hundred years old. The jaw turns out to be that of an orangutan, also about five hundred years old. Marston, the dentist, had been right. He had smelled out something that paleontology had been unable to.

Although anthropologists as a group were deeply relieved to have such an awkward character as Piltdown Man pruned from the family tree, they were also hideously embarrassed. The conclusions of the three examining experts were soberly reported by the British Museum in 1953, and they created an immediate furor.

Who had done this? And why?

The most likely candidate was obviously the original finder: the jolly amateur, Charles Dawson. Others involved by reason of having been present when fossils were dug up were Sir Arthur Smith Woodward and Teilhard de Chardin. Woodward was almost immediately exculpated by everybody; his scientific credentials – indeed, his entire scientific career – made it seem incredible that he would have done such a thing. A similar disclaimer was made for the French priest, whose long and equally distinguished career as a theologist/paleontologist seemed to lift him above any charge of guilt. However, there were doubters. Louis Leakey, who later succeeded Broom as the senior paleoanthropologist of Africa, felt so strongly that Teilhard de Chardin was guilty that in 1971 he refused to attend a symposium honoring the French priest. But most others believed that Teilhard was innocent.

That left Dawson. His character and rectitude were paraded by his supporters. Scientists recalled his good nature, his enthusiasm, his admiration for Woodward, his long devotion to an avocation from which he derived nothing but the respect of other scientists. A fellow member of the Piltdown Golf Club remembered him, in Ronald Millar's *The Piltdown Men*, as 'an insignificant little fellow who wore spectacles and a bowler hat. Certainly not the sort who would put over a fast one'.

But support for Dawson crumbled when it was learned that other archeological activities of his were pocked by fakery. Gradually the burden of guilt fell more and more squarely on him. It would have done so instantly and obliteratingly had it not been for one puzzling matter: how did he know enough to do what he had done? Where had he obtained a five-hundred-year-old cranium? How had he had the patience to produce bits of it over several years? How had he lured Woodward in so skillfully? How had he known enough to knock off a bit of the hinge that fastened the jaw to the skull so that the actual connection had to be assumed? How had he known enough to file down the molars? Where on earth had he gotten his hands on a five-hundred-year-old orangutan jaw? Those things were not exactly common in Sussex in 1912. Even more unlikely, where had he obtained the mastodon tooth he had put in the dig? A new uranium test of it indicated that it had come from Tunisia. Finally, what was there in it for him?

The arguments over Dawson buzzed and fizzed, and while they did, British science writhed in embarrassment. Efforts were made to explain the whole Piltdown matter as a triumph of scientific inquiry, but without much sympathy from the newspapers. One reported: 'Anthropologists refer . . . to the "persistence and skill of modern research." Persistence and skill indeed! When they have taken over forty years to discover the difference between an ancient fossil and a modern chimpanzee! A chimpanzee could have done it quicker.'

Published in 1972, *The Piltdown Men* gives a detailed review of the whole Piltdown story, running down many fascinating trails that are beyond the scope of this book to follow. It should be read by anyone interested in this extraordinary forgery. At its end the author, Ronald Millar, asked himself the same question that everybody else had been asking: who had done it, and why?

Generally, Millar is kindly disposed toward Dawson. He believes that the intricacy of the forgery – and it was considerably more intricate than is possible to explain here – required, at least, the work of a professional accomplice. But even the word 'accomplice' puts him off. Dawson simply was not that kind of man, says Millar; he admired Woodward too much and could not have betrayed him in that way. In the end, Millar suggests a surprising candidate: Sir Grafton Elliot Smith. He explains that Smith had the knowledge, the opportunity to get his hands on the proper fossils – having actually been in North Africa where the suspect elephant tooth came from – and finally the motive.

Smith could not abide Keith. For years the two had battled on

Principals in the Piltdown controversy are assembled in this group portrait. Seated in a laboratory coat is Sir Arthur Keith, who reconstructed a complete skull from some fossil pieces. Standing to his left is Charles Dawson, the amateur who found the fossils. Next to him, with a white goatee, is Sir Arthur Smith Woodward, who spent the last thirty years of his life studying Piltdown Man. Standing on the other side of Keith and pointing with a condescending finger is Sir Grafton Elliot Smith, who, some think, was responsible for the fraud because of his detestation of Keith and Woodward.

matters of scientific principle; he would have loved to see Keith made a fool of. He had no use for Woodward either. He had every opportunity to set the fumbling Woodward straight on his reconstruction of the skull, but did not. His own statements about it were always carefully hedged by the claim that they were based on an examination of casts and reconstructions, never on the original bones. Why, asks Millar, did Smith never bother to study them, when he could have at any time? Had he done so, being an expert anatomist, he probably would have quickly recognised the many small flaws in the entire Piltdown edifice. According to Millar, Piltdown was not so much a true forgery as a monstrous hoax intended only to embarrass Keith and Woodward.

The story of Piltdown Man does not end with Millar. Seven years after his book was published, an extraordinary story landed on the front pages of papers in England and the United States. In 1979, the English geologist James Douglas died at the age of ninety-three. Just before his death he set down on tape his recollections of the Piltdown affair, and he found a new culprit: William Sollas.

Sollas was the man who had written to Raymond Dart at the time of the Taung Baby controversy, excoriating Keith. He detested Keith and he detested Woodward. He had the Chair of Geology at Oxford for many years and was succeeded there by James Douglas in 1937.

In his taped recollections, Douglas noted a close thirty-year relationship with Sollas. He agreed that carrying out the Piltdown hoax would have been quite beyond the capabilities of the amateur Dawson, but well within those of the professional Sollas. He remembered having sent Sollas some mastodon teeth from Bolivia (similar to the one that turned up in the Piltdown dig) – and remembered vividly that Sollas had once received a packet of potassium bichromate (the chemical used by Dawson to stain the skull pieces brown). He remembered that Sollas had once borrowed several ape teeth from the Oxford collection (some loose ape teeth also figured in the Piltdown reconstruction).

Douglas also supplied Sollas' motive: professional revenge. He remembered Sollas' fury at Woodward's airy dismissal of his technique for making molds of fossils. Sollas waited patiently, according to Douglas, and finally got his innings. If Douglas' memory is right, how Sollas must have laughed, year after year, as the two great men, Woodward and Keith, wriggled helplessly in the bite of the phony anthropoid trap he had laid for them.

Douglas, of course, may not have been right, and he is no longer here for cross-examination. Others who have sifted through the evidence think the case against Sollas is a thin one, based on too few and too circumstantial clues. Sollas, they insist – and some of them knew him personally – was simply 'not that sort'. But that was the defense that Dawson's supporters put up for him. The perpetrator of the Piltdown hoax will probably never be known.

The third great event of the 1950s occurred in 1959 – and involved Louis Leakey.

Louis Seymour Bazett Leakey was born in Kenya in 1903, the son of an English missionary. He grew up speaking Kikuyu before he could speak English, and throughout his life was proud that he had been inducted into a Kikuyu tribe as a young man.

Sent away to school in England, the little boy from the African bush could not conform. He was a show-off – noisy, opinionated and generally detested by his schoolmates. Nevertheless, he did manage to get into Cambridge, only to be forced to leave as a result of a severe kick in the head during a rugby match. He returned a couple of years later and, despite having had a wretched and spotty education, graduated with highest honors. He was a brilliant young man with ferocious powers of concentration and volcanic energy.

A missionary's son is usually a poor boy, and Louis was one of the poorest. He had had no money as a schoolboy, had none at Cambridge, and continued to have none afterward. For many years he scraped along on fellowships and grants, cadging what he could where he could, lecturing for tiny fees, writing an occasional book or article. His trouble was that he had been infected with a passion for African prehistory almost from childhood. The past had lit such a hot fire in him that he could never thereafter put it out, and there was then no way of making an honest living from paleoanthropology or archeology in Kenya. He did what he could. He sold African ebony canes to his tailor in exchange for clothes. He turned a knowledge of handwriting analysis to account in identifying fraudulent documents. He assisted a man who was digging up dinosaur bones in Tanzania. He collected skeletons and pottery from burial sites in East Africa. He made an immense collection of stone tools. These things he took to London, where he was able to catalogue them and get grant extensions to return to Africa and collect more. Back and forth he seesawed during the 1920s. At one point Sir Arthur Smith Woodward got him a grant of four hundred pounds for two years' fieldwork. At another he was commissioned to write a prehistory of Africa. He turned out a 700,000-word work that went unpublished for thirty years because he refused to cut a paragraph of it.

During all this, with many false starts and collapses – physical, professional and financial – he gradually staked out the territory that would eventually lead him to world fame. He located fossil sites on the edge of Lake Victoria and in Olduvai Gorge in Tanzania; but along the way his career almost foundered.

As a young man he put two large blots on his copybook. While still teetering between scientific respectability and financial engulf-

Louis Leakey – at the height of his career, in the coveralls he always wore in the field, his glasses on a cord so that they would not fall off as he crawled and scrambled about, and with his hands full of fossil mammal teeth.

ment, he married a young Englishwoman, Frida Avern, who had come out to East Africa for a visit. At first she willingly accompanied him on his field trips, and returned home with him to a shabby shack overflowing with stone tools, bones, African artifacts and the skins of animals. But that life did not really suit her. She was only too glad to join him on his periodic trips to England. Even there she began spending more and more of her time alone while Louis worked at the museum, organising the mountains of material he had collected. Their life-styles could not seem to mesh, and a gap developed between them despite the birth of two children. Finally

Frida decided she had had enough. With her own money she bought a small house near Cambridge and settled there permanently.

Louis could visit there whenever he chose, but more and more he chose not to. He was engaged in ever more ambitious projects, and in 1933 had met Mary Nicol, a young archeology student and illustrator. She agreed to do some drawings of stone tools for a book he was writing. Not long after that, the two became lovers. Mary traveled with him to Olduvai the following year, and married him later, after he and Frida were formally divorced.

This whole affair was scandalous, confirming the general opinion of Louis' erratic character. His father was an Anglican minister, his mother a very proper minister's wife. Both were affronted, as were his sisters, both married to clergymen (one later became the Bishop of East Africa). In fact, both the Leakey and the Bazett (his mother's maiden name) families – a large and intricately related and ultraconservative group – were scandalised. Nairobi society too, unbending as colonial societies were apt to be, frowned on Leakey's unorthodox behaviour.

Louis not only had this scandal to live down. He had recently been buffeted by a more serious scientific one. In his work at Lake Victoria he had found two hominid fossils. The first consisted of several pieces of skull with a remarkably smooth forehead; it had none of the heavy brow ridges of Neanderthal or Peking Man. The second was a mandible – later to become infamous as the Kanam jaw – which was much more human-looking than the Piltdown jaw that still had everyone so bewitched. Both these finds had interesting animal fossils with them: extinct elephants and, in the case of the Kanam jaw, something even older, an elephantlike creature called *Deinotherium.* With these old mammal fossils and these strangely modern-looking hominid fragments, Leakey decided that he had unearthed something special. His mind took the kind of impetuous leap that would characterise his entire career. If Peking Man and Neanderthal Man exhibited primitive traits and his didn't, then his must represent the true line from which modern humans sprang. The others were split-offs, dead ends.

Throughout his career Leakey tended to identify his own finds as older and older representatives along the line of true man. All the hominid fossils found by others, whether in China, Java, South Africa or North European caves, he insisted were not ancestors. They were offshoots. It cannot be ascertained just when this belief seized him, but it was consistently held right up to his death. It is not too much to say that a mind tilting that way could easily have been thrown right over by what was found at Lake Victoria.

He made a jubilant and sweeping report to London, planning to follow there himself and make a personal representation of the fossils. He was warned by a friend that his geological evidence – the exact location of the specimens and the exact association with extinct mammals – had better be rock-solid. Leakey had been embarrassed before by a too-quick claim of great age for a fossil man from Olduvai that, humiliatingly, was later proved to be only fifteen thousand years old when the geologist Percy Boswell of the Imperial College in England visited Olduvai and pointed that out. But Leakey was a charger, a young bull, so full of surging belief in himself and the validity of his soaring ideas that he shook off these warnings as he would a gnat.

In London he again met his nemesis – Boswell – who was already suspicious of Leakey for his gaffe at Olduvai. Now Boswell picked and picked at the Lake Victoria evidence until Leakey in frustration asked that funds be raised by the Royal Society to send Boswell to Africa to see for himself. Again Leakey was warned to have airtight evidence.

He didn't have it. The iron pegs he had put down to mark the position of the fossils had been removed by native warriors to make spear tips. Heavy rains had altered the contours of the land. Leakey was hard put to locate exactly the spots where the finds had been made. Worst of all, some identifying photographs he had had taken proved to be of a different place.

Returning to England, Boswell demolished him in a devastating paper published in *Nature*.

These catastrophes haunted Leakey for years. He could have put them behind him more quickly if he had been willing to admit error. But he was stubbornness incarnate. He had written, 'The importance of the Kanam mandible lies in the fact that it can be dated geologically, paleontologically and archeologically, and that . . . it is not only the oldest known fragment from Africa, but the most ancient fragment of true *Homo* yet discovered anywhere in the world.'

Ideas of that magnitude are dangerously intoxicating. They are hard to give up, and Leakey never did give them up. He turned his attention more and more to Olduvai Gorge, where he and his new wife, Mary, would work as a team. Unlike Frida, Mary enjoyed life in the field. She understood Louis' work, complementing it with a growing expertise in the analysis of very early stone tools and in the archeology of ancient dwelling sites. Some of their best work was done in tandem, until they were driven apart by growing differences of opinion, by Louis' increasing absences on the lecture circuit, and by his interest in other women. It is ironic that Olduvai, where Louis' best-known work was done, was not visited by him at all in the last years of his life. Mary took charge of it, eventually moved in permanently, and lives there today in a small shack next to her laboratory.

What drew Louis and Mary Leakey to Olduvai was stone tools. Louis had been collecting them from all over East Africa for years, and had managed to match certain styles in toolmaking with similar 'industries' in Europe. Those at Olduvai were different. They were so primitive – single cobbles with a flake or two knocked off one end to provide a crude cutting or chopping edge – that to an untrained eye they would not have resembled shaped implements at all. Nevertheless, he and Mary recognised them for what they were, and gave the industry a name: Oldowan – another spelling for the Gorge itself.

Olduvai lies under the rain shadow of an enormous extinct volcano, the celebrated Ngorongoro Crater. Like Hadar, where I would find Lucy fifteen years later, it is a dried-up lake bed. And, like Hadar, the lake bed has been filled with sediments, some of them volcanic. Those sediments are displayed in the familiar layer-cake pattern of strata in the sides of deep gullies cut by rivers. The Gorge itself is Olduvai's principal gully, Y-shaped and running for several miles. It has now been cut to a depth of about three hundred feet. Its top stretches off flat to the westward and gradually merges with the Serengeti Plain. It was in the very bottom of the Gorge that the Leakeys began finding Oldowan tools. The lowest layer of sediment was named Bed I. It is covered by a volcanic layer that separates it from Bed II, and so on up, the sediments getting younger and younger and continuing to Bed IV at the top.

When the Leakeys first went to Olduvai in the 1930s, there were no roads for the latter part of the trip, and it took them four days of brutal travel to get there from Nairobi. Once settled, they would work as long as their hoard of supplies lasted, returning when they had scraped together enough money for another try. Over the years they went back again and again, reasoning that the makers of such extremely primitive tools were extremely primitive humans, and that sooner or later they would find their remains.

Olduvai Gorge is an eroded gully 300 feet deep. People first seeing it get the idea that early hominids lived at the bottom of it in a sort of trench, for that is where their fossils are found along with stone tools. In truth, there was no gorge here two million years ago, only a flat lake shore, the surface of the earth then being where the bottom of the gorge is now. All the sediments seen in this photograph were deposited later, covering the fossils. The gorge itself was cut still later by a river.

For nearly thirty years they failed, although they found a great number of other fossils – many of extinct animals, many new to science. They gradually got to know the geology of the Gorge very well, and began to get a sense of where the lake edge had been at various times as the lake shrank and grew. During the 1950s, with Louis' finances improved by his museum salary, the Leakeys were able to go there every year. Louis' respectability as a conscientious scientist began to solidify as he worked out the Gorge's geology and its animals with increasing detail and sureness. Young Clark Howell, just out of graduate school, visited him. They became friends, and Leakey helped Howell get started on a dig to the south of Olduvai. That was typical of Louis. He was a conspicuously generous and appreciative mentor to young colleagues whose ability and dedication he respected. He had a sense of the enormous amount that had to be done. The more worthwhile work dependable young men and women could do, the better.

In 1959, Howell, again with Leakey's assistance, made a survey trip from Nairobi into southern Ethiopia. When he got back, Louis invited him to dinner. At the end of the meal, Louis said he had a special dessert for Howell.

'With a funny little smile,' Clark told me later, 'he put a large biscuit tin on the table and watched me lift the cover. Inside was a magnificent fossil skull.'

Clark was flabbergasted. He recognised what it was instantly. He had been to South Africa and had made an extensive study of the australopithecine collections in Pretoria and Johannesburg. He knew the characteristics of the little *africanus* and the big *robustus* well. The skull staring at him out of the biscuit tin was a robust one, a finer specimen than any he had seen before. It was, in fact, super-robust; its molar teeth were enormous, big enough to crack walnuts.

'Well,' said Louis, 'we finally got one.'

Actually, it was Mary who found the skull. It was at the very end of their 1959 field season, and she had decided to go out alone because Louis was lying in his tent with a bad attack of malaria. Rugged as he was, he was prone to various illnesses, many of them coming from the reckless way he abused his body. He got bilharziasis, a tropical parasitical disease, several times from swimming in Lake Victoria. He had gall bladder trouble, which he ignored until his gall bladder had to be removed. He had occasional 'fits' – blackouts from exhaustion and possibly from the long-term effects of his rugby injury at Cambridge. His hip joint deteriorated until he could scarcely walk, and it had to be operated on. He was once attacked by a swarm of bees and suffered several hundred stings, which nearly proved fatal. He had heart trouble, suffered a coronary, injured himself in various falls, survived a head operation, and finally died of a stroke, brought on by a lecture tour he insisted on completing although he was already mortally ill.

So it is no surprise to hear that when Mary dashed into camp with the news that she had found what they had been vainly hunting for so many years, Louis bounded out of bed, chills and all, and ran to the spot, overcome with joy and nearly speechless at the sight of the skull just emerging from the sediment.

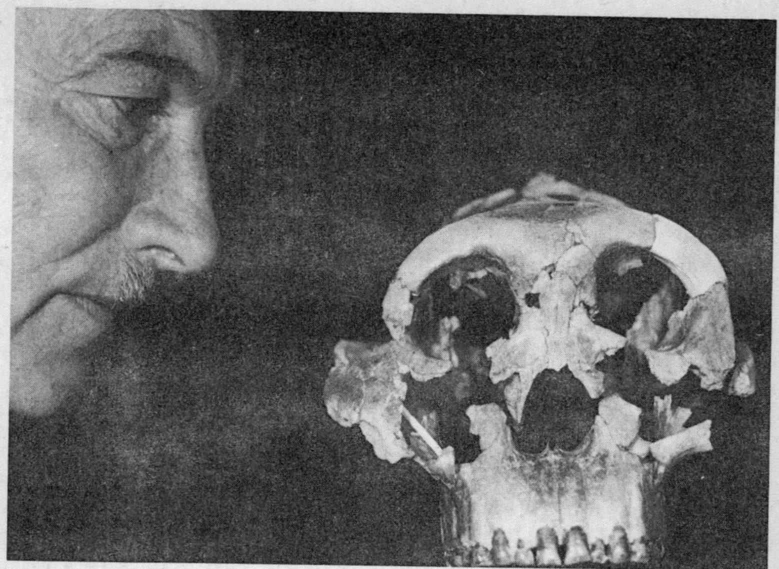

This is 'Zinj', found by Mary Leakey in 1959 and christened by her husband, Louis. It was the first australopithecine discovered outside South Africa, and the first anywhere to be reliably dated. Its age: 1.8 million years. It is a super-robust type, with the largest molars and jaw of any specimen found so far. Compare it with *A. robustus* (page 63); the two resemble each other very closely.

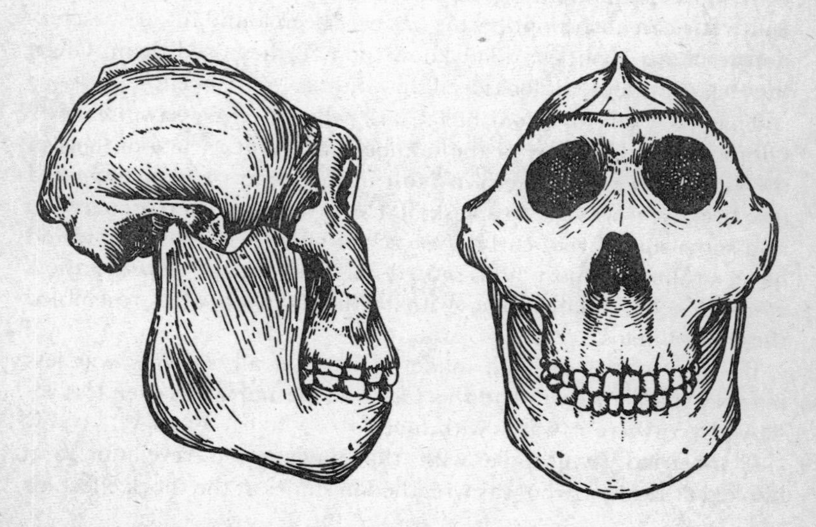

That is one story, the one most often repeated. Another is that he took one look at the fossil, growled, 'Why it's nothing but a god-damned robust australopithecine,' and went back to bed. Support for this second version comes from Louis' biographer Sonia Cole, who quotes Mary: 'When he saw the teeth he was disappointed, since he had hoped we would find a *Homo* and not an *Australopithecus*.'

Louis' fascination with *Homo* was obsessive. The australopithecines, he insisted, while interesting, were not humans; more, they were not ancestors. Peking Man and Neanderthal Man were not ancestors either, he believed. It was as if there stretched in Leakey's mind a straight line that went back from *Homo sapiens* to an ever cruder but always recognisably sapient forebear. And he was determined to find him.

His reservations about the ancestral credentials of the australopithecines were not unreasonable.

'Remember,' says Clark Howell, 'when Leakey found that first skull at Olduvai he didn't yet know how old it was. It obviously was not a man. It was even less manlike than the least manlike of those two South African types and don't forget that nobody knew how old *they* were either. By 1959 science had gotten around to acknowledging that they weren't apes. But it was still a big jump to turn them into our grandparents. We associate human beings with culture, with tools. And whatever Raymond Dart may have thought, no South African australopithecine has ever been found in sure association with any tool. We don't know today if they used them. Given the lack of evidence, it looks as if they didn't.'

That is what so astonished Clark when he first saw Leakey's Olduvai find. He knew of the number and great variety of tools in the Gorge. The presence of a skull in the same stratum staggered him because it was the wrong skull. He would have been prepared to find some sort of very early *Homo erectus* type there. He even would have swallowed hard and settled for the gracile *Australopithecus africanus* – the smaller one, with teeth that somewhat resembled those of humans.

But a robust one? That made no sense at all. *Robustus* was less manlike than *africanus*. And this Olduvai specimen was even less so.

And yet, there it was – with tools.

Clark tried to grapple with that unexpected revelation and blinked at Leakey, who was wreathed in smiles at the shock effect his 'dessert' dish had provoked. By this time, of course, Leakey had gotten over his initial disappointment that the skull was an australopithecine and not a *Homo*. It was a Leakey find, and therefore something special. He decided that its teeth were so much larger than those of *robustus* that it deserved a name of its own. He wrote, 'I am not in favor of creating too many generic names among the Hominidae but I believe that it is desirable to place the new find in a separate and distinct genus. I therefore propose to name the new skull *Zinjanthropus boisei*.' This statement was disingenuous in the extreme. Leakey was a notorious 'splitter', one who uses small differences as an excuse for naming new species, as opposed to a 'lumper', who tries to simplify by collapsing several types into one species.

The name he picked, 'Zinj', derives from an Arabic word that means East Africa. The species name *boisei,* honors Charles Boise, an early financial supporter of Leakey's work. That name did not last long. Others, noting the close resemblance to robust types, insisted that it be rechristened *Australopithecus boisei,* and it eventually was. Louis did not object too much. To him the skull was always 'Dear Boy'. The name 'Nutcracker Man' also got attached to it because of its outsize molar teeth. But to most anthropologists the fossil is familiarly known as Zinj. Now, more than twenty years after its discovery, it is still the best example of its kind ever found.

Zinj made Leakey famous overnight, although the fossil was really more Mary's than his. It was she who found it. It was she who fitted together all of its several hundred pieces, a job that took months. Clark Howell was one of the first to see the skull in its assembled form. It was typical of Louis to have chosen an obscure young anthropologist for that honor. He probably remembered what something like that would have meant to him at a similar stage in his own career.

Zinj was publicly unveiled at the Fourth Pan African Conference on Prehistory, held later that year in Leopoldville. It stole the show. Those present can still remember the shiver of astonishment that ran through them when Louis proudly displayed the skull. They marvelled that a being so primitive should have been responsible for making the stone tools that all knew littered the bottom of the Gorge. Whatever they may have thought, they had to be respectful of Leakey's public statement that Zinj – unlikely as it may have seemed – had been a tool maker.

There was more to come. Within a year Howell began to hear rumblings of a revolutionary dating technique being developed by Italian scientists to estimate the age of lava flows and volcanic ash deposits in the vicinity of Rome. This depended on measuring the amount of decay of a radioactive isotope of the element potassium into another element, argon. He paid little attention to it at first, admitting that his preoccupation with fossils and other geological and financial problems was so great that the potential of the new discovery had escaped him.

It didn't for long. Howell went to a meeting in Philadelphia to hear two geologists, Jack Evernden and Garniss Curtis, describe how they had been applying the method to get accurate dates from the volcanic deposits, or 'tuffs', from Olduvai. Clark was electrified. It hit him like a thunderclap that here at last was a way of finding the true age of something more than a few thousand years old – if the technique was reliable.

Apparently it was. Test after test was performed on tuffs at the bottom of Olduvai. Within a very small margin of error they kept giving the same result. Bed I was about 1.8 million years old. Therefore Zinj was also about 1.8 million years old – the first hominid fossil in the world to have a reliable age accorded to it.

The technique that pinpointed Zinj, now universally known as potassium-argon dating, has revolutionised geology and paleoanthropology. Its development was a stunning climax to the most productive decade yet in the long struggle to elucidate human origins.

To sum up the achievements of the 1950s:

First: Those erect, ape-brained australopithecines, thanks to the morphological studies of Le Gros Clark, had finally been recognised as hominids. That meant that they were members of the human family; but whether they were ancestors or cousins was still not determined. Despite the loud assertions of Louis Leakey to the contrary, most experts were willing, tentatively, to assign to *africanus* the role of ancestor – now, even more than before, on the basis of its presumed age. The reasoning went like this: If Zinj (a robust australopithecine) was nearly two million years old, then the robust type from South Africa should be of comparable age. That would mean that *africanus*, the more manlike gracile type, coming from presumably older caves in South Africa, would be older still. How much older, nobody was prepared to say. But there began to be a general

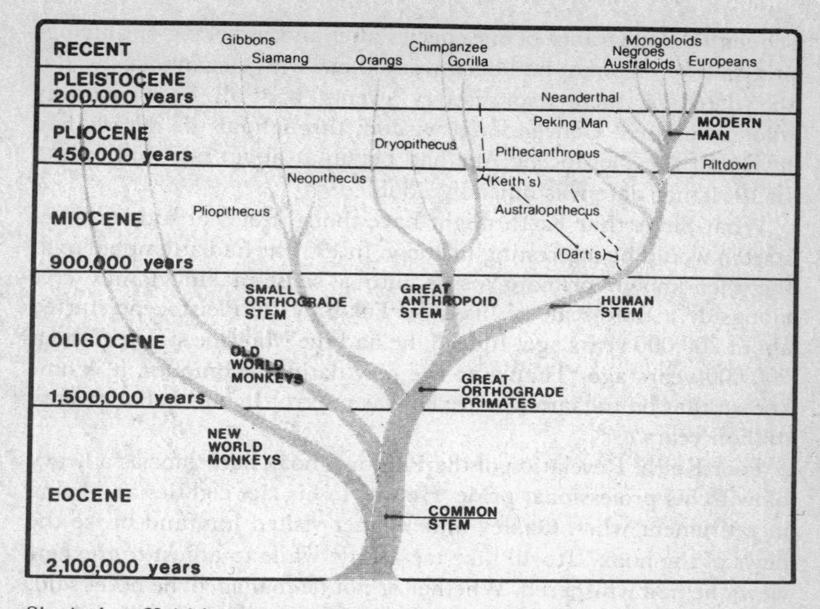

Sir Arthur Keith's primate family tree, drawn in 1931, reflects the shaky knowledge of hominid evolution which prevailed at that time. It suffered from the small number of fossils in collections; from poor understanding of their relationships; from a misplaced emphasis on differences between individual specimens; from failure to recognise Piltdown Man as a fake; and from a fatal misunderstanding of the true time scale. Radiometric dating now gives the Miocene about twenty million years; Keith gave it less than one million. Compare this arrangement with the one on page 369 made in 1979 by David Pilbeam.

consensus that *africanus* was at least two million years old, and might be as much as two and one-half million years old – plenty of time to evolve into a human.

Second: Piltdown Man was, thankfully, disposed of. Everyone was delighted to see the last of those fraudulent bones.

Third: Potassium-argon dating caused geologists and paleoanthropologists all over the world to reassess the ages they had been assigning to mineral deposits and fossils. The 1.8 million-year date for Zinj was responsible for some astonished second looks at the dates that paleontologists had been marking down alongside those trees of mammalian evolution they had been laboriously constructing. There was nothing wrong with the trees themselves. The sequential appearance of one species after another of pig or antelope or giraffe or elephant had been worked out with elegant clarity. But the dates of these evolutionary events had all been woefully underestimated. Conventional wisdom throughout the decade had held that the Pleistocene Age had begun at about one million B.C. In 1961 that date was officially doubled.

What Sir Arthur Keith might have thought of that sudden time-stretch would be interesting to know. In 1931 he had attempted to fit the then-known primate fossils into a sense-making family tree alongside a time scale of his own. For him the Pleistocene started about 200,000 years ago. Indeed, he had the Miocene starting about 900,000 years ago. Thanks to the new dating techniques, it is now known that on the latter figure he was in error by more than twenty million years.

Poor Keith. Revelation of the Piltdown hoax had come as a heavy blow to his professional pride. He was in his late eighties and living in retirement when Oakley and Weiner visited him and broke the news of the hoax. 'It will take me a little while to adjust to the new view,' he had whispered. Whether or not he managed, he never said. He was far from the anthropological scene, in the mists of old age, and died eighteen months later, well before he had a chance to learn that during his entire lifetime of estimating fossil ages he had so ridiculously miscalculated the age of every one.

4 East Africa: *Homo Habilis* – The Earliest Man?

'Homo habilis' has received a good deal of publicity since his sudden appearance was announced, and it is particularly unfortunate that he should have been announced long before a full and detailed study of all the relevant fossils can be completed . . . From the brief accounts that have been published, one is led to hope that he will disappear as rapidly as he came.

WILFRED LE GROS CLARK

I must protest against the use of quotation marks around Homo habilis, Zinjanthropus, *and* Kenyapithecus *in Sir Wilfred's letter, as well as the failure to put them in italics. All these are valid names in the terms of zoological nomenclature, and are not mere nicknames.*

LOUIS LEAKEY

Homo habilis *is an empty taxon inadequately proposed and should be formally sunk.*

C. LORING BRACE

The Zinj find of 1959 not only made Louis Leakey famous; it also made paleoanthropology fashionable. Not since the 1920s and 1930s, with the discoveries of the Taung Baby and Peking Man, had there been any new hominid fossils of their caliber to engage the public mind. Geology and paleontology were rated dull subjects in universities. Foundations turned their backs on requests for money to be spent digging for bones in obscure places. Newspapers ignored the subject.

But when Zinj fell into Leakey's hands, it was like a legacy from a forgotten uncle. Now he had a hominid, and that made all the difference. The National Geographic Society began funding his work and publishing articles about him in its magazine. Before Zinj, the Society had never given him a dollar.

Human fossils work a special magic. We have always been more interested in our own origins than in the origins of anything else. We trace our family roots and take pride in their length. We follow the histories of nations to their sources. We look behind recorded history to the beginnings of civilisations, and ultimately to the beginnings of humanity itself. Where *we* come from is where the interest lies. Fossil hominids have always had more clout than fossil clams.

I was still in high school when I read about Zinj in the *National Geographic*. The name Olduvai, with its hollow, exotic sound, rang in my head like a struck gong. I was about to graduate, and despite what my mentor Paul Leser had been telling me about the virtues of chemistry as a profession, I began thinking more and more about anthropology. Leakey's experience was proof that a man could make a career out of digging up fossils.

I went off to college, and Leakey promptly jolted me again. In 1962 there came a report that he had found another hominid fossil at Olduvai, this time not an australopithecine but a true human. With Leakey's well-known penchant for finding – and naming – things that were different from everybody else's, there was a period of waiting to see just what this new one might turn out to be. In 1964 a full report came. Leakey had formed a team consisting of John Napier from England, Phillip Tobias (Dart's student from South Africa) and himself. They had made a rigorous examination of some new fossils that had been recovered in the Gorge during the previous two years. These, they said, were larger-brained than australopithecines, and sufficiently unlike them in other respects to deserve placement in the genus *Homo*.

The shocker in the announcement was the age of this new *Homo*: about 1.75 million years, the same age as Zinj. At one stroke Leakey and his associates had tripled the known age of humans.

The name given the new find was *Homo habilis* (handy man, a name suggested by Raymond Dart), in honor of its having been the maker of the stone tools in the Gorge. That much, at least, about *habilis* was eminently satisfying: the idea of Zinj as a tool maker had never sat well. That such an extremely primitive-appearing creature with huge unmanlike molars, a small brain and a bony crest along the top of its skull should have been the tool maker – and thus, by inference, the ancestor of man – had been hard to digest. It was a relief to shove Zinj aside and accept this new larger-brained type as a more appropriate human ancestor.

From there on, however, *habilis* had heavy going. A principal reason was the scrappy condition of the evidence, for the four specimens found were badly preserved. As fossils so often are, they were given names. The first one found, a mandible with two cranial fragments, was named Johnny's Child after its finder, Leakey's son Jonathan. Cindy, the second one found, had a lower jaw and teeth, some bits of an upper jaw and a patch of skull. The third, George, had only his teeth and some very small skull fragments. The fourth, Twiggy, was represented by a crushed cranium and seven teeth. George's story was a frustrating one. He was found so late in the afternoon that the delicate job of removing him from the rock matrix was put off to the next day. During the night a herd of Masai cattle wandered up the Gorge and trampled on George, squashing him flat and grinding him into fragments, many of which were never recovered. Twiggy, who also was flattened – not by cattle but by the remorseless pressure of rock – got her name from a notably flat-chested English fashion model of the day. Cindy is short for Cinderella. Nobody seems to know where George's name came from.

Despite the fragmentary condition of these skulls, a preliminary survey suggested that they probably were larger than the typical gracile skull from South Africa. That was enough for Leakey. Always obsessed with finding human fossils, he insisted that these belonged to the genus *Homo* and should be so named. As his colleagues became more familiar with the fossils and began finding other things about them that seemed to connote *Homo*, they reluctantly went along with him. Tobias set himself to reconstructing the skulls and deriving from them an idea of their brain capacity. This turned out to be an extremely difficult task, because the skull pieces were so small that it was not always possible to decide with certainty at what angle they should be inserted in the reconstruction. Raise the angle slightly, and the brain would get larger; lower it slightly, and it would get smaller. Despite this difficulty, Tobias managed, by making inferences back and forth between the three skulls, to calculate that their mean brain capacity was 642 cubic centimeters (about 41 cubic inches). For Leakey, 642 cc was plenty. It was 200 cc larger than the mean for the gracile australopithecines, and to him clearly placed the three Olduvai hominids in another, more advanced species.

But should that species be *Homo*? Why not another kind of *Australopithecus*? How small a brain could a hominid have and still qualify as human? In fact, how did one even define a human?

It may seem ridiculous for science to have been talking about humans and prehumans and protohumans for more than a century without ever nailing down what a human was. Ridiculous or not, that was the situation. We do not have, even today, an agreed-on definition of humankind, a clear set of specifications that will enable any anthropologist in the world to say quickly and with confidence, 'This one is a human; that one isn't.'

It is not accurate to say that there were *no* standards for assessing humanness. Back in the days when Keith and Woodward were measuring Piltdown Man's skull capacity, a large brain was the *sine qua non*. The question was 'How small a brain can we accept in a human?' The answer: 'The smallest human brain that we have.' That amounts to circular reasoning, but lacking more and better fossils, it was the best that Keith and his contemporaries could do. His figure, then widely accepted, was 750 cc.

Le Gros Clark later shrank the human brain minimum to 700 cc. Again, this was an arbitrary figure: it measured the capacity of the smallest fossil human skull then in human possession. But Le Gros Clark, even as he set that new standard, realised that a newer find might force him to lower it again. He was also troubled by another problem that troubled many others. He knew that a species cannot be defined properly by one characteristic. There should be many, and that was what emerging *Homo* had always lacked. If there had been as much in the way of an arm and leg and foot and pelvic bones as there was of skulls and teeth, it might have been possible to arrive at some useful benchmarks among those parts of the skeleton too. But postcranial remains were horribly scarce. Therefore, other characteristics of the skull – its shape as well as its size, the nature of the jaws and teeth – would have to be considered.

Australopithecines fell into a multimillion-year gap between true humans and late Miocene apes. And it was Le Gros Clark's celebrated review of ape and human teeth that determined that australopithecines were not apes. But they were not humans either. Australopithecine teeth, in short, were their very own. So were their brains, which were in the 430–550-cc range – consistently larger than ape brains, notably smaller than *Homo erectus* brains. That in-between condition, in brain and teeth, was what made it possible to regard australopithecines as transitional on the line of descent from apes to humans – closer to *Homo*, but not yet *Homo*.

So, was *Homo habilis* a human or an australopithecine?

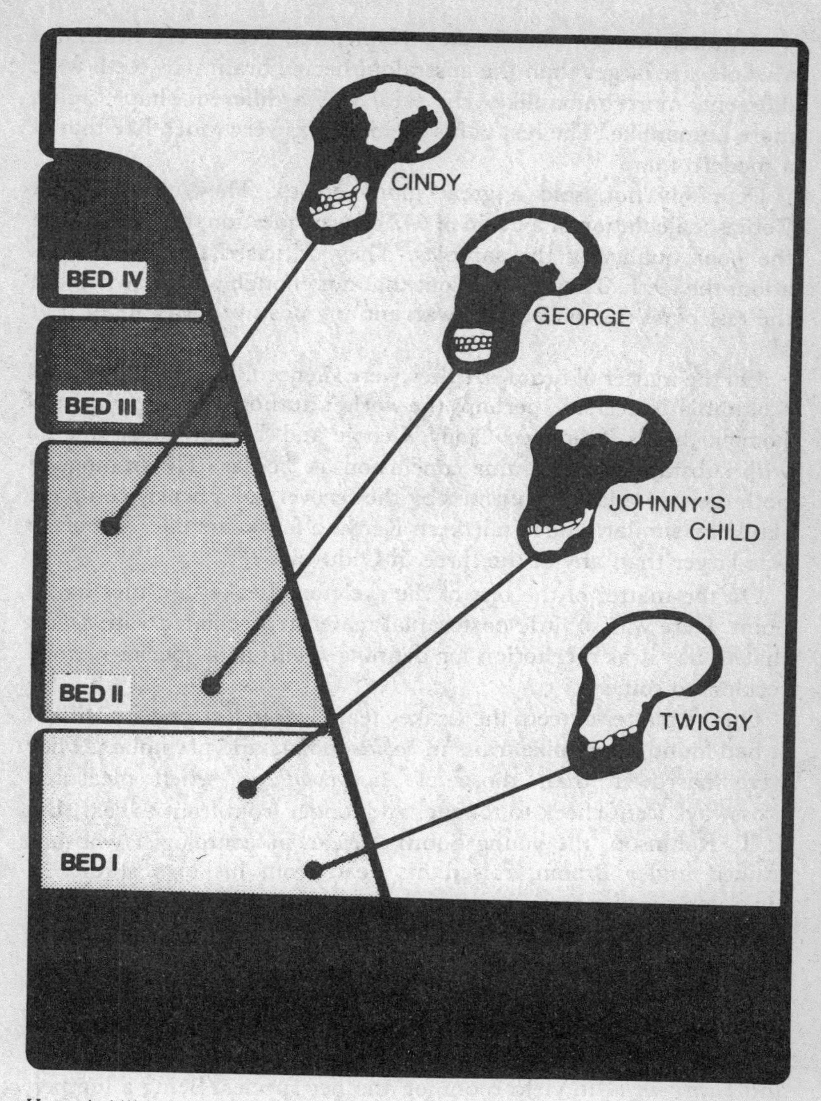

Homo habilis, the oldest known man, was identified and named by Louis Leakey in 1964 from these four specimens, all found in Olduvai Gorge, their preserved fossil parts shown against backgrounds of a complete *habilis* skull. Critics of the new species argued that the specimens were far too incomplete to justify their being given the name *Homo*, and that they should be regarded as gracile australopithecines.

Definitely human, said Leakey, Napier and Tobias. His brain was a whole size larger than the australopithecine brain. His teeth were different, more humanlike. His skull was a different shape, again more humanlike. The rest of his skeleton was very much like that of a modern man.

Definitely not, said a great many critics. They charged that Tobias' calculation of a brain of 642 cc was questionable, because of the poor quality of the samples. They criticised the conclusions about the teeth. They pointed out that not enough was known about the rest of *habilis*' skeleton to warrant any positive ideas about it at all.

On the matter of brain size they were silenced. Ralph Holloway of Columbia University, perhaps the world's authority on the insides of hominid heads, examined Cindy, George and Twiggy, and came up with substantially the same conclusion as Tobias. The findings of both men were later confirmed by the recovery of a nearly complete skull of a similar type in northern Kenya a few years later. Its brain was larger than any of the three in Olduvai.

On the matter of the rest of the skeleton, Leakey's critics had a point. Here was so little postcranial material associated with *habilis* that to use it as a criterion for defining *habilis* as a species seemed reckless to some.

On the matter of teeth the Leakey team was in a stronger position. It had found humanlike traits in *habilis* molars and premolars. They were narrower than those of *Australopithecus* when measured crossways from cheek to tongue, and longer from front to rear. But J. T. Robinson, the young South African paleontologist who had trained under Broom, raised his head from his own studies of australopithecine teeth to remark that those differences in shape were not significant. He said that one could find greater shape differences in a population of modern humans than Leakey had found between *habilis* and the australopithecines – or, in fact, between *habilis* and *Homo erectus*. Robinson's point was that on dental evidence alone there was too narrow a slot between *Australopithecus* and *Homo erectus* to yield room for another species. Being a lumper, he was not impressed by those somewhat narrower, longer teeth. Leakey, the archetypal splitter, was. Never one to back off from an argument, he threw himself into a rancorous exchange of letters in the pages of the scientific publication *Discovery* with Le Gros Clark, who tended to side with Robinson.

The argument ended inconclusively, the reason being that *habilis'* problem transcended mere measurement. It raised the deeper question of how species were determined, what criteria should be used, and how much variation could be permitted within a species. To take a single example: brain size alone is now recognised as a questionable index of species identification because of its variability. People today have brains that range in size between 1,000 and 1,800 cc, and in their lower range actually overlap the brains of *Homo erectus*, which ran from 700 to 1,250 cc. If the largest-brained *erectus* were to be rated against the smallest-brained *sapiens* – and all their other attributes ignored – their species names would have to be reversed. Similarly, the *habilis* brain (as would be learned later) overlaps that of *erectus*, varying between 500 and 800 cc.

It is clear that preoccupation with sheer brain size can be misleading. It becomes clearer when we learn that differences in brain size within our species appear to have no significant correlation with the intelligence of their owners. Rather, they reflect differences in body size. Big men have big brains, but they are no smarter than small men. Men are also larger than women and have consistently larger brains, but the two sexes are of equal intelligence. Since there has always been a high degree of sexual dimorphism in hominids, it must be accepted that there will be size differences in fossil skulls. If a large and a small skull are alike in every respect save size, the possibility cannot be ignored that the small one may be a female and the large one a male – and that they are the same species despite that size difference.

Add to this problem the one of differential rates of evolution in different parts of the body, and the difficulty of splitting a lineage of animals into species becomes even more evident. There is never a clean break in a line of evolutionary descent. An australopithecine mother never gives birth to a human son. There will be a period during which both mother and son will have such a blurry mixture of traits that assigning them to either species will be almost impossible. If one could assemble a complete series of mother–son–mother–son skeletons covering a couple of million years of time, and in the process going from something that unmistakably was not human to something that unmistakably was, one would be hard pressed to place a finger on the approximate spot – let alone the exact spot – where the crossover to humanness took place.

That was the trouble with *habilis*. He fell into that narrow shadow land between *Australopithecus* and *Homo erectus*, and perhaps there was not enough room for him there. Leakey continued to insist that *habilis* was a man, the oldest ever. Others claimed he was a gracile australopithecine, a slightly different version of the South African gracile type, but entitled to some differences because the two lived a couple of thousand miles apart.

What everyone agreed on was the need for better evidence, for more and older fossils. But there was no going deeper for them at Olduvai. The spade had struck rock bottom there at less than two million years. With the beginning of the Pleistocene now firmly set at two million, that date became a kind of doorway through which anthropologists began peering with growing intensity. Beyond that door lay the Pliocene, three million years of it. Somewhere in its depths were older hominids that would cast light on the origins of australopithecines and on their relationships with each other, and that might provide a better look at *Homo habilis* and a clue to what he really was. Perhaps, among those older hominids, there might even be one that walked erect three million years ago – or earlier. Those were big thoughts, but when one cannot see over the hill, one may think as big as one wishes.

PART TWO

The Golden Decade 1967–1977

5 Omo and Its Magic Ruler

I don't care what they look like. You can't nail down their relationships unless you know how old they are. The pigs will tell you that.

F. CLARK HOWELL

Ah – those wonderful pigs.

BASIL COOKE

There was never anything small about Louis Leakey's thinking. His mind ranged easily back through the Pleistocene, into the Pliocene, and beyond that to the Miocene. He had no way of determining the true dates of those ages, but he had a hard-earned smattering of geology himself, and he listened intently to the words of experts – always reserving the right to disagree with them. He hired Dr Richard Hay of Berkeley to decipher the stratigraphy at Olduvai. Hay spent more than fifteen years there, and as a result of his work, its canyon walls are nearly as well known to its students as are the walls of Fifth Avenue's canyon to students of New York City's architecture.

Like that of all paleoanthropologists before 1960 and before the introduction of potassium-argon dating, Leakey's knowledge of the various ages of the Tertiary (see diagram, page 11) was based on an interpretation of stratigraphy and an orderly arrangement of the animal fossils recovered from various strata. Thus it was possible for him to work comfortably around the shores and small islands of Lake Victoria in deposits that he recognised from the types of fossils he was finding as Miocene, without knowing exactly how old the Miocene was. Nevertheless, more than most anthropologists, because of the great variety of his professional skills, he had a finer sense of geological time. While others were speculating that the Miocene might have started five or ten million years ago, and whereas Keith had made his own crippled guess of 900,000 years, Louis grandly assumed that it had begun twenty or more million years ago – and the potassium-argon method has since proved him right.

His Miocene finds at Lake Victoria (recovered between 1932 and 1955), while not nearly as well known to the public as his extraordinary double strike of Zinj and *habilis* at Olduvai, are of comparable scientific importance. They include partial remains of many types of Miocene apes and give a rare glimpse into a world so distant that not one of its mammals survives today. The Miocene abounded in apes of various sizes, in contrast to the world of the present in which their descendants are reduced to only five types, all of extremely limited range and all threatened with extinction except one – man. It is sobering to note that for all practical purposes we are the sole survivors of the great blooming of apes in the Miocene; it is even more sobering to realise that we are responsible for the disasters that have recently overtaken all modern apes. Send a few more hunting parties into the mountain forests of Ruanda, Burundi and the Congo, and the gorilla will vanish. Continue to cut down the Budongo forest in Uganda, farm the Gombe stream, waste a couple of other fragile habitats, and the chimpanzee will vanish. Keep up the harassment in Borneo, kill a few hundred more mothers in order to get babies for zoos, and the orangutan will vanish. Turn the remaining hardwood forest tracts of Indochina and Malaya to lumber, and the gibbon will vanish. And that will be the end of the apes. It could happen in another fifty years – the winking out of twenty million years of evolution, evolution that the doomed apes themselves underwent in order to produce us, their nemesis. It is interesting that monkeys, not as closely related to humans as apes are, and presumably not as competitive with us, have fared better. Relatively rare in the Miocene, monkeys are much more abundant today.

Leakey's Miocene apes were excessively primitive. They bore only a working resemblance to the more specialised modern apes. While their fossils provided some tantalising hints about which ones might eventually evolve into chimps and which into gorillas, those evolutionary trends were still too dim to supply positive answers. To none of them could be assigned a hominid ancestry; hominid traits of any sort had yet to appear. For Leakey, always interested in the earliest men, the Miocene was simply too old.

He had poked into the Pliocene. South of Olduvai, in a spot called

Laetoli, he had extracted some Pliocene mammal fossils in 1935. But the recoveries were sparse and included no hominids – or so Louis thought. There was one hominid canine tooth that he did not recognise and that was not identified as such until years later by others. At any rate, he abandoned Laetoli and thereafter concentrated all his energies on Olduvai. But he kept the Pliocene in mind. When he offered to help Clark Howell make a survey west and north of Lake Turkana (formerly Lake Rudolf) in 1959, it was with the knowledge that extensive Plio-Pleistocene deposits lay on both sides of the lake and continued on up across the Kenya border into Ethiopia, particularly along the Omo River, the conduit through which water from the Ethiopian highlands runs down to keep Lake Turkana from drying up.

Clark raised some money, got a Land-Rover from Leakey in Nairobi, plus some advice on where and how to go, and headed north. He had previously obtained permission from the Ethiopian Embassy in Washington to enter Ethiopia. But when he arrived at the border, he was refused entry by the colonel in charge – obviously on someone's hate list, or he would never have been banished to that remote spot.

'I showed him my tourist visa,' said Clark. 'He said that didn't permit me to go roaming about, which was what I wanted to do. He was a long way from Addis Ababa, and he said he'd have to radio for instructions. Since his radio didn't work, I found myself hanging around his headquarters for several days doing nothing. My problem was that I had had to actually cross the border into Ethiopia in order to speak to him. I couldn't leave, and have any hope of returning, because I had a one-entry permit. That was ridiculous, because there I was, yards away from Kenya, with nothing within a hundred miles of me except that one police post, but I couldn't go back and forth.'

Clark finally persuaded the colonel to give him a written re-entry form. He crossed back into Kenya and spent several weeks driving around there and in the southern Sudan, surveying the emptier parts of one of the emptier regions of the earth. For most of the time he was alone. He had food, water and fuel with him, and he went where he pleased, crawling across an immense landscape whose silence seemed to summon him into the distant past. Clark remembers that trip as one of the great experiences of his life. He found some Early Stone Age tool sites near the Sudan border, but the Omo held more promise, and he went back to the colonel.

This time the colonel unbent a bit. He said that no word had yet come from Addis, but that as a special concession Clark would be allowed to walk around within sight of the police post, and in the company of a captain and a sergeant who would be assigned to him to make sure he did nothing out of order.

On his first walk, Clark came across an elephant skull poking up out of the ground. He photographed it and reported that evening to the colonel, who pulled a bottle of whiskey out of his desk and invited him to sit down for a drink. The colonel, who spoke perfect English, was starved for company, and over several evenings of drinking and fellowship, Clark was gradually allowed to expand his activities. He ended up going wherever he pleased.

The deposits at Omo nearly overwhelmed him. He had expected to find fossils, but he had not expected the complicated stratigraphy. A French paleontologist, Camille Arambourg, had been there nearly thirty years before and had collected a number of fossils. He later published an ambitious monograph on the Omo, describing it as a single moment in time with a very simple geological history. To Clark the geology was far from simple. It appeared to him to be incredibly complex – made up of innumerable layers of material, some of them extremely thick and apparently covering a huge span of time.

Clark took notes and photographs and made a small sample collection of bones, which he left with the colonel, asking that they be forwarded to Addis and eventually to the United States. He never saw them again.

'Did that bother you?' he was asked recently.

'Not at all. There were only mammals, no hominids. And we now have more than forty-five thousand mammal fossils from the Omo. We could spare those.'

Howell reported back to Leakey: Omo had a superb geological history and many layers containing fossils. Leakey asked him to stay for dinner, and at the end of the meal produced the biscuit tin containing the Zinj skull. The next year the first potassium-argon date was established for Olduvai and the start of the Pleistocene was shoved back to two million B.C. For Clark this meant that some of the strata he had been looking at in the Omo would probably have to be dated at three million years – squarely in the Pliocene.

But eager as he was to go back, Clark had other things to do. He was committed to working on the *Homo erectus* elephant-hunt site in Spain. Also, it was virtually impossible to mount an expedition in Ethiopia. It was not until 1965 that the subject of Omo came up again. That year Leakey was in the United States, and he and Clark discussed it.

'It should be an international expedition,' Leakey said.

'What does that mean?' Clark said.

'It means that we should ask the French; they've already been there. I would also like us to be a part of it.'

By 'us' Clark realised that Leakey meant only himself. 'That would be wonderful,' he said.

'Things are looking up,' Leakey went on. 'Give me a little time and I'll get word to you. You can go?'

'Oh, yes.'

'You're sure?'

'*Oh, yes!*'

Not long thereafter, Emperor Haile Selassie paid a state visit to President Jomo Kenyatta of Kenya. Leakey was summoned to describe for the Emperor the fossil wealth of Kenya and Tanzania.

'How is it,' asked Selassie, 'that there are no fossils in my country?'

'There are,' said Leakey.

'How do we find them?'

'That is not the problem, Your Majesty. I know of a young man who has been there and knows where they are. The problem is in getting permission to organise a proper expedition.'

Selassie waved his hand. 'By tomorrow there will be an order making this possible.'

A few weeks later Clark received a cable from Leakey: OMO OKAY SEE YOU SOON.

Leakey would have dearly loved to be the Kenya member of the tri-national expedition that was soon formed, but he was getting bad back trouble and his hip joint was in such serious condition that he could scarcely walk. Reluctantly, he decided to put the Kenya team under the leadership of his son Richard.

Richard Leakey was then twenty-three years old, a slender, alert young man with delicate features and a brittle, energetic manner. He bore no resemblance to his craggy father except that he had inherited a streak of do-it-your-own-way stubbornness. Although he and I are almost the same age, he had not reacted to his parents' Zinj and *habilis* discoveries in the same way I had. Instead of having his imagination inflamed by them, he was turned off. He had had a full dose of family anthropology since he was a small boy and wanted no more of it. When the time came for a decision about his university career, Richard said he didn't want one. His father, who had hoped that Richard would follow him as a paleoanthropologist, was understandably irritated. He said that if Richard did not want to be sent to a university he would have to look out for himself.

That was fine with him, Richard replied. He had had enough of bones. He liked the out-of-doors and the tracking and shooting of animals. He set himself up as a white hunter and began taking people out on hunting and photographic safaris. He proved to be extremely good at this. As a result, by the time he was in his early twenties he had learned a great deal about how to get about in rough country. It was that experience, and an obvious talent for organisation, which persuaded Louis that Richard should handle the Kenya end of the expedition, even though he had no training in anthropology or geology whatsoever. He *knew* quite a lot – how could he not, growing up with Louis and Mary Leakey as parents? – but his formal education had stopped in high school.

Haile Selassie kept his word, and the Omo Research Expedition was formed. Leakey's wisdom in suggesting that the French be included was sound. Not only would the French have objected to being excluded (and that would have made trouble because France, perhaps of all Western nations, enjoyed the best relations with Ethiopia), but under the unspoken rules of the paleoanthropological game, one does not move in uninvited and begin picking nuggets out of another prospector's claim. Arambourg had been the first to do any serious work at Omo, and the Ethiopian authorities knew him. Therefore, although his earlier conclusions about the geology of the place had been seriously in error and although he was old and ailing by this time, he was put in charge of the French team. To strengthen its geological expertise, he was given a second-in-command, a younger man named Yves Coppens. Howell headed the American contingent, Richard Leakey the Kenyan.

All the planning and provisioning for the expedition had to be done from Nairobi. Even though it was more than five hundred miles from there to Omo, and dreadful as the roads were, it was better than the shorter trip down from Addis Ababa. That way was virtually impossible because of the roughness of the country and the lack of roads. Furthermore, the simplest things, like toilet paper or glue, were unobtainable in Ethiopia and would have to be sent from other countries to a staging point in Addis. Nairobi was much more convenient and better provisioned, and it had the Leakeys right there to monitor the work.

The Omo venture, which had its first field season in 1967, was billed as the first truly cooperative international anthropological expedition, but it was not quite that. Each team was set up to act independently of the others. Clark Howell and Richard Leakey quickly found out that the French, with a prior lien on the place, had reserved for themselves the largest tract of deposits along a stretch of the Omo River a few miles above the point where it empties into Lake Turkana at the Kenya border. Clark was allotted a much smaller section just north of the French one. Richard Leakey found himself farther upstream around a bend in the river and on the far bank. With increasing frustration, he worked there for only a short time in deposits that clearly were too young to suit him. Not only was he disinclined by nature to work under anybody, but he quickly realised that he was in danger of becoming the expedition's 'Esau', the untutored hewer of wood and drawer of water, and that its 'Jacobs', the scientists Arambourg, Coppens and Howell, would get all the credit for any discoveries that were made. Having turned his back earlier on paleoanthropology, he decided that if he was going to reverse himself and become a paleoanthropologist after all, he would be wiser to do it on his own terms and in a better place. Taking a calculated gamble that he could find a better place, he borrowed a helicopter that the expedition had jointly leased and flew southeast back into Kenya, down the eastern shore of Lake Turkana and over a large stretch of eroding deposits that looked promising. He landed, and found himself in a desolate Plio-Pleistocene landscape that was littered with fossils. Even he, a nonprofessional, was stunned. He returned to Omo and announced that he was leaving the expedition. Henceforth he would conduct his own investigations in his own country – a bold decision for a twenty-three-year-old with an irascible father to face.

That helicopter ride was the most fateful ever taken by a fossil hunter. Richard Leakey organised a team of professionals and set up a permanent camp at Koobi Fora on a promontory jutting out into Lake Turkana. What he found there over the next few years (to be discussed in the next chapter) would have a profound effect on how we look at our origins.

The French and Americans stayed where they were. Arambourg died, and Coppens took over for the French. Work continued every year at Omo until 1974. Thereafter, Ethiopian politics became so turbulent that large-scale field seasons were rendered impractical. But even before 1967 the methodical Howell had suggested, and Louis Leakey had agreed, that a preliminary survey by a trained geologist would be a good idea.

'I'm not a professional geologist,' Howell said. 'I've had to learn a lot of geology, but before sending forty or fifty people out to Nowheresville I thought it would be prudent to have Omo checked out. I was fairly sure that the things I had found on my survey in 1959 were Plio-Pleistocene, but I wanted to make dead sure.'

A geologist, Frank Brown, retraced Clark's steps up to the lonely command post on the border. When he came back to the University of California to report in December 1966, he greeted Howell with a broad smile. 'The deposits are Plio-Pleistocene, as you expected,' he said. 'You will be pleased with them.'

That turned out to be the understatement of the decade. The great triumph of Omo is its geology, coupled with its animal sequences. Together they shine like a bright star in an otherwise dark and cloudy firmament of paleoanthropology – a field that is swept constantly by storms of disagreement over how old things are and, thus, how they are related. South Africa remains a nearly impenetrable murk. Other sites have had dubious dates attached to them because of the difficulty of interpreting confused stratigraphy or because of the impurity of the volcanic samples chosen to get dates.

But Omo is a marvel. It preserves a continuous record of events that is unique in the study of fossil hominid evolution. Its uniqueness results from a combination of special qualities.

'First, it's fairly big,' said Howell recently. 'It's not one gorge like Olduvai, but a jumble of eroded outcroppings that cover well over a hundred square miles. Second, it's deep. The deposits are more than three thousand feet thick, which means that the bottom ones are extremely old; you don't get three thousand feet of material deposited overnight. Third, it's faulted. That means that you don't have to dig down three thousand feet to get to the bottom layer. The

The deposits at Omo are notable for their depth (over half a mile), their accessibility, and their good dating. They consist of several hundred strata tilted upward at an angle by movements of the earth's crust so that even the older, deeper layers appear on the surface as long parallel waves of differentiated material. The ones in the foreground are older than the ones in the background. Their total age runs from four million to one million, with datable volcanic marker tuffs occurring throughout the section.

earth's crust has tilted or buckled, and the layers are sticking up at an angle, so that even the oldest are exposed here and there on the surface of the ground. In fact, you don't have to dig at all. You just walk along. As you go, each step you take carries you forward or back in time.'

The exposed sequence of strata at Omo constituted an immense geological ruler, with sections of time marked off by about two hundred separate and distinct deposits of mud, clay, sand, gravel and volcanic ejecta. More than a hundred of these layers contained fossils. All together, they spanned about three million years in time, running roughly from four million years ago to one million years ago.

'The volcanic activity – lava flows and tuffs – was the fourth extraordinary thing about Omo,' said Clark. 'They were the measuring markers that could be dated. Omo must have been a pretty hot spot back then, because there are dozens of tuffs. Some of them are more than forty feet thick – really big puffs from a volcano that has since disappeared; we're still looking for it. The point of contact between the sediments and the tuffs is razor-sharp in many places. That has been a godsend to the geologists. They can collect extremely pure samples of ash to be sent back to the States for potassium-argon dating. When the lab begins returning dates to us we can get a very close fix on the ages of the animal fossils, because some of the tuffs are extremely close together. They are as little as fifty thousand years apart. That means that for a fossil sandwiched between two of those tuffs at two and a half million years, say, you can get a date with an error of only about twenty-five thousand years either way. In the time ranges we're dealing with, that's nothing. It's as if you asked an extremely aged man if he was a hundred years old, and he replied, "No, I'm only ninety-nine".'

For dating, there had never been anything like Omo in the entire history of paleoanthropology. With a scale of that precision, it was possible to organise the multitude of mammal fossils that littered the Omo landscape, for they, like the deposits themselves, were exposed on the surface of the ground. In the eight years that the Omo expedition operated, it collected close to fifty thousand specimens, representing more than 140 species of mammals. Some of the little ones – rats, mice and shrews – were valuable because such small animals do not move about much and are extremely sensitive to environmental change. Thus the presence of a certain kind of mouse in a particular sediment says something precise about what the local climate was like when that sediment was laid down. The same is generally true of larger mammals, but the information is not as precise because many of them – antelopes, for example – range widely with the wet and dry seasons. Nevertheless, there are groups of mammals that together support a dry climate; others, like hippos, indicate swamps, lakes and rivers. Through the combining of information about animals, plants and geology, an increasingly accurate picture of the geography and climate at Omo through long stretches of time began to emerge.

Equally important, Clark's practice of carefully mapping every fossil made it possible to follow the evolutionary development of various types through time. Beautiful sequences of antelopes, giraffes and elephants were obtained; new species evolving out of old ones and appearing in younger strata, then dying out as they were replaced by still others in still younger strata. Evolution, in short, was taking place before the eyes of the Omo surveyors, and they could time it. The finest examples of this process were in several lines of pigs which had been extremely common at Omo and had evolved rapidly. Unsnarling the pig story was turned over to the paleontologist Basil Cooke. He produced family trees for pigs whose various types were so accurately dated that pigs themselves became measuring sticks that could be applied to finds of questionable age in other places that had similar pigs.

This work is called biostratigraphy. It is time-consuming, repetitious and dull. It goes on and on in sequestered laboratories and is never heard of outside professional circles. But it is the template to which every grand dream of paleoanthropology must be fitted. A gorgeous hypothesis about East African hominid evolution, if it does not fit one of those pedestrian pig sequences, must shrivel. If it does fit, it must command respect.

What did Omo say about the evolution of australopithecines, or about the respectability of *Homo habilis*? Directly, it did not appear to say much. The hominid fossils found there were fragmentary and of poor quality.

'I must confess,' Clark said to me once, 'that eight years of fieldwork should have produced more. We did find hominids. We found them in eighty-six different localities. But most were damaged by abrasion, and a great majority of them were teeth. You can't tell a great deal from a single tooth.'

Clark understates the value of his hominid finds. What he hoped to find was The Perfect Skull. He never did. I suspect he never will. Omo is more a place of rivers, of faster-moving water, than of the still conditions you find on the edges of lakes or in marshes. The grain size of sand and pebbles that are carried along is directly related to the speed at which the water is flowing. Teeth aren't any bigger than small pebbles, and if the water is moving pebbles it will move teeth – maybe miles from the jaw they were loosened from. The jaw itself gets banged around and worn down. Its finer features disappear. That seems to be particularly true of Omo.

'A lot of trouble with quality at Omo,' Clark explains, 'is, I admit, due to bone movement. But I think there are ways of getting around that. Toward the end of our work at Omo we had begun to pinpoint places where fossils were particularly numerous, and we had identified particularly rich strata. We were planning, in succeeding field seasons, to concentrate on those places. Instead of surface collecting, we were going to dig. The work would have been slower,

Camille Arambourg (*standing*) explored the Omo for fossils in the 1930s. He went back with Clark Howell and Richard Leakey in 1967 as the French leader of a tri-national expedition, but died soon thereafter. He was replaced by Yves Coppens, shown here excavating an elephant jaw.

Clark Howell, a pioneer in rigorous site development, is the man ultimately responsible for the orderly development of the Omo deposits. He is working here on a horizon containing a number of small fossil pieces. To preserve their exact spatial relationship, the finder cleared away the material around them and left each one on its own little pinnacle.

but I think the quality of the recoveries would have improved. Unfortunately, because of politics, we haven't been able to go back. It's often that way in this business. Just when you learn the most, something happens to prevent you from cashing in on it.'

I still feel that Clark cashed in better than he usually will admit. Nobody, of course, underestimates the value of his fantastic dating and animal sequences. But they are *so* fantastic that the poor old hominids tend to be overlooked. Clark found a lot of them. And one thing you can say about a couple of teeth – which by themselves don't say much – is that if in a similar time horizon somewhere else you find a better fossil with those same teeth in place in a jaw, then Clark's teeth snap into focus. That's what happened at Koobi Fora and Hadar. Both places have produced some beautiful fossils that throw considerable light on what Clark has found.

People keep comparing paleoanthropology to a jigsaw puzzle. It is a very good analogy, because in both you are faced with the problem of fitting individual things together one at a time to form a complete picture. But the jigsaw-puzzle solver has it easier. He has all the pieces. He just dumps them out on the table and goes to work. If he has any sense, he doesn't take one piece and begin sorting through all the others until he finds one that will fit it. He sets aside all the blue pieces and all the purple ones and tries to fit them together. He also finds all the pieces with straight sides. They are the edge of the puzzle, and it's pretty easy to begin stringing them together to get a frame. Then he can start fitting in the blue and purple hunks that he is gradually assembling.

The anthropologist can't quite do that. The pieces don't come to him in any orderly way. With that first Neanderthal skull there was only one piece, nothing to fit it to. Later, with the Java ape-man, there was another, but no clue to how they connected. The South African caves produced quite a few pieces – blue, let us say, for gracile, and purple for robust. Dart, Broom and Robinson managed to get some small pieces together, but there was still no frame for the puzzle. Then along came Louis Leakey with another piece, Zinj, *and a date* – the first piece of the edge of the puzzle.

What Clark Howell did was produce a whole strip of edge with his Omo dates. The hominid finds there, although their quality was not exceptional, did fit the frame. He found four different kinds. Since he knew how old they were, he could arrange them on the table – which had not been possible before – even if he could not fit them together.

Equally important, some of his finds were Pliocene – more than two million years old. He had broken out of the Pleistocene, just as he had hoped.

Clark's four kinds, according to a review he made in 1976, consisted of:

> A small hominid resembling the South African gracile australopithecine. It began to appear at Omo at three million, went on certainly to 2.5 million, and might have persisted to about two million, although the quality of the fossils makes that further persistence debatable.
>
> A robust hominid. It more closely resembled the super-robust Zinj than it did the South African *robustus*. It appeared at two million and disappeared at one million. Of the four types found at Omo, this one was by far the most abundant.
>
> Traces of *Homo habilis*. These occurred in a few seemingly manlike teeth and are dated at about 1.85 million.
>
> Finally, *Homo erectus*. First seen at about 1.1 million.

When those four were sorted out and set down on paper according to their ages, ideas about possible relationships began to assert themselves. For example, with all the gracile types occurring before two million and all the robust ones after two million, the suspicion arose that one might be descended from the other. The suspicion became stronger when one turned to South Africa to find gracile older than robust there also. Was that just a very queer coincidence, or was it shaping up as a true pattern in the jigsaw puzzle?

Second, with *Homo habilis* occurring at Omo as well as at Olduvai *after* two million, it began to be possible to speculate that it too may have had a gracile ancestor. With all that evidence falling so neatly and suggestively into place, it was tempting to turn the suggestion into fact by pushing the four lines together (see diagram, far right, page 285). But that would have been premature. 'Suggestive is not conclusive,' as Clark Howell has said to me more than once. Before we could accept that diagram, widely agreed upon though it was, we felt we should wait for more evidence. That, as the next chapters tell, would come from Koobi Fora, from Laetoli and from Hadar.

When I transferred from the University of Illinois to Howell's graduate program at Chicago, Omo was on everybody's lips. Howell's other graduate students were talking about nothing else. A deluge of evidence – maps, statistics and fossils – was flowing steadily into the laboratory. Hypotheses about how it all should be interpreted were bouncing around wildly. As usual, Clark himself was very cautious.

He admits to having been dubious about the reality of *habilis* at that time. 'I rode the fence on *habilis*,' he says. 'I wobbled. Louis Leakey's insistence that *habilis* was the earliest known type of *Homo* seemed logical as far as could be deduced from the fossil fragments. But those fragments didn't go far enough. I was given many chances to study all the Olduvai material. I know it well. It was extraordinarily interesting, but not conclusive. It was not until later, when we began finding similar *Homo*-like material at Omo at 1.85 million, that I began to feel strongly that Louis' original assertion was probably right and that there *was* room for another species between an australopithecine ancestor and *Homo erectus*. But we were still a long way from proving it.'

I was tremendously excited at finding myself in the centre of that forward rush of knowledge and debate, at being on hand when new ideas were in the process of being tested. I was more sanguine about *Homo habilis* than Clark was. Being younger, less experienced, and having far less professional responsibility, I could afford to be. At the time I was just another voice in the postgraduate babble of Clark's lab.

The first time I looked at some of those teeth, I thought: *Homo*. It was a hunch. I had no right to think it. I had never really studied the australopithecine collections in South Africa. My only familiarity with them was from reading about them and looking at some casts that Clark had. What I had been studying was chimpanzees. I had decided that my doctoral dissertation would be a complete review of chimp dentition. The total morphology: sexual differences, population variability, infant development, every bump on every tooth. Anything anybody ever wanted to know about how a chimp's jaw developed, how it chewed and why, I would be able to tell him.

With access to a sample of three hundred chimpanzee skulls in the United States, I worked out a file-card system based on IBM punchcard techniques for collating all the data I was taking from individual teeth. In the process I became so familiar with chimpanzee teeth that when I began comparing them with other ape teeth – gorillas' and orangs' – and then with human teeth and casts of australopithecines', I began to get a powerful intuitive sense of what made an ape, what made an australopithecine, and what made a human.

Most anthropologists, sooner or later, begin to settle more and more comfortably in a particular corner of their business. My corner was turning out to be teeth. It was lucky I hit on them. Your doctor will say to you, 'You are what you eat.' That is more true than even he realises. Your teeth betray you – what you are, where you come from. They can't help it.

When I took a look at some of the Omo teeth that Clark was tentatively assigning to *Homo*, I thought we should come right out and call them *Homo*. I repeat, I *thought* so. In the very process of having that thought, I realised that I was going to have to know more. I had to see the South African collections myself – the original fossils, not just casts. For the first time in my life I realised there was something I just had to know more about. I was also beginning to feel hungry about getting out into the field and finding things myself.

I also needed more chimpanzee skulls to complete my dissertation: There were two collections in Europe that, added to those in the United States, made a total of 826 specimens, a large enough sample to be statistically valid. (These, incidentally, were all of modern chimpanzees. No fossil chimpanzee skull has ever been found.)

At that point, I stuck my neck out. Clark had helped me get a small grant to go to Europe to finish up the chimpanzee skulls. I learned that he was planning to go to South Africa that same summer to study the australopithecine collections. So I asked if I could go with him. I explained the recording system I had worked out for chimpanzee teeth and showed how it could be used for a more detailed and orderly description of australopithecine teeth than had ever been made before. Clark fell for that. There were certain cusp patterns and dental grooves that had never been analyzed. He said, 'Okay. Write a little more money into your grant application to cover your air fare to South Africa from Europe. We'll meet in Nairobi.'

Then I *really* stuck my neck out. I said that if I was going to get as far as Nairobi it would be a shame if I didn't go just a little farther, up to Omo, and work there for a while that summer. Clark gave me a funny look, but again he said okay. He spoke to Gerry

Eck, his field coordinator, about me. Eck knew me as a graduate student, and he groaned. He said, 'I'm not so sure about that guy. He strikes me as the kind who likes big cities and bright lights. Do we need him?'

I have to confess that Eck was right about me. I do like cities and bright lights. But Clark explained that I had some training in archeology and that I might be useful if they did any digging that summer – and that it was going to be only a few weeks anyway. Eck rather grudgingly gave in.

In the spring of 1970 I went to Europe, assembled my chimpanzee data and flew to Nairobi, where Clark and I spent several days together. It was a time of great excitement for me. I felt myself opening up, my career beginning. At last I was in Africa. I was getting a firsthand look at fabulous fossils – Zinj, Twiggy, Cindy and George – about which I had read so much but never seen. Here they were, lined up on a table. Reaching out, I could actually touch them. I picked up the Twiggy skull, noticing how heavy it was – like stone. I wondered if I would ever find anything like that, wondered if I would have had the skill and patience to reconstruct a rounded skull from the pancake-flat Twiggy that had been found.

I was introduced to Mary Leakey and renewed an earlier acquaintanceship with Richard. Richard had already become a glamorous figure, well known to the world's press. He was just down from Koobi Fora with some sensational new robust fossils from there. I had a chance to look at them – another thrill, for papers on them had not yet been published.

I was filled with a sense of being on the cutting edge of a science, of talking to people who were shaping it, who had found famous skulls and had themselves become famous. The feeling stayed with me when Clark and I flew to Omo in a light plane that the expedition used to ferry people and small supplies back and forth. I looked out of the plane as it went north over the green rolls of the Abardare highlands, past the snowy spike of Mt Kenya, on up to the dustier, drier north. The country began to spread out immensely, with gray-and-lavender mountains standing up in a waste of flat brownish landscape. Lake Turkana came into view, a hundred miles of it, and at northern tip a long coil of green that wriggled away still farther north to disappear in the heat haze. That green snake was a riverine forest, a narrow strip of trees clinging for life to the banks of the Omo River.

Gerry Eck was at the bumpy gravel airstrip when we landed – not enthusiastic, but not unpleasant either. He showed me the tent I would be sleeping in. I unpacked, got into a pair of shorts and charged out, determined to prove my usefulness to Eck immediately. Within half an hour I had changed my mind; I was near collapse from the heat, scarcely able to stand up. Eck was surveying, walking rapidly around in the blazing sun, looking for fossil fragments. Not only could I scarcely keep up with him, but I realised with a lurching sense of inadequacy that I could not identify a single one of the fossils that we were finding. That night, sweltering alone in my tent, I realised that I would have to behave differently. I confessed my uselessness to Clark the next day. He advised me to take things more easily until I could get better acclimated to the intense heat. He invited me to sit with him in the afternoons while he sorted out the bags of bones that were coming in from each morning's survey. After about a week of labelling and numbering, and being told over and over again that this one was a pig and that one an antelope, I began to recognise a good many of them, and Howell said, 'From now on, *you* can begin telling *me* what they are.'

I liked camp life very much and found ways of making myself useful; I checked supplies, learned to repair broken vehicles, did mapping and a great deal of paperwork. I got a grounding in Plio-Pleistocene mammal fossils. More than that, the pattern of the expedition itself – Clark's role and style as leader, Eck's as coordinator, all the details that had to be attended to – began to come into focus. I watched Eck at work, and came to admire him greatly. Eck was able to mesh a variety of practical skills that had nothing to do with his professional expertise – fossil monkeys – and use those skills to keep the camp supplied with food, its native labor cheerful and productive, and the work of the scientists coordinated. I also watched Clark intently. When my three weeks were up and it came time to leave, I realised that I had not dug a single shovelful of dirt as an archeologist – what I had supposedly been engaged to do.

'You did okay,' said Clark on the airplane as we headed south to Pretoria and Johannesburg. 'You learned a few things – didn't you?'

'Boy, did I ever.'

In South Africa, as in Nairobi, and again thanks to Clark, I got a look at the fossil collections, at specimens that Broom had himself pried out of the breccia in the Sterkfontein and Kromdraai caves. I approached one skull with what amounted almost to reverence. It was an *africanus* specimen known as 'Mrs Ples', one of Broom's most famous because of its fine state of preservation. I ran my fingers over it, over the cracks between the little pieces that Broom's own clever old fingers had so skillfully put together a quarter of a century before. Again I wondered if I would ever have a chance to do something like that. I looked at other fossils recovered by Robinson, at Dart's big collection from Makapansgat: the legendary discoveries of legendary men. There they all were, hundreds of them. I had not realised until that moment how large the South African collections were. True, they were mostly teeth and jaws, but that was what I felt most at home with. I settled down with Howell for an analysis of them. My recording system worked very well. Together we compiled an immense amount of data, and planned to put out a paper on it the following year. We have never done so; we have been so busy since that we have not found the time.

I left South Africa with a vivid sense of the reality of australopithecines, something I had never had before. I had seen dozens of jaws and hundreds of teeth, and their owners had gradually begun to come alive to me. The wear patterns – the small flat facets that had been worn in the teeth by actual chewing a million or two million years ago – were clearly visible, faithfully preserved by the fossilisation process with a precision that no plaster cast could ever match. The sight of those wear facets in those ancient jaws turned their owners into flesh-and-blood chewers, into digesters.

It was as clear to me as it was to Howell, and had been to Broom and Robinson, that there were indeed two kinds of australopithecines. I began to get a feeling for the many features of tooth, jaw, skull shape and muscular attachment that had to be subtly blended to produce a robust or a gracile type. One tooth did not necessarily do it; there was even some overlap in size and shape among some of the teeth. But overall, the two types were clearly distinct. *Robustus* had a thicker, more massive jaw; bigger, broader, more heavily enameled molars; distinctive premolars; stronger chewing muscles running up the sides of the head; a crest of bone along the top for anchoring those larger muscles. All those features, taken together, said one thing: powerful chewing. *Robustus* quite obviously was an animal adapted to a diet of coarse vegetable matter: roots, buds, stalks, The gracile australopithecine was much less so.

That was the conclusion Robinson had come to more than a decade earlier in a celebrated monograph on australopithecine teeth. Now I could see with my own eyes that Robinson had been right. I departed with a strong admiration for Robinson, not only for the logic of his analysis but also for the rigor of his descriptions of individual fossils. Growing up professionally under Howell, I was developing a lasting respect for accuracy in description. Of all the fossils I have examined since, I think Robinson's are as well described as any. I made up my mind in Africa that if I were ever lucky enough to get fossils of my own to describe, I would use Robinson's as a model.

Back at Omo, I tried out some of my impressions on Clark. 'You know, those robust molars look to be an extension of gracile molars. They're the same kind of teeth, just more so.'

'Yep.'

'They get bigger and bigger, more specialised, over time.'

'Yep, yep.'

'Why couldn't it be an adaptation that got started in gracile types and just kept going?'

'Could be.'

'Just gradually got more and more robust until it *was* robust.'

'Maybe.'

'Well, if *africanus* went on to *robustus*, then how could it be an ancestor of *Homo habilis* at the same time?'

Clark was hard to nail down. I persisted. 'How will we work that one out?'

'Takes time.'

'From the data we've collected?'

'From more fossils.'

'But we've just been through a big collection of them down there.'

'No good dates from down there.'

'But we know *robustus* is younger.'

'How much younger?' said Clark. 'Can you tell me that? I'd like to know that. I'd like to have some better fossils, some better-dated ones from somewhere else.'

'It seems to me we always need better fossils.'

'Yep.'

I spent the winter continuing my postgraduate studies and working on my dissertation. In the summer of 1971 I returned to Omo. By the end of that field season I had a thorough familiarity with Plio-Pleistocene mammals and a good sense of the problems involved in unraveling the stratigraphy of a geologically complex site and the difficulty of getting accurate dating. I had become a competent surveyor and had even found a couple of hominid fossils myself.

I also learned something of the prickliness of multinational expeditions. There was scarcely anything that the French did in the same way as the Americans. Their funding was different. The support that their embassies gave them was different. Their attitude toward work was less urgent.

Originally the two camps were physically separated by a considerable distance. But after Arambourg's death Coppens invited Howell's group to move down to the French concession and work a small section at its northern end. This was separated from the part that the French were working by a watering road that ran down to the river and that came to be known as the DMZ, across which there would be no professional trespassing.

Before I got to Omo, according to Eck, the American coordinator, the Americans were treated almost as second-class citizens by the French. That used to infuriate him. He remembers that the first job on arrival at camp each year was to construct an airstrip. This meant cutting away bunch grass by hand with bush knives in a fifty-foot swath for half a mile, an exhausting three-day job. If this was not done, the grass would stick up in tufts as the dust around it was blown away by propellor blast, and the strip would become so bumpy that it might shake a wing right off a plane. Each year the Americans made the strip, and each year the French immediately

Three stages in the geology of Hadar, which has been affected by climatic change, earthquakes and volcanism. The top drawing shows its condition 4.0–3.6 million years ago, with a small lake developing in a low spot. Material from the plateau and from mountains farther to the west is already beginning to flow into the depression, and has created several alluvial fans. Basalt is welling up from a volcano in the lower right corner. In the second drawing (3.6–3.3 million), rivers have cut through the plateau and are carrying more silt and gravel into a lake that had grown much larger as a result of a rainier climate. In the third drawing (3.3–2.6 million), the region is drying out again and the lake has shrunk. Hominids lived along the lake edges throughout this entire period (4.0–2.6 million). The Kadada Moumou basalt (the one dated by James Aronson) has been buried beneath deep alluvial deposits. Today the area is a near desert. There is no lake, only a small river, the Awash, on whose bank the present campsite is located. The basalt has been laid bare once again.

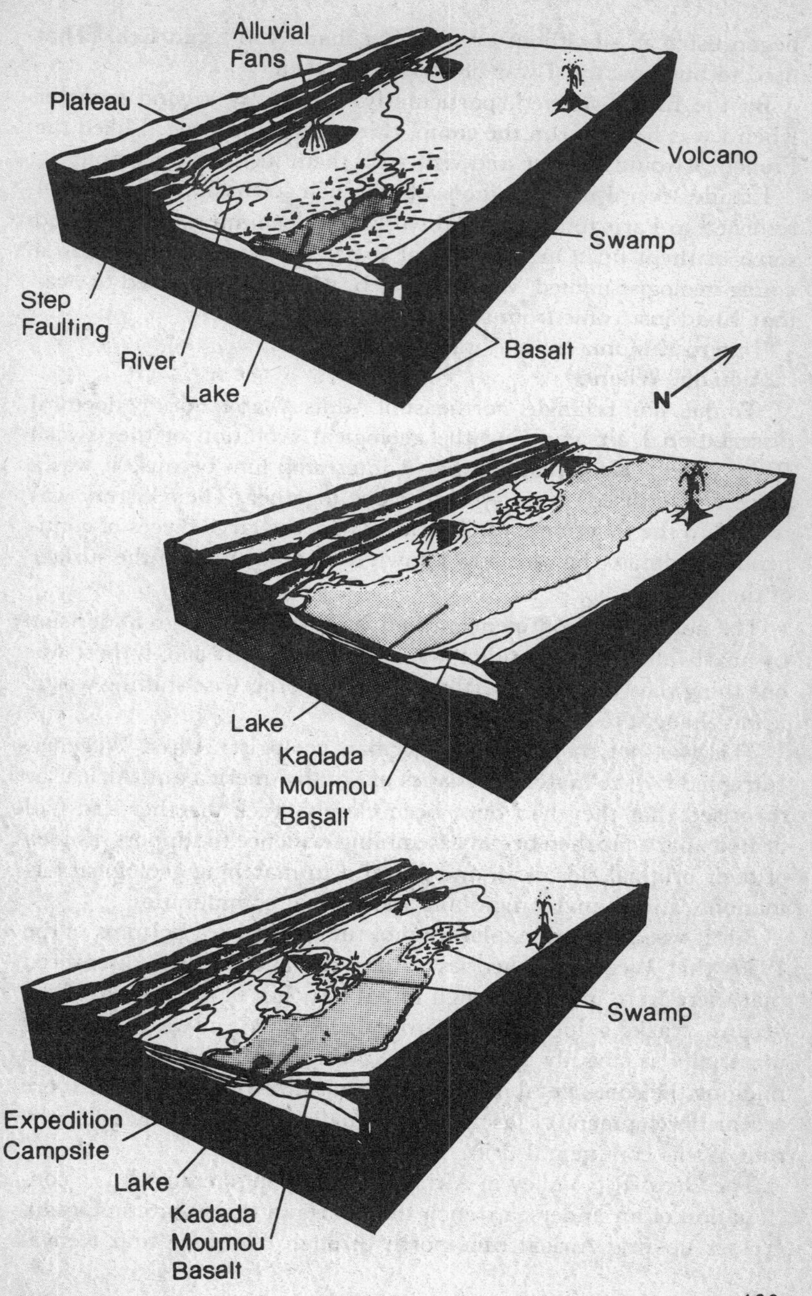
Alluvial
Fans
Plateau
Volcano
Swamp
Step
Faulting
River
Lake
Basalt
N
Lake
Kadada
Moumou
Basalt
Swamp
Expedition
Campsite
Lake
Kadada
Moumou
Basalt

began using it – 'without even saying thank you', said Eck. 'That used to burn me up. I was hot-tempered then.'

By the time I arrived, particularly during my second summer when I was helping run the camp, things had improved. I liked the French. I would go over and visit with them just to be sociable.

I made several friends among the French scientific staff. When I returned to Paris to do my final work on the chimp skulls, I looked some of them up. One evening at a party I was introduced to a young geologist named Maurice Taieb, who was interested to hear that I had just come from southern Ethiopia.

'I go to Ethiopia myself,' said Taieb.

'You do? Where?

'To the Afar Triangle, northeast of Addis Ababa. For my doctoral dissertation I am studying the geological evolution of the Awash River valley.' He explained that it interested him because it was a place where three rifting systems came together. Their history was related to the movement of tectonic plates, the large pieces of continental landmass that are now known to float around on the surface of the earth.

The notion that continents moved at all was held in great derision by nearly all scientists until about twenty-five years ago. If there was one thing that was reliably stationary in an otherwise shifting world, it was thought to be the continents.

This was not so, insisted a German geologist, Alfred Wegener. Intrigued by the matching shapes of South America and Africa, he theorised that they had once been closely stuck together and had drifted apart. He then began assembling evidence to support his idea of their original closeness, and found it in matching geological formations, and in matching plant and animal communities.

But it was not until exploration of the world's sea bottoms in the 1950s that Wegener's theories gained credence. It was discovered that there were immense longitudinal trenches in the middle of the oceans, cracks going halfway around the earth, in which volcanic material was steadily spewing up, flowing out from the cracks and nudging the continental masses in various directions. With the more recent development of lasers, it is actually possible to measure the rate of this continental drift.

The Great Rift Valley in Africa is a continental crack. It is a continuation of an undersea trench that starts in the Antarctic Ocean, crosses up into Africa, runs north through Tanzania and Kenya, across Ethiopia, and ends in the Dead Sea, between Israel and Jordan. The rift is a region of great instability in the earth's crust. Whatever stews are bubbling deep below the surface, they express themselves in a series of volcanoes that have been building and dying along its course for millions of years. Olduvai Gorge, its vanished lake, and the volcanoes past and present that surround it are all part of the Rift. Lake Turkana is a skinny little Rift puddle; the Red Sea, a big one. Africa is splitting up along that seam – one piece of it skidding off toward Arabia at the rate of a few inches every century, the other part swinging up against Europe. It is the movement of that second, northwestern plate that has thrust part of the European plate into a series of wrinkles known as the Alps.

The Afar Triangle was one of the rare spots on the earth's surface where three rifting systems were bumping and grinding together. Taieb had been out there to do some mapping, and one of the things that caught his eye was a huge area of eroded deposits containing fossils. Their age, by his guess, was Plio-Pleistocene. They were lying about all over the place, and appeared to him to be in remarkably good condition. Not being a paleontologist, he hadn't the slightest idea what they were.

'You should go out there with me and examine them,' Taieb said.

The allure of that invitation was tremendous. I had spent two summers at Omo. I was beginning to suspect that it would never produce exceptional hominid fossils. I knew, however, that there were places along the Northern Rift that did. Richard Leakey's Koobi Fora site was one. What if the Afar turned out to be like it, rich in quality as well as quantity? Richard had taken a chance and struck gold. Shouldn't I?

On the other hand, my personal situation was in knots. I had not finished my dissertation, had no teaching job lined up and no money. I had committed myself to another season's work at Omo. My marriage, only a couple of years old, was already collapsing as a result of my heavy work schedule and increasing absences on fieldwork.

But if I didn't go, Taieb would almost certainly ask somebody else. Taieb showed me some photographs he had taken of the Afar fossils. I recognised them instantly. A couple were Pliocene pigs' teeth; another was an unusually well-preserved elephant's skull. That skull pushed me over the line. For someone truly hooked on the past, bait like that is as irresistible as a U.S. airmail invert would be to a stamp collector – well worth a big risk. 'All right,' I said, 'I'll go.' Somehow I would fit a trip to the Afar into all the other things I had to do. I made a tentative agreement to be Taieb's partner on a short survey trip in April and May of 1972, before the Omo field season started. I would use my Omo air ticket – with an added stopover at Addis Ababa – to get me there. The other matters I would work out somehow.

When I got back to the United States and explained all this to Clark Howell, I was told that I would be acting very unwisely.

'You should finish your dissertation,' Clark said. 'Without it, you can't get your Ph.D. degree. Without a degree you're nothing. You can't get a job. You have no career.'

I thought that over.

'You can't ever raise any money either,' Clark added.

I left that interview feeling very subdued. But a few weeks later the job prospect brightened. Case Western Reserve in Cleveland had heard about me and offered me an instructorship in anthropology. I went to Cleveland to discuss it, and flabbergasted myself by saying that if I could be advanced $1,000 for fieldwork that summer I would join the faculty in the fall.

A young man who hasn't even finished his doctoral programme just doesn't ask for extra money before he's hired. I don't know what made me bold enough to say that. But you have to press your luck. I had been lucky in making that contact with Taieb. I was lucky again in getting the Case offer so fast. The people at Case gulped when I mentioned the thousand dollars. But I had come well recommended. They wanted me, and they gave it to me.

With that and with my entire savings of $600, I could just manage. I met Taieb in Addis Ababa, and together we scrounged for a week, borrowing a Primus stove from an Addis-based American couple who liked to go camping, and a tent from a Dutchman Taieb knew. We found some canned goods. I was able to buy a sleeping bag. Taieb provided a couple of elderly Land-Rovers from the CNRS – the Centre National de la Recherche Scientifique – an agency that funds French scientific activity worldwide and maintains a small headquarters in Ethiopia. Taieb's priceless contribution was an exploration permit from the Ethiopian Ministry of Culture. With that we headed northeast, dropping down from the central highlands into a hotter and hotter land of arid flatness in which almost nothing stirred except groups of Afar tribesmen with their cattle, their goats and their camels.

A German construction company, TRAPP, was building a road from Addis down to the city of Assab. We made that road a sort of running headquarters, surveying out from it for a day or two at a time, doing 'blitz' anthropology, trying to cover as much territory as possible in the shortest possible time. We made friends with the construction crews working along the road. These were lonely men, glad to give us bunks and extra food and water in return for our company and an occasional bottle of liquor. Taieb bowled over one crew one night by cooking up a dish of flaming bananas well laced with brandy over a roadside campfire.

We picked up an Afar guide named Ali Axinum to help us follow disappearing tribal trails, to steer us down into deep washes, or wadis, and out again, and to provide names for riverbeds, mountains and tribal headquarters, for we were doing mapping as we went. We managed to locate some of the deposits Taieb had seen before. As he had promised, they were rich in fossils of remarkable quality. In one of them I found the nearly complete skeleton of a colobus monkey – heartening because it was a primate.

'How does this compare with Omo?' Taieb kept asking.

'Better, much better,' I kept answering.

Two places in particular were better: Houna and Leadu. By Omo standards they were big, a couple of hundred square miles of eroded badlands in the 2.0-to-2.5-million-year range. I wanted to linger there, but Taieb insisted there was a still better place. All one day we looked for it, looping aimlessly over a landscape that was as featureless as a billiard table. Our guide, Ali, who was almost stupefied by a large wad of a narcotic leaf called *qat* that he was chewing, kept getting lost. He got us stuck in the beds of sand rivers and we would laboriously haul ourselves out. Just at sundown, and with no warning whatsoever, we arrived at the lip of a huge expanse of badlands stretched below us. In the low, slanted light, the bands of varicoloured sediments that had been exposed by erosion shone brilliantly. A small river crawled through the bottom of one of the gulches.

'This is it,' said Taieb.

'Hadar,' said Ali.

I had never seen anything like it. We camped on the rim, and were nearly blown away during the night by a roaring wind that swept over the tableland. Unfortunately, as I would learn, this did not penetrate to the bottom of the wadi, which was as hot as Tophet. We were up at daylight. It took more than an hour of driving back and forth to find a way down into the deposits for the Land-Rover. Once there, we were able to cruise about almost at will, through one dry gulch after another, all of them well stratified, all of them seeming to ooze fossils. It was a place paleoanthropologists see only in their dreams.

We stayed there three days, making a small collection of some of the best-preserved mammal skulls and long bones that we found. I also collected some pig teeth to compare with Omo pigs for dating. Thanks to my familiarity with the Omo mammals, I was pretty sure we were somewhere in the three-million-year range, but I wanted hard evidence. My belief was strengthened by Yves Coppens, who had hooked up with us a few days earlier. Coppens had asked Taieb if he could join the Afar enterprise, and had been accepted. Having been the French leader at Omo, he too knew its pigs well. 'These are the same,' he said. 'This is a better place.'

Back at Addis, I made an agreement with Taieb and Coppens to organise a formal expedition. This would be a truly cooperative international affair and would avoid the separatist difficulties of Omo. Funds would be raised and personnel recruited individually, but the work in the deposits would be jointly conducted. Taieb would be chief geologist, Coppens chief paleontologist, and I would be chief paleoanthropologist. We drew up a formal agreement specifying our various responsibilities, and the International Afar Research Expedition came into being. I went to Omo, finished out the summer there, returned to the United States, started teaching at Case, worked frenziedly over nights and weekends to complete a mammoth dissertation on chimpanzee dentition which I ultimately boiled down to 450 pages. I defended it and was awarded my Ph.D. I wrote up a grant proposal which I submitted to the National Science Foundation, asking for $130,000 for two years' fieldwork. I received $43,000.

I recruited two men: Gene Dole, who came highly recommended, and Tom Gray, who did not. Gray was a graduate student in archeology at Case. The faculty said he was unreliable and lazy and a poor choice. I saw him differently. I knew that he was extremely intelligent, and had a hunch that he would do well in the field.

Field appointments of untried men are a terrible gamble. One never knows about a person who has not been out there until that person goes. Dole was a serious young man, but camp life proved difficult for him. The heat incapacitated him. The threat of war along the Ethiopian border upset him. He was sick off and on. After the second year he did not return. But Gray, the near-dropout in archeology, was a gem. Apparently, Hadar was the catalyst he needed to energise him. He became my coordinator and most valuable associate.

Meanwhile Taieb had been gathering his own crew. They included an artist/scientist named Guillemot and two young women who in succeeding seasons would turn out to be invaluable. They were Raymonde Bonnefille, a botanist and fossil-pollen specialist, and Nicole Page, a geologist. The French group was larger than the American one. It was underfunded by CNRS, and before the first field season was over I found myself dipping into my own grant money to assist the French.

It was agreed that the first field season of the Afar expedition would start in the fall of 1973, just after an important anthropological meeting held in Nairobi. This was the Circum-Rudolf Conference, and had been convened to discuss the findings to date from fossil sites around Lake Rudolf (Turkana). The two most important sites, of course, were Omo and Koobi Fora. Howell and Coppens held forth at length about the Omo work. The star of the show, however, was Richard Leakey, who for the first time gave a detailed public review of his Koobi Fora finds. They were sensational. Not only had he recovered a large number of very fine robust fossils representing various parts of the body, but he had a magnificent *Homo* skull that his potassium-argon dating team had given an age of 2.9 million years. That skull and that date were the shockers of the conference.

I attended, but kept a very low profile. I had no fossils, only a stakeout and a hope. Faced with the glittering finds of Richard Leakey, I felt rather insignificant. When Richard asked me what I was doing, I described my hopes for Hadar.

'Do you really expect to find hominids there?' Richard asked.

'Older than yours,' I replied bravely. 'I'll bet you a bottle of wine on that.'

'Done,' said Richard.

6 Koobi Fora: The Triumph of *Homo Habilis*

I just find the fossils. I'll leave it to the experts to name them.

RICHARD LEAKEY

The aim of science is to seek the simplest explanation of complex facts. We are apt to fall into the error of thinking that the facts are simple because simplicity is the goal of our quest. The guiding motto in the life of every natural philosopher should be, 'Seek simplicity and distrust it.'

ALFRED NORTH WHITEHEAD

In the years after he abandoned the Omo expedition and set up on his own at Koobi Fora, Richard Leakey grew from being an obscure bushwhacker into one of the most prominent young field anthropologists in the world. The Kenya Government supported his work, later putting him in charge of all paleoanthropological activity in that country. The National Geographic Society stepped in to help, just as it had done with his father. Its magazine articles about Richard were read by millions.

Koobi Fora is not quite as hard to get to as Omo, but it is still far from accessible. Marsabit, the only town in Kenya's Northern Frontier District, is the jumping-off place for Koobi Fora. A good road runs to it from Nairobi, but from there one must traverse about two hundred rough miles, following a poorly marked track that crosses an alkali desert and then wanders generally northward through a wasteland of volcanic boulders and gravel that breaks the axles of all but the most rugged vehicles. There are romantic photographs of Richard Leakey whipping a camel along this desolate route, but the best way in or out is still by truck or Land-Rover, or by light plane. Leakey has since learned to fly, and can now land at an airstrip that is handy to the Koobi Fora promontory where the camp is located. It was on this point, jutting out into the lake, that Leakey established his headquarters. By 1969 he was deep in the problem of deciphering the extremely intricate stratigraphy of the East Rudolf deposits. The lake had risen and fallen a good many times in the past, and the strata at any one spot were not easily matched with those at another, being interrupted by erosion and by later incursions of different types of material – from a more recent river, say, its signature left in depositions of new gravel or sand before it dried up and vanished. Only those swirls of gravel remain to mark its existence, and to baffle geologists.

A number of geologists have worked at Koobi Fora, and have gradually reconstructed a stratigraphic 'column' of deposits for the area that run from about three million years ago to about one million years ago. There are a couple of 'marker layers' of volcanic tuff in this column that provide potassium-argon dates. One of them was discovered by a young woman from Yale University, Kay Behrensmeyer, who has achieved geological immortality by having the tuff named after her. It is called the KBS tuff for short, and is of critical importance because it falls between a marker tuff at the bottom and another higher up in the column, thereby supplying an in-between date for calculating the ages of a great many fossils that have been found just above it or just below it in the sequence.

In addition to his concentration on geology, Richard Leakey took another important organisational step at Koobi Fora by training a team of native Kenyans to do much of the surveying for bones. These men have become expert at finding and identifying fossils. Under the leadership of Kamoya Kimeu, they have located most of the notable ones recovered from the Koobi Fora area. Kimeu is regarded today as the best hominid-fossil field expert in East Africa.

Hominid bones began to appear almost immediately in recoveries from three search areas, one of them starting almost at the doorstep of the camp and running inland from the lakefront for a distance of ten miles or more. A second lies about twelve miles north of the camp; a third – and very productive one – at Ileret, about twenty-five miles north. There has been considerable local faulting throughout the area, dislocations upward or downward of the strata in one place as compared with those in another, with erosional washouts between sites. Much of the work of the geologists has been to unscramble this mess, and find correlations between sites – that is, certain sequences of strata that are recognisable in one spot as being identical with those in another. From such match-ups, even though the volcanic marker layer at the first site may have eroded away, it is still possible to date the fossils there because of its geological conformity with the second site.

The hominids the Koobi Fora team began finding in the greatest abundance were robust australopithecines. Also found were some smaller specimens which, to Leakey, seemed to fall comfortably within the range of Cindy and George and Twiggy, those elusive *habilis* specimens from Olduvai. This heartened him mightily. Like his father, Leakey had a fix on Old *Homo*. He hoped not only to strengthen his father's claim that *Homo habilis* existed as a valid species, but also to extend the known life of that species farther back in time. The prospects for doing both at Koobi Fora were good. The deposits were larger and richer than those at Olduvai, they went deeper into time, and many of the fossils being yielded were of better quality.

In 1972, Richard Leakey announced an utterly dazzling find: a superb skull from below the KBS tuff. Inasmuch as the tuff had been dated at 2.6 million years by a pair of English potassium–argon specialists, that meant that this new skull was older than that. Leakey gave it a tentative date of 2.9 million.

'What do you call it?' he was asked in an interview in 1972.

'I call it *Homo*. I won't give it a species name because that is not my speciality. For me there are two kinds of hominids at Lake Turkana. All the robust ones I call *Australopithecus*. All the gracile ones I call *Homo*.'

The new find was designated KNM-ER 1470. The letters stood for Kenya National Museum–East Rudolf. The number was its accession number in the collection. For some reason it has never been given a nickname. It is known universally as 1470 and is one of the notable fossil finds of the century. If one were to blurt out the number 1470 in any group of paleoanthropologists anywhere in the world, all would know what he was talking about.

Homo it was, with a vengeance. Its skull was thinner, higher and rounder than any australopithecine skull. Most remarkable of all, it had a brain capacity of 775 cc. This was a firm figure; it had none of the iffyness about it that the brain measurements of the Olduvai fossils had because 1470's skull was virtually complete. It had been assembled in an inspired burst of jigsaw-puzzle work by Richard's wife, Meave, after he and a couple of others had given up on it; there were too many pieces, they were too small, they could be fitted

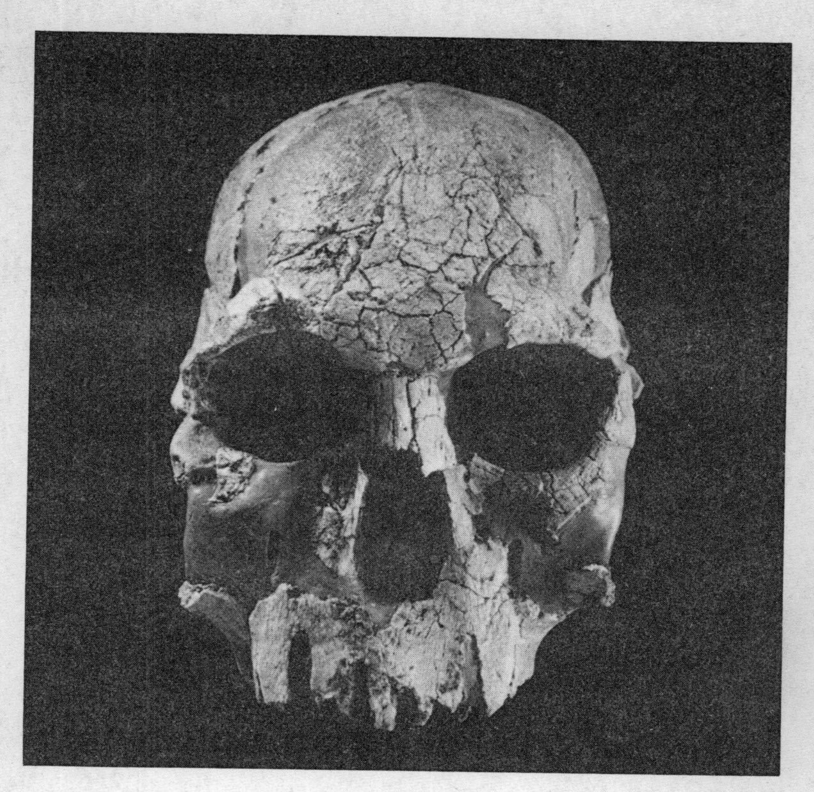

Skull 1470 is Richard Leakey's most dramatic find at Koobi Fora, and one of the most important hominid fossils found anywhere. It is definitely *Homo*. It is in far better condition than any of the *habilis* specimens from Olduvai and has a larger brain. It was first thought to be nearly three million years old – an awkward date indeed, for that made it older than many of its presumed australopithecine ancestors. More recent dating from purer ash samples has convinced most authorities that it is less than two million years old.

together in too many different ways. But Meave, working slowly and with exquisite care, managed a reconstruction, each piece in place reducing the perplexity caused by the remaining ones.

The 1470 find was a shocker to everyone but Louis Leakey. He was immensely pleased with it. It confirmed what he had been asserting all along: that humanity was old, very old; proof would be found. That was one of the few bright spots of the year for Louis. He was old himself, seriously ill, and harassed by a load of responsibilities that he refused to relinquish. It was an unflawed triumph for the father that the cloud of uncertainty hanging over his last and perhaps most cherished hominid claim – acceptance of *Homo habilis* as a valid species – should be dispelled by the son with a truly smashing find of his own. What a grand moment for Louis, and how appropriate that it happened in 1972 – the year in which he died.

What was such a shock about 1470 was its remarkable humanness coupled with its remarkable age. Those other *habilis* specimens, Cindy and George and Twiggy, were about 1.5 to 1.8 million years old. Their skull capacities averaged 640 cc. Here was a skull with a brain that was a quantum jump larger, *and a million years older!*

Sudden revelations of this sort leave paleoanthropologists breathing very hard. They have to sit down by themselves for a while to do some heavy thinking while they adjust to the realisation that the family trees they have been sketching may be totally askew, for 1470 posed some profound challenges to ideas about early hominid relationships.

To begin with, there could no longer be any doubt that robust australopithecines were not ancestral to *Homo*. The oldest robust fossils found at Omo were about two million years old. Now the same type was popping up at Koobi Fora, and the oldest specimens there were two million years old also. If the date of 2.9 million for 1470 held up, then by simple logic, 1470 could not be descended from a type that was at least three-quarters of a million years younger.

By the same logic, was *any* australopithecine ancestral to 1470? Suddenly Louis Leakey's claim that none was seemed to have improved. There had been a growing belief that the gracile type, *africanus*, would fit that role, but now even that idea seemed less likely. Again, it was a matter of dates. The age of *africanus* in South Africa was still uncertain, although more and more evidence was building to suggest that it was between 2.0 and 2.5 million years. At Omo a few gracile teeth had been recovered from the 2.0-3.0-million-year range, but their quality was so poor that Clark Howell was not able to do better than pin a general australopithecine label on them. Whether they were *africanus*, or ancestral to *africanus*, or exactly what, he could not say for certain. In any event, there was no good fossil in either the north or the south that was sufficiently older than a 2.9-million-year-old 1470 to stake out a confident claim that *africanus* was 1470's ancestor.

On a visit to Koobi Fora just before his death, Louis Leakey examines a fossil cranium found by his son Richard. Although the two had not gotten on professionally for a number of years, they were reunited at the end, in some measure by Richard's finds, which seemed to confirm his father's view of the separateness and great age of the *Homo* line.

In fact, the only older hominids from anywhere were two from northern Kenya – an arm bone from Kanapoi that may be four million years old, and a jaw piece containing one molar from Lothagam that is about five and one-half million years old. Both are too badly worn and fragmentary to say much of anything about themselves except that they are probably hominid. But whether they are ancestors, cousins, or australopithecines, no one really knows. They lie far out in an ancestral suburb where hominid traits themselves begin to get dim.

To summarise: when the australopithecine collections were hastily re-examined after Richard Leakey's astonishing announcement of 1470, it became disturbingly clear that here was a remarkably advanced *Homo* that was about as old as any of them. Or, to put it the other way round, one of the best-preserved and oldest bona fide hominid fossils in the world was a human and not an australopithecine. And yet australopithecines were clearly more primitive in a number of characteristics. That just did not make sense.

What did make sense was *Homo habilis*. His credentials were suddenly and dramatically improved. As noted, Richard Leakey made a point of not pinning a species label on 1470. 'I will merely call it *Homo*,' he reiterated. Others, however, had no difficulty in associating 1470 with Olduvai's three *habilis* specimens. Now the narrow time slot that had been so confining to Cindy's and George's and Twiggy's reputations was broadened to a million years. That provided more than enough room for a species, and *Homo habilis* thereafter was generally accepted as one.

There were dissenters. The principal one was C. Loring Brace of the University of Michigan. For years Brace had favored an extremely simple family tree. A lumper among lumpers, he would reduce the number of species labels to the very minimum, preferring to explain differences between individual fossils as expressions of extreme variability *within* species, rather than to keep creating new species to accommodate those differences. Brace and his followers have for a number of years espoused a family tree that is essentially a straight line. It goes: *Australopithecus—Homo erectus—Homo sapiens*. This means lumping all the australopithecines and *habilis* in one species. Brace's framework has the virtue of simplicity, but – with the presence of 1470 – forces one to stretch to accommodate them under one label. Recently Brace has dealt with this dilemma by splitting off the robust type as a separate species, but he continues to regard the gracile and *habilis* types as conspecific.

Richard Leakey disposed of the problem in a different way. Until recently he refused to recognise the existence of a gracile type in East Africa. He called all gracile specimens *Homo* and all robust ones *Australopithecus*. But either way, whether one followed Leakey or Brace, *africanus* and *habilis* ended up in the same basket. Leakey and Brace just put different labels on the basket.

In dealing with such matters we arrive back at the old question:

what makes a human? More broadly stated: what makes a species? Even more broadly: should we care? To answer the last question first, yes we should. Basic to all scientific inquiry is the organisation of data into sense-making arrangements. Evolution could not be understood or traced without sequences of fossils to demonstrate that it has happened. Those fossils cannot be put into sequences and given names unless their relationships are worked out by their being carefully described and their differences noted. We can arrive at no insights into where man came from – when, how or why – without going through that process. In the process it is inevitable that species will be named, because it is no longer possible to argue that extinct humans are the same as modern humans. The question reduces itself to a decision as to how many human species will eventually be identified.

As to what makes a species, Ernst Mayr puts it this way: a species is 'a group of interbreeding natural populations that are reproductively isolated from other such groups'. That is the classic description of a biological species, or biospecies. Le Gros Clark goes on to say that a biospecies is 'a genetic entity, genetically isolated from other species'.

Both men stress isolation, for it is only in isolation – without the opportunity of exchanging genes – that groups of animals or plants will gradually begin to develop behavioral and structural differences. The isolation can be physical – a desert or an ocean that effectively keeps apart two populations that once were one. It can be behavioral – a refusal on the part of two creatures to mate because they no longer recognise each other as suitable partners. Dogs, for example, do not mate with cats. Not only does it not occur to them to try, but because of the large genetic differences that have built up in them, it would be useless to. It is impossible today to produce, even by artificial insemination, a litter of puttens or kippies – or whatever one might wish to call them. Horses and donkeys, however, being more closely related than dogs and cats, can still be mated to produce mules. But the mules are sterile, and the definition of a species holds up. Horse and donkey have become reproductively isolated; their mating cannot produce fertile offspring.

These examples are of living creatures. They might be called 'horizontal' examples, in the sense that they deal with a particular moment in time – the present – to demonstrate the appearance of species through ongoing evolution in two directions from a common ancestor. If that ancestor is very close, as it apparently is in horse and donkey, then the possibility of a successful breeding is also very close: the mule just misses; it gets born, but cannot reproduce itself. Go back a million years, and we might find a horse and donkey that could mate to produce fertile 'mules'. Go back another million or two, and we certainly would. Modern horses and modern donkeys had not yet evolved. Only their common ancestor was on the scene, and it mated freely with all others of its kind that it came into contact with. It was only through separation of populations over time, and the gradual appearance of differences in those populations, that two species, horses and donkeys, would emerge where one had been before.

Isolation in single populations is also achieved through time. Here the separation is not so much horizontal as vertical. We are concerned with descending lineages, and the changes that take place over many generations within groups, not between them. *Homo erectus* did not produce several types that now, a million years later, confront each other as biological strangers. Humans the world over can and do interbreed today. What *Homo erectus* did was sire descendants who might not recognise *him* as a sexual partner. It would be interesting to know if a modern man and a million-year-old *Homo erectus* woman could together produce a fertile child. The strong hunch is that they could; such evolution as has taken place is probably not of the kind that would prevent a successful mating. But that does not flaw the validity of the species definition given above, because the two cannot mate. They are reproductively isolated by time. Therefore, somewhere along the evolutionary path that leads from one to the other a species line may be drawn if, in the opinion of anatomists, the differences between them are significant.

Since the word 'significant' means different things to different people, there will always be disagreement about where – or whether – to draw species lines in extinct lineages. Loring Brace concedes measurable differences between *africanus* and *habilis*, but he does not regard them as important enough to justify pasting another label on the latter. He would prefer waiting until evidence of evolution has progressed a little further, until a still larger brain, greater differences in skull proportion, a shorter face and a more humanlike set of teeth show up – as they ultimately do in *Homo erectus*. Brace has no trouble drawing a species line there: the earlier types are all australopithecines; the later ones are all humans. Most others disagree. They would insert *habilis*. On the evidence of 1470, that appears to be an appropriate insertion.

Not everyone was happy with the 2.9-million-year date for 1470. Nevertheless, it was the one given, and it was tentatively accepted. That made 1470 one of the oldest of all decently preserved hominid fossils, and meant that a search for its ancestors would have to be conducted in deposits that were older yet – three million years or more. When I heard about 1470 at the Circum-Rudolf Conference, I recognised a splendid opportunity: the deposits I would be working at Hadar in a few weeks' time *were* three million years old.

'If we can find hominids there,' I said to a colleague in great excitement, 'they've got to be ancestors.'

'Good luck. I hope you find them.'

'They might even be *Homo*.'

'I doubt that.'

'Why not?' I said. 'If 1470 is 2.9 million years old, they could be.'

'Don't quote me, but I don't think 1470 *is* 2.9 million years old.'

'But it's below the KBS tuff. The tuff is 2.6. That's a potassium–argon date.'

'Then there's something fishy about the date. Look – we're back the other side of two million, right? And there's nothing back there except a bunch of small-brained primitive-looking australopithecine-type things with funny teeth. Then we go back another half-million years and suddenly out of nowhere we hit 1470 – something that's so modern it makes your hair stand up.'

'You're suggesting that date is *wrong*?' I asked.

'I wouldn't be surprised.'

I had no comment to make on this, because I had been given no chance to study 1470. However, I had visited Koobi Fora in 1971 and knew that its stratigraphy was extremely complex and difficult to interpret. The deposits were shallow, and hence easily eroded away or contaminated by other equally shallow ones. Moreover, the volcanic marker tuffs were not continuous from site to site. They too were interrupted by erosion and faulting and were hard to correlate with one another. But what struck me most strongly was that the animal fossils found just above and below the 2.6-million-year-old KBS marker tuff did not look like the 2.6-million-year-old fossils from Omo. Theoretically, animals of that age should have been identical in the two places, which were only about fifty miles apart.

A wide-ranging animal like an elephant could have – and very well may have – walked from one to the other. The climate in the two places was the same.

It was about then that Clark Howell remarked to Basil Cooke, the pig specialist, 'The Koobi Fora pigs don't look like our pigs from the same time range. Their pigs are too young.'

'I agree,' said Cooke. His interest in pigs, already very large, grew.

By this time I was back in Cleveland, teaching. When my students asked me about 1470, I replied by giving them an examination paper on it:

> Assume a new skull is found. All you know about it is that it is large-brained, with an estimated cranial capacity approaching 800 cc; that the vault of the skull is not thick; that the brow ridges are not prominent; that the fossil is truly bipedal because it has associated leg bones identical with those of modern man; that the skull comes in several hundred pieces; that the pieces were found below a 2.6-million-year-old marker tuff. Fit this into your model of the early Plio-Pleistocene, and consider how the specimen might influence contemporary ideas about early hominid evolution, with specific reference to the single-species concept as well as a two-species interpretation of australopithecines. What questions would you be most interested in asking about this specimen to help solve the problems posed by its discovery?

In other words, what did *they* think? What I wanted them to say was that 1470 was *Homo*. I'd given them a photograph of it along with the description, and it should have been clear to them that it was. But *if* it was, then that raised some relationship problems. I would have given an 'A' to anybody who raised questions about australopithecines' being ancestral to it on account of its age. I'd have given an 'A' to anybody who questioned the date. I'd have given an 'A+' to anybody who went on to say he would like to know more about how the skull had been reconstructed. That would have shown that he was thinking about the brain size and that it might be wrong. Students tend to believe what you tell them. *Africanus*, an ancestor, was rattling around in all the literature. I wanted them to wonder if even that might be wrong. I would have given a poor mark to anybody who stuck to the single-species theory. A *Homo* that is more than two million years old and a robust australopithecine that is only one million years old just can't be in the same species.

In deference to the one-species theory, it should be pointed out that its adherents recognise this anomaly. It should also be pointed out that the theory looked more plausible when first formulated than it does today.

To start with, some of the most sophisticated paleoanthropologists working between 1930 and 1960 had come to the realisation that variability within a species was apparently a good deal greater than had initially been thought to be the case. As more and more fossils were found, this became more and more obvious. As a counter to the tendency to name a new species each time a somewhat different fossil was discovered, the single-species theory was welcomed as a healthy simplification. It was only when the South African caves began to supply what were quite obviously two kinds of australopithecines that the theory first got into trouble. There was an explanation: males and females. That was plausible, since there are marked differences in the skulls and teeth of many male and female primates, notably gorillas and baboons. But when it was pointed out that all the male australopithecines would have been living at Swartkrans and Kromdraai (where the robust type is found) and all the females at Sterkfontein and Makapansgat (where the gracile ones are) – a kind of sexual segregation that is unknown anywhere in the primate world – that explanation quickly crumbled.

But that crumbling did not necessarily crumble the single-species theory. It was pointed out that robust types were younger than gracile ones and could be descended from them, their differences explained by evolution within the species over time. Looking at australopithecines that way – strung out, so to speak – it would be reasonable to expect some differences at either end.

The most important concept that the single-species theory had going for it was an idea that had prevailed about human evolution since the beginning: there never had been and never would be room on the planet for more than one kind of man-becoming biped at a time. Erect ape descendants are unique and always have been. There is only one kind today: *Homo sapiens*. There was only one kind yesterday: *Homo erectus*. So, why should there have been more than one kind the day before yesterday? The emerging ape, wherever and whenever he began to emerge, would alone carry the torch toward humanity. He would pre-empt the niche that his bipedalism and his growing brain and culture were helping him carve out. Any other invader who came along later and tried to compete would be handicapped by being a less able walker, a less intelligent animal, less adept as an emerging tool user, and so on. It would fail to compete and would fall by the wayside. More likely, it would never get started.

Stated as simply as it is here, this is a powerful argument. It goes to the heart of a basic tenet of biology which says that indeed no two species can occupy the same niche at the same time. The less well-adapted one will either disappear or edge its way into another niche.

Among birds, for example, the fifty-odd species of warblers that summer in North America might seem to contradict this precept, but the fact is that they do not compete. Each has its own niche. Different species breed in different parts of the continent. Where they overlap, it turns out that they choose different habitats – bushy thickets, pinewoods, tall deciduous trees, even different levels in the trees – and eat different kinds of insects, for insects too have their own habitats. One of the marvels of evolution is how so many seemingly similar plants and animals have managed to find ways of making a living cheek by jowl with each other. This seems to have been the case for early hominids as well. The gracile/*Homo* type begins to emerge as a smaller-jawed, smaller-toothed omnivore; the robust type as an increasingly specialised eater of coarse vegetable matter. The divergence in diet alone would be enough to permit the coexistence of two types of erect hominid; and thanks to Richard Leakey, we know today that two did coexist. Among his most significant finds at Koobi Fora is an excellent *Homo erectus* skull. It is dated at 1.5 million. It lived alongside robust australopithecines – and no anthropologist in the world would argue today that those two are conspecific.

So, if two kinds of erect hominids coexisted 1.5 million years ago, then it is possible that two kinds – or even more than two – existed earlier. More and better fossils in the two-to-three-million-year range would have to be found to clear up that matter. About to embark on my first full-scale field season at Hadar in 1973, I hoped to be one who would find them.

I had talked about Hadar as enthusiastically as I could at the Circum-Rudolf Conference. Present were dozens of Plio-Pleistocene men of great renown, several of whom I hoped ultimately to entice into working with me. Getting them committed, however, was not easy. All had full schedules. They listened politely to my rosy descriptions of deposits bulging with fossils of superior quality, and to my confident predictions that hominids would be found. They listened because I was Howell's man. I had worked three seasons in the field with Howell and came well recommended.

But I had never headed an expedition myself. What assurance was there that I was capable of running one? What was the caliber of others on my team? How reliable would our stratigraphic work be? Would I succeed in persuading the Ethiopians to let me take fossils out of the country? If so, would their dating and mapping be pinpoint-accurate? Would they be made available for study promptly, or would they languish in museum drawers for a decade or so until someone remembered to do something about them? All these things can go wrong, and often do. The greatest frustration for a busy anthropologist is to find himself involved in some aspect of a project that fizzles, that just fades away from lack of attention, lack of funding, lack of resolve – worst of all, from lack of confidence in the reliability of its data.

I desperately needed assistance in half a dozen specialised areas. But my only stock in trade at Circum-Rudolf was a show of confidence.

I had to lobby those people, persuade them that if I sent them a bunch of antelopes to study they could be sure of the dating. I explained that our deposits at Hadar were comparable to the lower time horizons at Omo – where the fossils aren't so good – and that this would be a great opportunity to fill that gap. I pushed Alan Gentry hard on that. Alan is from the British Museum. He is a world authority on antelopes; I had to have him interested. I talked to John Harris about giraffes. I talked to Basil Cooke about pigs. I did a great deal of talking.

I left for Hadar with many expressions of interest, but very few firm commitments.

7 The First Hadar Field Season: The Knee Joint

Don was squatting in that blazing sun, fitting those two bones together, and suddenly it hit him – that's not a monkey, it's a human being.

MAURICE TAIEB

Addis Ababa, although the capital of Ethiopia, is more like a mixture of casbah and boomtown than a national center. Its royal palaces, a few business buildings and some of its ministries are splendid, but its markets are shabby. Most of the city has the flavor of an African village grown very big and sprawled very wide. Under Emperor Haile Selassie all energy trickled down from the top, through a bureaucracy answerable to him. For a foreigner, getting anything done from a distance was impossible. One had to be on the spot.

Maurice Taieb knew all about that. When I arrived in Addis in the fall of 1973 with Dole and Gray, I found my French partner already scurrying from ministry to ministry getting the necessary permits: for vehicles, for entry into the Afar region, for the removal of fossils. Everything had to have its paper; every paper had to be stamped by the proper official – in some cases by several officials. I just tagged along.

There were endless rounds, endless little cups of coffee. The Ethiopians are extremely cordial, but they aren't always on the same wavelength as you. You just have to plod your way through.

In a week I learned a great deal about Ethiopian bureaucracy, and which people in which ministry were important to archeologists. I also made a valuable connection with an Englishman who taught history at the national university. This was Richard Wilding, who would gradually become the expedition's man in Addis, its confidant, adviser, troubleshooter and supplier when it was in the field. Finally, with the last paper stamped and the last vehicle repaired and the last sack of flour loaded, a small caravan set out for the Afar. There were four Americans, seven in the French party, plus some Amhara workmen recruited in Addis, a cook, Kabete; and a representative from the Ministry of Culture, Alemayehu Asfaw. My $43,000 had already dwindled rather alarmingly. I had had to buy mosquito nets and other equipment for some of the French, who had arrived without them and without funds.

Camp was established at Hadar on a low bluff overlooking the Awash River. It was hot on the bluff, but not nearly as hot as in the windless little gullies and stifling canyons where the surveying would take place. Also, the river was handy for bathing and, after its water had been filtered, for cooking and drinking. But when there had been heavy rains in the far-off mountains, the campers would know it; the level of the Awash would rise and its water would turn coffee-colored. At those times it was useless for anything. According to Tom Gray: 'I'm not much for swimming, but one day I realised I was so dirty that I could no longer stand myself in my own sleeping bag. So I went down to the river and stuck my leg in. It came out dark brown. So I decided to be smelly for another week.'

Some members of the expedition would not swim at all because of the infestation of the river by crocodiles. But these were smaller than the man-eating monsters of Kenya and Uganda, and did not seem to be consuming any of the local Afar people, who were in and out of the water constantly. After a couple of weeks most of the scientists were bathing daily. Luckily, the flow of the river was rapid enough to eliminate the threat of bilharziasis, a debilitating disease carried by freshwater snails that afflicts thousands of people who wade or bathe in slack-water places like Lake Victoria.

Camp was plain and practical. It consisted of one large tent for dining, geology and paleontology, plus mosquito nets for sleeping. The large tent was actually a fly with open sides for coolness and small flaps hanging down to keep out the sun. It had long tables in it. All of the classifying, sorting, cleaning, numbering and paperwork took place there in the afternoons when it was too hot for surveying and while the light was still good enough for delicate picking at fossils, which was difficult in the evening because of the feebleness of the camp's butane lamps.

Once the camp was set up, scientific work could start, and Taieb immediately threw himself into the complex task of interpreting the geology of Hadar. Unlike the Omo deposits, Hadar's are not tilted. There has been little faulting there, and thus a minimum of upward slanting of buried deposits to the surface of the ground. Instead, the layers are revealed in the sides of gullies by erosion. Some gullies are high and shallow. Others start low and run deeper. Some areas are almost flat, and are contaminated by material that has been dropped on top of them much later, replacing other material that may have been washed away, and thus confusing the original stratigraphy. In other spots the strata are clean and undisturbed, as neatly differentiated as the layers in a cake. All together, Hadar is a maze of eroded beds of different depths and different stratigraphic patterns. It was Taieb's job to interpret those differences and fit the whole area into one coherent geological scheme from top to bottom – a so-called stratigraphic column.

He proceeded at a crawl at first, getting himself familiar with a few of the commonest and most easily identifiable strata, and learning to recognise which ones always lay on top of which other ones. With that fixed in his mind, he could pick up a similar pattern across a gully – or in another gully some distance off – and establish a relationship between them. In some places those commonest and most striking strata were absent; they were either totally eroded away or still buried. It was disconcerting to walk into an area and find a pattern of deposits that resembled no other. Were they older or younger?

To answer that question, Taieb had to find linkup spots whose top layers seemed to match the bottom layers of others, and vice versa. In that way he was able to begin putting together a number of short sequences, and once in a while to triumphantly hook two of them together into a much longer one. Then would come the crushing letdown when cross-checking revealed that they did not hook after all; the key layer that supposedly connected them had turned up somewhere else in the middle of some layers that did not match anything. Maurice would have to dismantle his sequences and begin fitting their sections together in other ways.

One difficulty was the similarity of many sequences, undistinguished alternations of mud or clay and sand. It was not easy, looking at a small slice of two or three such layers, to determine just how high or low in the overall sequence they belonged. One way would have been by measuring their thickness, but even this varied according to the slope of the land and the depth of the water in lakes.

To balance that confusion, there were detectable differences in grain size between strata. The faster water moves, the larger the objects it can carry with it. At the feet of the mountain range off to the west, and flowing out onto the plains, is a sorting of constantly smaller and smaller rocks carried by streams that roar down the mountain flanks in wet-season floods. By the time this water reaches Hadar, a hundred miles away, it is moving slowly, and can carry only gravel, sand, fine silt and clay. As it approaches a lake, its movement may be extremely languid. Fine sand is dropped on lake-shores and river deltas. The finest stuff of all, clay, drifts very slowly out into the centre of a lake and finally settles there. One way of identifying an old lake bottom at Hadar is to find deposits of ex-remely fine material with no fossils in them. Muds or clays that do contain bones are usually interpreted as being the ghosts of swamps or of marshy bottomlands along the margins of bygone lakes.

As Taieb and his associates wrestled with the geology of Hadar, the paleontologists went to work. They had no topographic maps of the region or aerial photographs, but they kept careful track of their surveys. Whenever they picked up a fossil they would establish a 'locality', give it a number and paint that number on a rock. Ultimately all fossil information would be coordinated on a single master map, so that all temporal and distributional relationships would be known.

Without that, we would have been totally bewildered. You have to know exactly where each fossil comes from. I put Tom Gray in charge of this work, and he laid down Gray's Law: 'If you're not going to mark it down carefully, *don't collect it.*' Tom got in a rage one afternoon at one of the scientists who came into camp with some interesting mammal fossils that had not been pinpointed. They were useless; he threw them all in the river.

Gray was also concerned about the reliability of the overall fossil sample that was beginning to accumulate. Did a person who had a good eye for antelope horn cores tend to overlook pig teeth, and thus give a false idea of the relative abundance of each species? Was there a tendency to collect only the rare and unusual fossils in a given area and ignore the common ones, again skewing the sample?

Gray was at a symposium in San Francisco not long ago, and he explained the surveyor's constant dilemma very well to a group of people to whom he was describing the work at Hadar. 'You're out in all that heat,' he said, 'dragging your butt along. You've collected a hundred horse teeth. You see another one over there. Big deal. Do you bother to go over and collect it? That means you have to establish a new locality, write it up on the master map, send a geologist out later to do the geology on that spot – all so that you can date that one horse tooth.'

'So do you collect it?' he was asked.

'Ideally, you do. If you had infinite time and infinite resources, you would collect every scrap of bone in every locality you marked. That wasn't practical, of course, so that left me worrying about sample error. We tried to do our surveying according to a controlled plan, working the deposits systematically. But some of the people didn't always do this. I found I had to retrace their steps and do what I called "census" sampling – to see if the percentage of what I found matched what they picked up.'

'Did it?'

Gray grudgingly admitted that it usually did. Like Gerry Eck at Omo, he was not overly admiring of the collecting done by some of the French in the field.

I, on the other hand, have always gotten on well with the French. The differences in the French and American ways of doing things have never bothered me. My partner Maurice Taieb and I have somewhat different temperaments. He is an ebullient, mercurial, loquacious and friendly man who is sometimes impatient and hot-tempered. Rather than being upset by his occasional outbursts, I have recognised that they have been useful in clearing the air at several critical moments when Ethiopian lethargy and red tape have threatened to bring the expedition's activities to a standstill. On the whole, after four joint field seasons, I feel that the pluses of a two-nation effort far outweigh the minuses.

What bothered me that first year was not international relations but fossils. The season was half gone and not one hominid had been found. I had not exactly *promised* hominids in my request for funds from the National Science Foundation, but I knew when I wrote up my grant proposal that if I did not include a strong pitch for hominids I would get no money at all; the likelihood of my being sent to Ethiopia to collect pigs' teeth was remote. Even so, the $43,000 I was given was only a third of what I had asked for. Some of it had already gone to the French, and it was becoming clear to me that if they were going to be kept in the field through the end of the season, I would have to part with more.

What does a young man do on his first expedition, when he is given a two-year grant and has exhausted most of it the first year and has not found what he went out to look for? He wonders what he will do the second year. He wonders if he may not crash, if he may not get a reputation for irresponsibility before his career gets properly started. He sweats.

I did my share of sweating and hoping, and dogged surveying. I kept wondering how I was going to explain that start-up costs had been big. I had had to buy all those tents. I had had to put up $10,000 for a Land-Rover. Those were nonrecurring expenses; there was no avoiding them. Out in the deposits, I realised every day that all my money would be gone by the end of the year, whisked away on one spin of the wheel. Would I ever have a chance at another? Would it have been prudent to have started smaller, to have planned on a field force of four or five scientists instead of eleven?

Those thoughts so preoccupied me that when I was out surveying late one afternoon I idly kicked at what looked like a hippo rib sticking up in the sand. It came loose and revealed itself, not as a hippo rib but as a proximal tibia – the upper end of the shinbone – of a small primate.

A monkey, I thought, and decided to collect it. I marked the spot in my notebook and gave it a locality number. As I was writing that down, I noticed another piece of bone a few yards away. This was a distal femur – the lower end of a thighbone – also very small. It was split up the middle so that only one of its condyles, or bony bumps that fitted into the shinbone to make a knee joint, was attached. Lying in the sand next to it was the other condyle. I fitted the two together and then tried to join them to the shinbone. They were the same size and the same color. All three fitted perfectly. A rare find.

As I studied it, I realised that I had joined the femur and the tibia at an angle. I had not done it deliberately. They had gone together that way naturally; that was the way they *had* to go. Then I remembered that a monkey's tibia and femur joined in a straight line. Almost against my will I began to picture in my mind the skeleton of a human being, and recall the outward slant from knee to thigh that was peculiar to upright walkers.

I tried to refit the bones together to bring them into line. They would not go. It dawned on me that this was a hominid fossil.

Gray walked up. I showed him the tibia. 'What do you think?'

'A monkey?' said Gray. 'A little primate of some kind? You can't tell much from the tibia alone.'

'How about if I add this?' I said, producing the two pieces of the femur. I laid them on the sand, the angle between femur and tibia

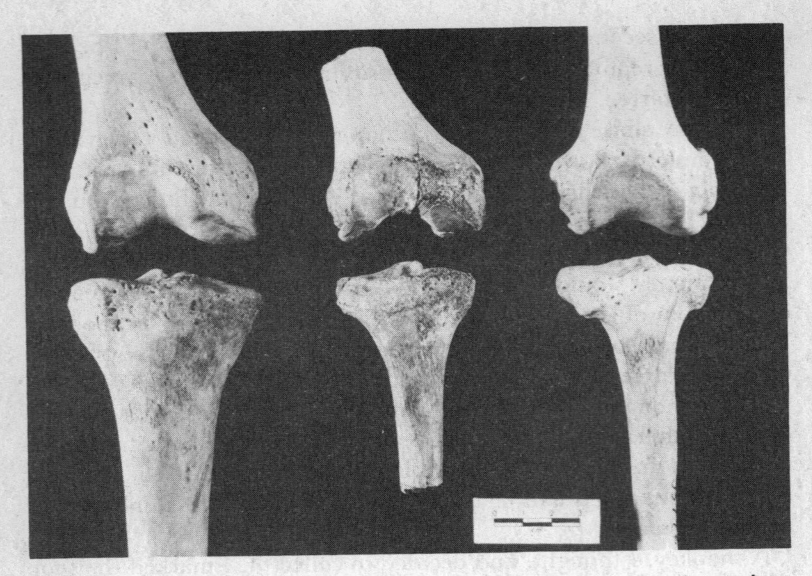

Johanson's first hominid find at Hadar was a knee joint, shown here between a modern human knee (left) and an ape knee (right). At more than three million years, it is the oldest such fossil on record. The angle at which the two bones connect is what convinced him that it was a hominid. In apes the connection would be straight. Viewed from the side (opposite three drawings) all three joints look remarkably alike. When they are examined in cross section, however, the difference between the two hominids and the ape is clear. Their bones are oval; that of the ape is round.

unmistakable.

Gray stared at the fossils, and then at me. 'One of *them*?' he finally said.

I nodded, and at that moment became fully aware of the burden that had lain on me all month long. Suddenly it was lifted. I felt light-headed, floating with relief – and at the same time incredulous at my luck. As I wrapped up the bones to take them back to camp, I wondered if I weren't dreaming. No one in the history of anthropology had ever seen the knee joint of a three-million-year-old hominid before. If this was indeed one, it was unique in the world. Could I – *I* – really be the finder of this extraordinary rarity? I had always imagined myself as finding teeth and jaws, or, with great luck, a skull. But something hitherto unknown to science was too much. By the next day I was beginning to have doubts about my original flash of recognition. I had had time overnight to think about some of

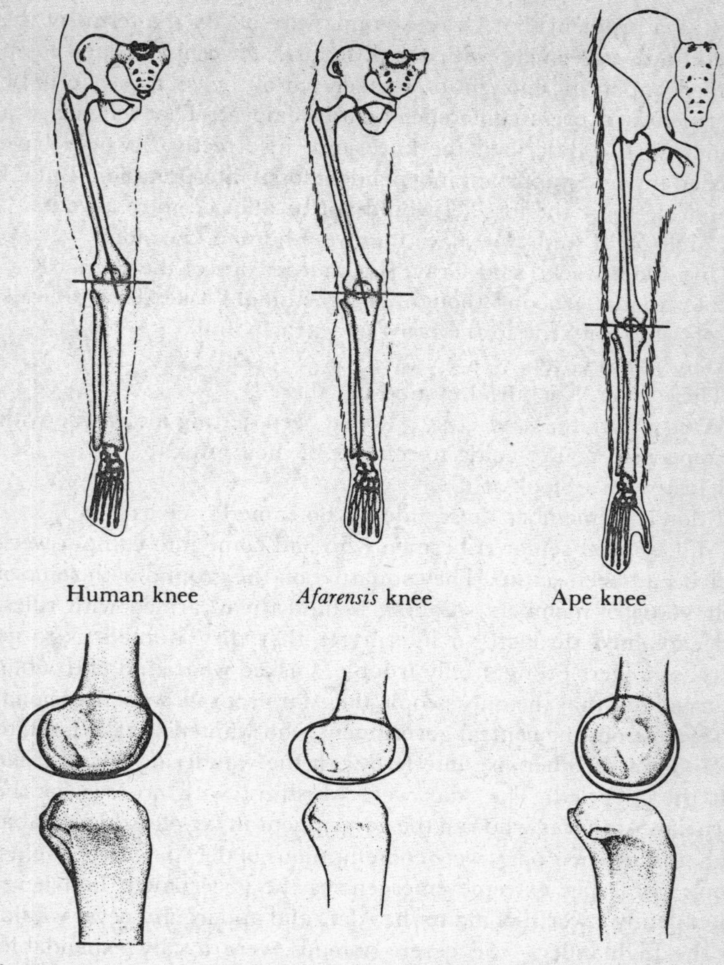

the implications of this find, and had realised that a manlike knee joint meant manlike walking. This would be the first evidence from anywhere that anything had walked upright three million years ago.

'We're going to have to have something to compare this with,' I said to Gray a day later. 'I don't trust my anatomy well enough to be absolutely sure. We need a human knee joint.'

'Who's the biggest goof-off on the camp staff?' said Gray. 'Maybe we can spare him.'

'Be serious,' I said. 'I have to know.'

I had a problem that Gray was unaware of. By the terms of the expedition's agreement with the Ethiopian government, any fossil deemed important enough to be taken out of the country had to be described at a press conference before removal. That left me in a quandary. If I described the knee joint incorrectly, I would have botched my first important independent fossil interpretation. But if I did not describe the fossil, I would not be allowed to remove it.

'Catch-22,' I said. 'We have to have a human knee joint.'

'Give me a week,' said Gray. 'I'll murder one of the French.'

Later that afternoon I thought of something. I asked Gray to walk out of camp with me to the brow of a nearby hill.

'Why?' said Gray.

'There's an Afar burial mound out there.'

'Wait a minute,' said Gray. 'We've been getting along fine with these people. You're going to rob one of their graves?'

'I just want to look at it.'

'I don't. Remember those elders who came to see us?'

I did: several senior tribesmen who had come into camp a week after it had been set up. They squatted on the ground with some of their younger members standing behind them, armed with rifles, and explained through an interpreter that they wanted us to go away; strangers brought only trouble. I asked what kind of trouble. It turned out that the only people the Afar ever saw were occasional surveyors from the central government who wanted to take land, to make dams or otherwise interfere with the wandering pastoral life that the Afar led. The Afar were Muslims, with no love for the Christian Amharas who ran the government in far-off Addis Ababa. Amharas, on their part, were contemptuous of the Afar. In droughts, famines or other extreme emergencies the government had never sent a penny's worth of aid to the Afar, and apparently never would. To the highlanders, the desert nomads were totally expendable. Worse, they might even be sympathetic to other Muslims from Eritrea and Somalia, with whom the Ethiopians were already on such bad terms over border disputes that war was threatened. For all those reasons, Amhara–Afar relations were chronically and admittedly terrible.

Taieb and I had talked patiently for most of a day to the Afar elders, explaining that we were there only to collect old bones and old rocks. The elders, of course, did not believe a word of this. The land was nothing *but* rocks – what earthly use were rocks? As for bones, they were all over the place too. They were not good to eat. They could not be burned. They and their ancestors had been kicking them aside for hundreds of years. But after an exchange of some small gifts and an invitation to have a couple of tribesmen posted to the camp to see for themselves what the Europeans were up to, we were allowed to stay. When, after a few weeks, it became clear that the crazy strangers actually were collecting stones and bones – and nothing else – the inspectors drifted off. Others drifted in. They sold goats to the expedition cook, Kabete. Some of them signed on as workmen. Wives of the workmen arrived with their babies. A camp of native huts sprouted a couple of hundred yards from the main one. Rather quickly the stagnant local economy heated up.

'Let's not bust all that,' said Gray. 'What if they catch us?'

'It's late. They won't notice.'

'They'll shoot us. They'll run us off.'

'I have to have a femur.'

By that time we had arrived at the burial mound. It was a loosely made dome of boulders and was probably a good many years old, because one side had fallen away. I looked in. There was a large heap of bones inside – a family burial place. Lying on the top, almost asking to be taken, was a femur. Tom took it. We looked around. There was no one in sight. Tom put the bone in his shirt and carried it back to camp. That night I compared it with the fossil. Except for size, they were virtually identical.

The knee joint was found near the end of the first field season, and gave an extraordinary lift to everybody in camp. Although more – and better – fossils would be found in succeeding field seasons, none would affect me quite as deeply as this one. There is something special about your first nugget. I am sure the gold prospector has that same feeling. He is a gambler too. He is living on hope. He has nothing to sustain him except the belief that the next pan of dirt, or the one after that, will contain gold. Once he has found it, his whole reason for being there, the core of his own most secret motivation, is justified. You have the same feeling when you are surveying. The landscape is so monotonous. You are so hot. You have come such a long way. So many people have bet on you. The bones are interesting, certainly, but you soon get jaded by them. As Tom Gray said, you can hardly bring yourself to collect one more horse tooth. But you do. You are establishing a framework, putting together the jigsaw puzzle. Each piece helps. Then, suddenly, you hit The Piece. You never forget that moment as long as you live.

The knee joint was the great find of that first field season. With the proximal (upper) end of the femur gone, the joint did not in itself say anything about how it might have fitted into a pelvis. But two other bones found near it did. They were proximal femurs, the broken-off upper ends of the thighbone. One lay almost next to the knee joint. The other was about fifty feet away; it had carnivore tooth marks on it, suggesting that it had been dragged off to be eaten. These bones were from the same horizon as the knee joint. They were the same color and size. I believe, although I cannot prove it, that they represent the upper legs of one individual.

The field season wound down. Taieb and his team went back to France. Gray and I were left to close down the camp. We packed everything in the trucks, finished our paperwork, ate the last odds and ends of our supplies. We paid off the local workers, who, watching us go, were as mystified by our sudden departure as they had been by our arrival.

By this time the Afar had come to take us wholly for granted. They had been paid well for goats – up to $15 per goat, far above the going price before we came. They had cash wages, and a large supply of tin cans and other odds and ends that we, with our incomprehensible ways, had thrown out in favor of worthless bones and stones. Some of the men who had worked around the camp had picked up a few words of English. They were delighted to learn that we would be back the next year, and they promised to be on hand to work again.

But the next year, only one or two turned up. Although the Afar are remarkably nice people, one cannot really pin them down. But to call them unreliable would be to judge them by a Western standard, which would be unfair. Nomads are different. Their sense of time is different. They have no feeling for progress or history because they have none themselves. Their lives have not changed in a thousand years. They have no concept of the shape or the size of the world. Almost none of them have ever been outside the Afar Triangle.

The strangers who come in are the weird ones, the ones who do not fit. They cannot live off the desert the way the Afar do. They would die of thirst or starvation in a week. Nearly everything we had, we brought in; everything but the water we got from the Awash

River and the goats the Afar sold us. Even the goats were not plentiful. Although the Afar get a little milk from their goats, they use them mostly for trading. They seldom eat one themselves.

I worry sometimes about the effect of our intrusion into their lives. We caused a terrible inflation in goats by our constant purchases of them. People were bringing them from miles away because they knew that Kabete, our cook, would buy one every couple of days.

For all their nomadic ways the Afar are good workers in the sense that they are intelligent and obliging. They have incredibly sharp eyes. They are good surveyors. When they learned what we were looking for – particularly when they learned that it was hominids we were looking for – they became very good at recognising them. But to expect an Afar to show up on a regular basis to sift gravel on a hot day when he feels like doing something else is not to understand the Afar way of life.

The best steady workmen were the Tigre, from an area to the north of the Afar Triangle. We had a few of them in camp, and some Amharas, members of the ruling tribe from the Ethiopian highlands. Amharas are the best educated and consider themselves superior to others. They are more accustomed to giving orders than to taking them. The Afar are independent in a different way. They live in a world of their own – out of touch.

I sat down once with one of our Amhara workmen who could speak Afar, and tried to explain through him to several Afar what we were doing. They were polite, but it was obvious to them that we were far from being normal people. That did not bother them. They enjoyed looking at us and speculating about what we were going to do next. They amused themselves a great deal by mimicking us, although the greatest camp mimic was a Tigrean. He decided that anybody who walked around holding a little black box in front of his eye and going click-click-click was crazy. He would do that with a cigarette box or a biscuit tin, running around in a kind of crouch and saying click-click-click, and crack up the whole camp. When he had everybody's attention, he would then strut around, sticking out his chest, with his hands on his hips and his chin in the air. Then everybody would really crack up – he had Taieb down to a T.

Back in Addis, I left all the camp equipment with my new friend the archeologist and historian Richard Wilding, who agreed to store it until the next season. The Land-Rover I left with an American, Jon

Kalb, who had been living in Addis and who had joined the expedition as a geologist. Although Taieb had returned to Paris and could not share in any publicity about the knee joint, I was obliged to make a press announcement of it in order to remove it from the country. Members of the Department of Antiquities in the Ministry of Culture were thrilled to learn that an extremely old hominid fossil had been found in Ethiopia, but I persuaded them to limit press coverage to Ethiopian papers and prepared a short release for them.

The next day I flew to Nairobi to show the knee joint and some mammal fossils to the Leakeys. I was met at the airport by Richard Leakey's wife, Meave, who said, 'We heard all about your bones on the radio last night.'

I was astonished. I wanted to know how.

'I don't know how. A foreign stringer must have got hold of your press release. Anyway, the radio said that you had found a knee joint that was between three and four million years old. It said that this was the oldest evidence of hominid bipedalism yet known. Is that true?'

As far as bipedalism went, I said, it was true – realising that my professional judgment was now on the line. As for the date, I was not sure. Taieb had found a lava flow – a layer of basalt – in part of the deposits. Samples had been sent away for potassium-argon dating, but the results had not come back yet. Meanwhile, on biostratigraphic evidence, we had decided that the basalt layer was about three million years old. The knee joint was much deeper in the deposits than the basalt layer, and therefore considerably older. There had been some discussion about how much deeper. Gray and I felt that the basalt layer should have been placed lower in the stratigraphic column, with fewer deposits below it, which would have meant less time represented by those deposits. But Taieb was the chief geologist, and we deferred to him. Furthermore, like everyone else fingering the first threads of the hominid story, we were inclined to lean toward the old. The higher the basalt layer was placed in the column, the more strata there would be beneath it, and the older the underlying fossils would be. Gray and I had settled on 'between three and four million' as a reasonable figure.

Mary and Richard Leakey confirmed my diagnosis of the knee joint: hominid. That was a vast relief to me. I took it to Paris, where Taieb and Coppens made their own formal press release 'to the world', thereby gaining their share of the credit – important to them for future fund-raising and continued government support. That done, I continued on to the United States. I went straight to Kent, Ohio, where my friend C. Owen Lovejoy, a world authority on locomotion, teaches anthropology at Kent State University. While we were talking, I casually opened the box containing the three knee joint pieces. 'Take a look at this, Owen.'

'They're adult,' he said, after examining them carefully.

'That's what I thought.'

'But they're so small.'

'They are, aren't they.'

'How old did you say they were?' he asked.

'Three million years.'

'That can't be. This is like a modern knee joint. This little midget was fully bipedal.'

'*About* three million,' I amended.

'What's this "about" stuff? Don't you have a date?'

'I've got biostratigraphic evidence. A lot of it. The animal fossils say three million. I'll have a potassium date in a few weeks.'

'It had better be right,' said Lovejoy. He is a gray-eyed man with a Franz Liszt haircut and a swooping, maniacal sense of humour. He threw back his head and let out a bray of laughter.

'What's wrong?'

'What's *wrong*? The whole goddamned thing is wrong. I'm thinking of what some people are going to think when they first see this little fellow. He could run on his hind legs. He could really scamper. But I'll bet his brain wasn't much bigger than a peanut. How could it be? He was scarcely more than three feet tall himself.'

'So?'

'So, have your pals got a place on the family tree for something like this? A little three-foot ape running around on his hind legs? He'll mess up everything. Where are they going to put him?'

'I don't know.'

'They won't believe you. You'd better go back and find a whole one.'

'But he could walk upright?' I persisted.

'My friend, he could walk upright. Explain to him what a hamburger was, and he'd beat you to the nearest McDonald's nine times out of ten.'

That was what I wanted to hear. At last I had the confirmation of an expert.

8 The Second Hadar Field Season: Some Hominid Jaws and Lucy

I have something strange here: a baboon with very big back teeth.

ALEMAYEHU ASFAW

Life is the art of drawing sufficient conclusions from insufficient evidence.

SAMUEL BUTLER

Look at all the sentences which seem true and question them.

DAVID RIESMAN

My first public description of the knee joint was at a meeting of anthropologists held early in 1974 in New York by the Wenner-Gren Foundation, an organisation founded by a Swedish industrialist to further anthropological research.

Everyone asked what on earth it was. I said I was not sure. All I could tell them was that it was bipedal and extremely small; if it was an australopithecine, it was smaller than any collected so far.

I thought I had been convincing; but when I was having lunch afterward with Mary Leakey, she said, 'I can't tell you who, but there are a couple of people at this conference who are saying those are monkey bones.'

'But they're not,' I said.

'Of course they're not. You know it; I know it. But it's always the way when you find something new. All those anatomists think they're so smart. They don't want to admit that somebody has found something different'.

'They'll say that?'

'Of course they will. But you stick to your guns. You know what you've found. They're just jealous. Snickering away, saying, "We'll wait until he publishes; then we'll carve him up".'

I was not able to publish right away. I had to get back to

Cleveland and my teaching job. I also had to begin raising money for the next field season. My original grant of $43,000 for two years had been almost entirely spent the first year; I could not go back to the National Science Foundation for more. I decided, at the end of a lecture I was giving at the Cleveland Museum of Natural History, to make a blatant plea for funds. Somewhat to my surprise, this resulted in local contributions of $25,000, most of it from one Cleveland couple.

That was a delicate moment for me. It was my first attempt at money-raising. It was not like writing up grant proposals for a foundation. I had my hat in my hand and I was holding it out to real people, not a committee.

I had another reason for being nervous. I had had a run-in over money the year before that had turned out badly. An American scientist, hearing about the Afar expedition, had explained that he would be in Africa that fall and had asked if he could join the field team for two weeks. When I said yes, he had then stipulated that he be paid a per diem not only for his time at Hadar but for his other time in Africa also. I had told him bluntly, 'I will pay you only for the days you are with us. You can't expect me to be responsible for work you do somewhere else in Africa.' The scientist was offended and refused to go to Hadar.

I had made a mistake, an enemy. I should have found some way of explaining politely that my funds were so short that I could not afford him. But I did not think. I was irritated by the attempted rip-off. I have since learned to try to be more diplomatic, which is one reason why Maurice Taieb and I have gotten on so well. We understand the need to be flexible. When there is some sort of Franco-American disagreement we sit down and talk it out. With Maurice that is easy, because we respect each other. With some others one has to watch one's step. I was learning that being an anthropologist isn't just bones.

But bones help, especially the right bones. If there had been any carping or mumbling about the knee joint at the Wenner-Gren conference, I did not hear it. Most participants had been fascinated by it. I rather quickly realised that it had brought me up a step; in my dealings with other scientists I was standing taller. I now had a unique hominid fossil of my own and some superb mammal specimens. Scientists who had only listened politely the year before when I was trying to talk up Hadar were now seriously interested in the work being done there. Basil Cooke, the pig specialist, was one. He agreed to work on the Hadar pigs. Several French experts were so enticed by the first year's mammal collection that they asked to go to the field for the second year. Michel Beden was caught by the quality and abundance of elephant fossils; he signed up. So did Vera Eisenmann (horses), Germaine Petter (carnivores), Jean-Jacques Jaeger (rodents), Claude Guérin (rhinos). An expedition that the year before had been mostly geology and surveying, and notably thin on paleontological expertise, was now rich in all three. Things were shaping up beautifully. With money in our pockets and a full field crew of specialists signed up, Taieb and I were about to leave for Ethiopia in August to do the field preparation when we got a horrible shock. A letter came from the Ministry of Culture informing us that our right to work at Hadar – indeed, our competence – had been challenged by another scientist.

The challenger was Jon Kalb, the American who had been living in Addis Ababa. He was known to have done some work in geology and had been recommended to me by Taieb. Kalb had gone with us for the first field season at Hadar, but the relationship had not worked out. When Taieb and I arrived in Addis, we found Kalb in the Ministry with a list of our incompetences and transgressions: poor scientific work, breach of contract, bribing natives, playing on tribal fears. The Minister read this document and informed us that until the charges could be disposed of no permit would be issued.

Taieb was stunned. He nearly collapsed. It is the only time that I have seen that energetic, sanguine man nearly wiped out. There was nothing to do but start talking; so I began. Kalb had asked for an outside referee to adjudicate the case. I agreed, and suggested Clark Howell. Kalb did not want Howell. It turned out that he already had a candidate: the man with whom I had had a run-in the year before over his per diem. I knew that that man was still angry at me for refusing to give him that extra money, so I refused to accept him. I explained to the Minister that the man and I had never worked in the field together and that he had no way of judging my competence; therefore he was useless as a referee. The Minister listened to that argument. I took heart and kept talking. I threw names around. I would have liked to get the National Science Foundation on the phone, but they are a government agency and will not make statements about their scientists. I wrote to Howell anyway, and the Minister got a strong letter back from him.

We were in the Ministry every day for nearly two weeks. Gradually we began to prevail. I think the Minister finally realised that we were competent scientists, and that this was an argument that he should not have to find himself listening to. Eventually he decided in our favour and awarded us the all-important permit.

During the last sessions of argument I had felt myself getting sick. The day the permit was issued I definitely had a fever. I had just enough strength left to phone Kalb for the return of the Land-Rover that had been put into his safekeeping at the end of the previous season. Kalb said he no longer had it, and hung up. I collapsed.

Taieb came roaring to life. He extracted from Kalb the information that there had been a fee due on the Land-Rover that Kalb had been informed of but had not attended to; instead he had simply turned the Land-Rover over to the Customs Office and mislaid its logbook. Now there was no record of its entry into the country and no way of tracing it. Luckily, I had kept a Xerox copy of the logbook. With that Taieb was able to trace the Land-Rover. Then with a huge sheaf of papers he rumbled through the bureaucracy like a tank, past desk after desk, waving his arms, waving his papers, getting people to stamp them. At the very end he encountered a man who told him that the Xerox was no good. For release of the Land-Rover, only the original logbook would do.

Taieb exploded.

Getting angry with bureaucrats is usually a poor idea in any country. With Ethiopians, who are naturally quiet and dignified, it is particularly counterproductive. But Maurice was at the end of his rope. He caught the man by surprise. He bowled him over. He pounded the desk. He shouted, 'I'm just a stupid foreigner. I do everything you ask. I get all the papers you tell me to. I go here. I go there. I run around this building for an entire day. Now you tell me it's all for nothing. I won't accept that. Here – sign!' The man was so astonished that he signed.

The next day my fever left. I went to the Ministry to thank the Director General, a sophisticated and understanding man, for issuing the field permit. The Director General opened his desk drawer and took out a letter.

Apparently the per diem man had been nursing an even bigger grudge than I had realised. He had written a letter on the stationery of the American university where he held a professorship. It was addressed to the Ministry, and explained in some detail that I was an incompetent scientist, that Taieb's and my reports about the geology of Hadar were not to be trusted, and that the fossils I had taken out of Ethiopia were not really headed for the United States for study and return, but were in the Kenya National Museum, where they would stay.

I was appalled. 'You had that letter during our – our argument with Jon Kalb?'

'Your permit discussions? Yes.'

'I want a copy of it.'

'I won't give it to you. I deliberately withheld it during the discussion because I knew it would prejudice your case. Now I am going to destroy it' – which he did. The whole episode left me badly shaken. I had been introduced to a seamy, vengeful side of anthropology that I had not known existed.

Some months later Taieb and I were made aware of Kalb again. He had started a field organisation of his own, the Rift Valley Research Mission, written to a number of American scientists in an effort to get them to join him, and persuaded an Ethiopian official to sign over to him a big section of the Hadar deposits covered by our permit. Taieb and I went back to the Ministry. The atmosphere there had changed rather markedly since our previous visit. It was more guarded now, reflecting a growing uneasiness among all Ethiopian officials over a looming crisis in the government. Haile Selassie's hold on the country, ironbound for decades, had loosened. There were bloody student riots and constant rumors about an army take-over. The minister to whom we talked seemed wary and weary.

'The official who signed that paper for Mr Kalb,' he said, 'has been transferred. The political situation is not comfortable here right now, as I am sure you know.'

We nodded.

'You can prosecute this official if you wish. If we find that he has acted improperly in any way, he may be condemned to death. Do you wish to do that?'

We shook our heads.

'Under the circumstances, I think it would be prudent to wait and let this matter work itself out. Do you agree?'

That minister was a wise man. Again we nodded, and left.

It was a relief to get back to the simpler and more direct problems of the field. Camp was set up as before on the bank of the Awash, but farther upstream and on a larger and more efficient scale. Surveying activity was stepped up, with dozens of new locality numbers marked on the master map. Soon they passed the two-hundred mark.

We had started with locality number 1. Each time we found a fossil on a new place we would establish another locality – 2, 3, 4, and so on. Some localities are only forty or fifty feet apart. Others are half a mile from one another. Each has to be described. If I were to find a fossil in my living room, I would give it a number and then go on to say something like this: 'Locality 300 is characterised by a sofa on the north and a fireplace on the south. It is about twenty feet across, and its depositional surface is a Persian rug.' In the field, a man would say, 'Locality 199 is characterised by DD3 sands [a stratum we had come to recognise and to which we had given a name] on the north, deeply eroded with steep sides on the south and east.' He would take a Polaroid picture of the site, circle the area of the locality on the picture, and spot the location of the fossil. Then he would mark that information on an aerial photograph and finally enter it on our master topographic map.

As the season went on, geologists would follow in the wake of the surveyors and work up the geology on each locality. In that way, the physical nature of the entire area was slowly filled in and fitted together. I have already mentioned the problem of getting Hadar organised into a single stratigraphic column. During the 1974 field season, that all began to come together. We had already learned that certain horizons were much more productive of fossils than others, and we concentrated on them. But suddenly we would lose one. We would be working our way along it, and it would disappear abruptly. Taieb would help with that. He would come out, take a look, and say, 'Forget it. There's a fault here. What you're looking for is buried forty feet down.' His assistant, Nicole Page, was a marvel. She is a very conscientious and accurate worker who helped us trace horizons across gullies and into other parts of the deposits. Nicole, along with Taieb and possibly one other exception, has a better overall picture of Hadar than any other geologist who has been out there.

The exception is James Aronson, an American from Case Western Reserve University and a potassium-argon expert. He is the man to whom I had sent samples of the basalt layer after Taieb had found it. Basalt, being volcanic, can be dated by potassium-argon methods.

Recently Aronson had sent us his first test results. They indicated that the basalt was at least three million years old, with a possible error of two hundred thousand years either way. Since the knee joint of the year before had been found below the basalt layer, our tentative figure for its age – 'between three and four million' – now seemed more secure.

'Thank God,' I said. 'We're off the hook on that one.'

'What do you mean, off the hook?' said Gray. 'We were never on the hook. Don't you trust your own knowledge of animal fossils? You've been to Omo. You know what they look like at three million there. We've got the same things here.'

'Okay, okay. I'm just glad to have it confirmed by a potassium-argon date.'

That was reassuring, but there was still something that bothered us. As we wandered about on our daily surveys, we began to learn an increasing amount about the geology of Hadar ourselves. Independently we came to the conclusion that our last-year's suspicions about the position of the basalt layer in the stratigraphic column were justified. Taieb had placed it too high.

'You think so too?' said Gray.

'I'm beginning to.'

'What do we do about that?'

'We get Aronson over here. We can't afford any mistakes in our geology.'

I was sensitive about this. In a paper describing the first field season's work at Hadar, I had drawn a stratigraphic column that placed the basalt layer high. I had covered myself in a footnote saying that all geological data were tentative and subject to revision; still, the stratigraphic column was in print with my name on it, and it made me very uncomfortable. Now, with a better sense of the geology at Hadar, I was more so.

I was also acutely aware of a controversy that had been ballooning over the age of the KBS tuff at Koobi Fora. Although Richard Leakey still staunchly defended the date of 2.6 million that had been given it by two English scientists, Frank J. Fitch and Jack Miller, rumblings about it were growing louder and louder. A few months before, Basil Cooke had published a devastating paper on pig evolution. It showed that the Omo pigs from two million were identical with Koobi Fora pigs from the KBS tuff at 2.6 million. With a six-hundred-thousand-year discrepancy, there was something wrong, either with Cooke's pig analysis or with the Fitch–Miller potassium argon date. Since Cooke knew how rigorous a collector Clark Howell was and how precise the dating at Omo was because of its many clear volcanic marker tuffs, he stood by his pigs. Leakey stood by Fitch and Miller.

I had begun to suspect that Leakey might be wrong, and was acutely aware of the enormously complex rearrangements that he would have to make in his thinking about hominid evolution if it turned out that he was indeed wrong. On the strength of the 2.6-million-year date for the KBS tuff, Leakey had asserted that his *Homo habilis* skull 1470 was 2.9 million years old, which made it difficult to accept any known australopithecine as an ancestor of *Homo*. Leakey therefore, and following his father's belief, had logically anchored himself in the view that australopithecines were collateral relatives, and that the *Homo* ancestor, when discovered, would turn out to be more *Homo*-like and less australopithecine-like than any fossil found so far. To rearrange all that in his mind on the basis of some pig fossils would be shockingly difficult. 'Either we toss out this skull [1470],' he wrote in 1973, 'or we toss out the theories of early man. It simply fits no previous models of human beginnings.' He preferred to toss out the theory that australopithecines were ancestors.

I was determined not to get into dating dilemmas; the Hadar figures would have to be rock-solid. I began to fret about the samples that had been sent to Aronson for potassium-argon dating; maybe they were bad samples. I realised that the only way I could reassure myself on that was to get Aronson to collect his own. I asked Taieb if he thought it would be a good idea to get Aronson to come to Hadar.

'Yes, if he'll come,' said Taieb.

Aronson was a hard man to get. He had been asked the year before but had turned me down: he had a great backlog of dating tests to run on his machine in Cleveland, the work was extremely difficult, and he was in the process of training an assistant. He would have liked to go to Hadar, because he is an almost pathologically careful man who pushes a show-me attitude to the limit and he was somewhat suspicious of the basalt samples he had. Instead, he stayed in Cleveland and put a strong qualifier on his test results: 'I get a date of three million, but I have no assurance of its accuracy because the sample may have suffered some weathering. If so, you can forget about that date; it's no good. I'd be a lot happier with a sample I collected myself.'

With Taieb's concurrence, I pounced on that and called Aronson: 'Okay, come and collect one.'

Aronson agreed to. He said he would arrive in December 1974, toward the end of the second field season.

One of the reasons for the step-up in surveying, as the 1974 season got under way in early October, was that the Ethiopian representative from the Ministry, Alemayehu Asfaw, was turning out to be extraordinarily good at it. His ability to recognise different fossils grew daily. Soon I was trusting him to go out by himself, collect his own specimens and make his own notes on them. This suited him admirably. He was a quiet man, exceptionally observant, who worked best alone.

One day Alemayehu found a small piece of a lower jaw with a couple of molars in it. They were bigger than human molars, and he told me that he had a baboon jaw with funny big teeth.

'You think this is a baboon?' I asked him.

'Well, with unusually large molars.'

'It's a hominid.'

The knee joint of the year before had proved the existence of hominids at Hadar. Everyone had been sanguine about finding more of them in 1974. In fact, the French had been so eager that they had gone rushing out to survey on the very first day, leaving it to the Americans to put up the tents. But after weeks of searching without results, that ardor had dimmed somewhat. Now it flared again, but in no one more than Alemayehu himself.

It is impossible to describe what it feels like to find something like that. It fills you right up. That is what you are there for. You have been working and working, and suddenly you score. When I told Alemayehu that he had a hominid, his face lit up and his chest went way out. Energised to an extraordinary degree, and with nothing better to do in the late afternoons, Alemayehu formed the habit of poking quietly about for an hour or so before dark. He chose areas close to camp because, without the use of a Land-Rover, they were easy to get to. He refrained from saying – although I feel sure that this was a factor in his choice of places to survey – that he had begun to realise that he was a more thorough and more observant surveyor than some of the others who were doing that work.

The day after he found the hominid jaw, Alemayehu turned up a complete baboon skull. I had it on the table for a detailed description the next afternoon when Alemayehu burst into camp.

His eyes were popping. He said he had found another of those things. After having seen one, he was sure this was another human jaw. I dropped the baboon skull and ran after Alemayehu, forgetting that I was barefoot. I began to cut my feet so badly on the gravel that I was forced to limp back to my tent to put on shoes. Guillemot and Petter, who were with me, kept going. When I rejoined them, it was in a little depression just a few hundred yards beyond the Afar settlement. Guillemot and Petter were crouching down to look at a beautiful fossil jaw sticking out of the ground. Guillemot ruefully pointed out his own footprints, not ten feet away, where he had gone out surveying that first morning in camp and seen nothing.

A crowd of others arrived and began to hunt around feverishly. One of the French let out a yell – he had a jaw. It turned out to be a hyena, an excellent find because carnivores are always rare. But after that, interest dwindled. It began to get dark. The others drifted back to camp. I stopped surveying and was about to collect Alemayehu's jaw when I spotted Alemayehu struggling up a nearby slope, waving his arms, completely winded.

'I have another,' Alemayehu gasped. 'I think, two.'

I raced over to him. The two turned out to be two halves. When I put them together, they fitted perfectly to make a complete palate (upper jaw) with every one of its teeth in position: a superb find. Within an hour Alemayehu had turned up two of the oldest and finest hominid jaws ever seen. With the addition of the partial jaw of a few days before, he has earned a listing in the *Guinness Book of World Records* as the finder of the most hominid fossils in the shortest time.

The little depression is now known as Hominid Valley. All together, it has yielded three jaws: Alemayehu's two, and another a few days later, found by one of the camp's Tigrean workers. These sensational finds were reported to Addis Ababa, where politics was now in a state of great tumult. Emperor Haile Selassie had been deposed and put under house arrest. A military junta, known as the Dergue, had taken power. Many prominent members of the old regime had fled the country.

The expedition kept in touch with the capital via an unpredictable airplane that delivered supplies and news of the political situation in Addis on an irregular schedule. As far as Taieb and I could tell, the Dergue was not harassing the bureaucrats in the various ministries;

the expedition's contacts seemed secure. In fact, the announcement of Alemayehu's jaw finds caused a great flurry of excitement at the Ministry of Culture, several of whose members expressed an interest in visiting Hadar. Arrangements were made to welcome them. We began playing host to a stream of officials who would fly out, buzz the camp as a signal, and then go on to a small airstrip that had been cleared about forty-five minutes' drive away. A Land-Rover would pick up the officials, who usually stayed a few hours, looked with interest but incomprehension at the fossils, inspected the premises and then flew back to Addis Ababa, relieved to get out of the intense heat and back to the cool highlands.

Those weeks were ones of great pressure for me. My first examination of the jaws had revealed some very peculiar things about them. I was desperate to find time for a long and careful description of them, followed by some undisturbed thinking. But this was impossible. I had had to prepare a press release, go to Addis and deliver it, then talk to all the visitors – official and reportorial – who came to camp. Worst of all, I had to face the fact that the expedition was again running out of money. My two-year grant from the National Science Foundation was now entirely gone, as were most of the funds I had raised in Cleveland.

I had discussed my financial affairs with Clark Howell, who knew from his own field experience that the Afar expedition was underfinanced. When I told Clark about Alemayehu's jaws, he realised their importance and made up his mind to do what he could to keep me in the field as long as possible. He cabled for more details about the jaws, then used that information to try and raise some money. His target was the Leakey Foundation, set up in honour of Louis Leakey, who had died two years before. The Leakey Foundation sent $10,000, which turned out to be just enough to get the expedition through the 1974 season, relieving me of the embarrassment of going back to my Cleveland backers for a second contribution.

With my financial worries eased temporarily, and with officialdom gone, I could at last turn to the jaws. They were extremely puzzling. My first conclusion was that they were australopithecine; the long sessions I had spent in South Africa with Howell two years before, studying and measuring the fossils there, had burned in my mind a strong sense of the australopithecine dental condition. The teeth in

Alemayehu's jaws, overall, had very much the same character. But they also had unsettling differences. One that I noticed instantly was the size relationship between certain teeth. In australopithecines, the molars are very large and the incisors – the front teeth – are very small. In humans the reverse is true. Our molars are small, and our incisors comparatively large. In this respect, Alemayehu's jaws were more human than australopithecine. The canines, on the other hand, seemed neither human nor australopithecine, but held a hint of a more apelike condition than either.

Strange, strange jaws. Neither one thing nor another. What on earth were they? The longer I studied them, the more I wondered. Their peculiar blend of *Homo* and australopithecine traits, witrh a whiff of something more primitive, was utterly baffling. Adding to the perplexity was a size problem: although both of Alemayehu's specimens were adults, one was much larger than the other. Were there two species of hominid at Hadar?

I realised I had reached a point where I should be discussing these strange fossils with another Plio-Pleistocene fossil hunter. I wanted the back-and-forth of argument, and finally decided to ask Richard Leakey to visit Hadar. I owed Richard the courtesy of a visit. I had been to Koobi Fora two years before, had had a chance to see the work being done there and to talk to the scientists doing it. Now that I had something to show Richard, I should ask him back. But most of all I wanted to talk, to compare, to listen, to bounce ideas around. I sent a letter off to Richard, inviting him; his mother, Mary; his wife, Meave, and whomever else Richard cared to bring along.

All three came, and with them John Harris, a giraffe specialist who was married to Meave Leakey's sister and who served as a paleontologist on the Koobi Fora team. Richard flew them all up to Addis in his own light plane. There he visited the Ministry of Culture to put to rest the accusation that the bones I had been taking out of Ethiopia were now permanently locked in the museum at Nairobi. That charge, said Richard with all the force he could muster, was false. The fossils were in the United States for study, and would be returned within the prescribed five-year loan period. He then flew his passengers on to Hadar, getting an aerial view of the deposits that we members of the Hadar expedition had never gotten, always having gone in and out by truck. The Leakeys were overwhelmed by their magnitude – thousands and thousands of square miles of eroded landscape, enough to keep a dozen fossil-hunting teams in the field for a couple of centuries.

Mary and Richard Leakey were as anxious to see the jaws as I was to show them. They examined them with great care. 'They are certainly not robust australopithecines,' said Richard. 'They are nothing like the *boisei* specimens we have been finding at Koobi Fora. Their jaws are too delicate and their molars much too small to be robust. Don't you agree?'

'I do,' I said.

'All in all, I'd have to call them *Homo*.'

'So would I,' said Mary.

I heard that. I had half-expected to hear it, half-anticipated it, and half-believed it myself. If that diagnosis stood up, it meant that these were the oldest human fossils in the world.

I then took the Leakeys on a tour of the deposits and showed them the basalt layer that Taieb had found, to whose estimated three-million-year age all the Hadar dating had been fitted. We then returned to camp for an examination of the mammal fossils, of which there was a large collection by this time, many of them in exceptional condition. The Leakeys subjected these to the closest scrutiny, particularly a collection of teeth from an ancestral horse, *Hipparion*, that had become extinct sometime after three million B.C. The true horse – *Equus* – is a migrant from Asia and appears in East Africa at two million. John Harris turned the *Hipparion* teeth over and over in his fingers.

'What are you looking for?' I said to him.

'The ectostylid,' said Harris. The ectostylid is a small cusp that appears on the lower teeth of *Hipparion* but not on *Equus*. It is one way of telling the two apart. 'There doesn't seem to be one.'

'It's there,' I said. 'It's on all those teeth. It has to be.'

'I don't see it on this one,' said Harris.

'Right there,' I said. 'The crown's just not worn down enough for it to show.'

'Hard to see at all,' persisted Harris. 'It might be absent, you know.'

'It's not absent. If you want me to get a hacksaw and cut the tooth in half to prove it, I'll do so. Every one of those teeth that you're looking at has an ectostylid.' Harris reluctantly agreed that that was so.

'What was that all about?' asked Gray a little later.

'I think it's dating,' I said. 'They've got troubles. They have some

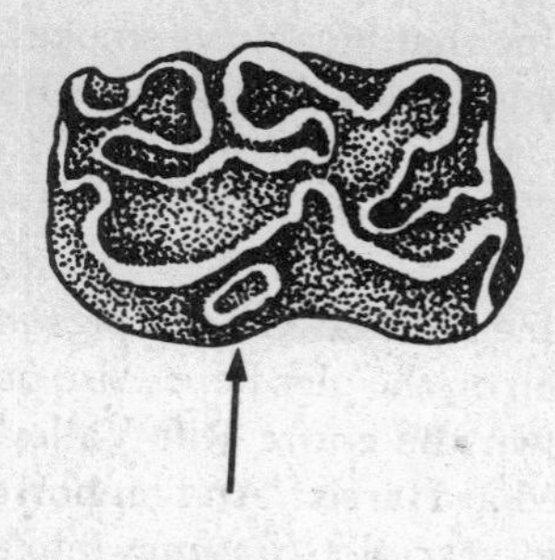

Hipparion

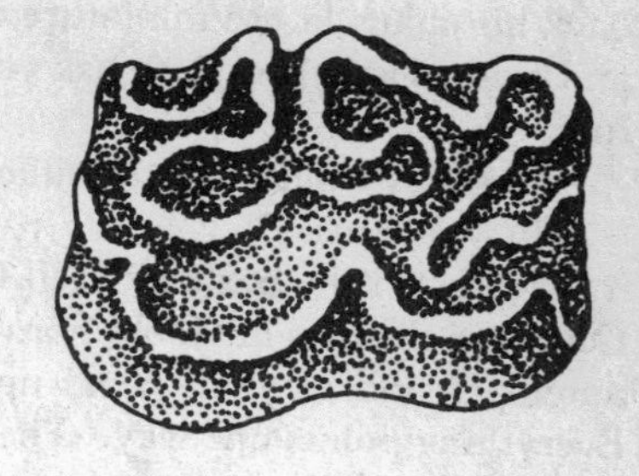

Equus

The ectostylid that John Harris was searching for in the collections of Hadar horse teeth is the small circular bump (arrow) on the biting surface of the tooth. *Hipparion*, which normally has the bump, became extinct after three million and was replaced around two million by a true horse, *Equus*, whose tooth lacks the ectostylid. The presence of *Equus* teeth that cannot be more than two million years old in the same strata as *Homo* fossils at Koobi Fora was one reason that doubt was cast by Harris and Tim White on the claimed 2.9-million-year age of skull 1470.

Equus teeth from below the KBS tuff at Koobi Fora. By their dating, that would make them nearly three million years old. That's awkward, because there aren't any *Equus* teeth in eastern Africa that are three million years old. The oldest that anybody has found anywhere are only about two million. Nobody's buying three-million-year-old *Equus* teeth. So Harris feels he has to hunt around and find some. I guess he hoped to find them here.'

'The ectostylid?'

'Right. They saw our basalt layer today. Maurice explained all the geology. They know our horses are three million years old. All they have to do is find one without an ectostylid and, presto, they have a three-million-year-old *Equus*. And boy, do they need one. John was just trying to help Richard over a bump.'

'Why didn't he come out with it?'

'Too touchy.'

It was indeed touchy – the most sensitive unresolved paleoanthropological question of the decade: for on the age of the KBS tuff hung the age of Richard's skull 1470, and on the age of 1470 hung a network of conclusions about human–australopithecine relationships.

Richard brought the matter up himself after dinner. He pushed back his chair, lit the miniature curved pipe that he always smokes, and said, 'Don, I want to ask you a question. How old is the KBS tuff?'

'My guess would be 1.8 million years,' I replied.

'On what basis?'

'Mostly on the basis of Basil Cooke's pig studies. On the pigs at Omo and the pigs at Koobi Fora. And even on the pigs here. Also on antelopes and elephants. All up and down the entire Rift Valley. Everything points one way.' Then I looked at Harris: 'And on horse teeth, John. You can't turn a three-million-year-old *Hipparion* into a two-million-year-old *Equus* just because you want to.'

Harris conceded that point, but went on to speculate that the rate of evolution might have been different for certain species because of differences in the environment. He pointed out that Omo had a river and forest environment that was different from the dry savanna at Koobi Fora, and that that might well have affected the speed of evolutionary change enough to account for some apparent discrepancies in the fossil record.'

'Let's keep it simple,' I said. 'We don't need those complex speculations when we have an overall fossil record that's as plain as this.'

Richard cut in: 'But what about John's point, Don? Do you think that conditions at Omo, being different from those at Koobi Fora, could have brought on a differential rate of evolution?'

I said that there might have been slight differences in the environment, over very short periods of time, which could have increased the pressure to evolve.

'That's right,' said Mary Leakey. 'Don't let him get you confused. He's talking just like Louis. He'll argue with you, try to trap you. You stick to your guns.'

Encouraged by her comment, I then said that none of the fossil collections was refined enough in its dating to reflect that. 'We can't pinpoint our fossils over tens of dozens of years,' I continued. 'We're talking of tens of thousands of years, or even hundreds of thousands, and there the record is overwhelmingly clear – all over the place.'

'Very interesting evening,' said Richard. He knocked out his pipe and went to bed. Skull 1470 – the heart of the matter – had not been mentioned. The next morning the Leakeys left.

They left me with a great deal to think about. They had confirmed my suspicion that the newly found jaws had *Homo* traits. More than that, they had promoted my suspicion into something like belief. I would have to write all that up in a paper. I had been putting it off, hesitant to commit myself. But when I woke the next morning I knew what I was going to do: I would assign the jaws tentatively to *Homo*. At the same time I would stress the excessively primitive features they showed. I reasoned that if they were *Homo*, they would be the oldest *Homo* fossils known, old enough to retain some ancestral apelike characteristics. That was all right. Indeed, it was to be expected. If one were to poke back three million years, it would be peculiar if apelike traits were not encountered. With a kind of mental blink I acknowledged that I had arrived at the point where other anthropologists before me had arrived, where each had had to ask the old question: where did one draw the line?

I had never had to ask myself that question. I had had no fossils of my own to force it on me. Now, for the first time, I became aware of what a problem it might have been for others to draw the ancestral human thread thinner and thinner, back farther and farther, until it was not a human thread at all. This could be an emotional problem as well as a scientific one. The investigator would find himself pressing ever deeper to pass beyond the origins of man, and yet not be able to give man up. There would always be a human ancestor: a little older, a little more primitive – but still human.

I shook myself out of that. I would never permit myself to become sidetracked in sentiment, realising at the same instant that this might be a moment when sentiment was affecting my judgment. Were these jaws *really* human?

Were they? Ultimately, it would be I who made the announcement. And it might well end up being a judgement call, in which a single human tendril in a bush of apishness decided. I looked once more at the jaws. I slid by a disturbingly primitive-looking premolar and fastened on a manlike molar. It was rather lightly enamelled, small in comparison with the front teeth. It was not like the South African australopithecine molars; that was certain. And that was enough; my waking decision had been the right one. I would go with *Homo*.

My mind made up, I looked at my worktable. It was a blanket of papers, of letters to be answered (which had not been because of the visitors from the Ethiopian Ministry), of bills to be paid (which now could be because of the Leakey Foundation contribution). All this would have to be attended to immediately, along with a description of the jaws. I would start after breakfast.

I went to the dining tent and got myself a cup of coffee. Drinking it, I found myself under a strong compulsion to put off paperwork and go out surveying instead. I knew I shouldn't. But at that moment Gray came in and began asking where Locality 162 was. Combined with the powerful feeling that this should be a day spent hunting fossils, that decided me. The papers could wait. Gray and I drove out of camp. Two hours later we found Lucy.

Lucy was utterly mind-boggling; there was no other way to describe her. She left the entire camp reeling. Everything about her was sensational. That nearly half of a complete skeleton should appear on the table of the anthropology tent, as her various parts were sorted out and laid in place, seemed incredible to the scientists even as they saw the evidence accumulating before their eyes.

Just as astonishing was the creature that was being assembled. It was not more than three and a half feet tall, had a tiny brain, and yet walked erect. Its jaw was V-shaped, not as rounded in front as some of the other mandibles, and smaller than any of them. Furthermore, its first premolar had only a single cusp. The larger jaws had two-cusped premolars. Since the one-cusp condition is the more primitive and the two-cusp condition the more human, I came to the tentative conclusion that Lucy was different from the larger-jawed type. As I studied the fossils, it seemed to me that Alemayehu's jaws represented something very early on the *Homo* line, as the Leakeys had said, and that Lucy represented something else – possibly a very early representative of the australopithecines.

When bones are scarce, speculation about them can be as daring as one cares to make it, and no one can contradict the speculator. When bones become more numerous – when a single fossil is augmented by a large sample of fragments from a number of individuals – those fragments begin to make assertions about themselves that forbid some earlier speculations. The sheer increase of information cuts off a number of possibilities as to what they might be or what they might do. On the other hand, a good assemblage of bones increases the respectability of certain other speculations. On better evidence they improve themselves from merely hopeful guesses to logical probabilities. Once in a great while, a set of bones provides a certainty.

Lucy provided a certainty. Previous guesses about the presence or absence of early bipedalism – those old arguments about whether *Australopithecus africanus* walked with a waddle, a scuffle or a shuffle – were cut through with a brutal finality. Here was an ape-brained little creature with a pelvis and leg bones almost identical in function with those of modern humans. I remembered my timid conjectures about the knee joint of the year before, and my relief when Owen Lovejoy had reassured me about its resemblance to a modern human knee. Now I knew, with the certainty provided by this extraordinary fossil, that hominids had walked erect at three million B.C More surprising yet, they had walked *before* their brains had begun to enlarge. There could no longer be any argument about that, or any conjecture over whether a certain leg bone and a certain skull did or did not belong to the same individual. Here they were, together, in one unbelievable skeleton.

But like all fossils that provide certainties, this one also posed some new questions. One of them virtually howled for an answer: If erect walking had been perfected before brain enlargement had taken place, what had caused erect walking? A popular guess for years had been that a combination of manual dexterity, increased tool-using and brain development had forced certain apes up on their hind legs as a growing daily reliance on manipulation of objects encouraged them to stand erect so that they could carry more and more things around with them. One eloquent spokesman for this view was Sherwood Washburn of the University of California. He had argued persuasively in the 1960s that tool-using and brain enlargement came before bipedality, and probably had been responsible for its development.

Lucy destroyed that argument. I caught myself wondering what would replace it. That would be something to really tie up Owen Lovejoy for the next couple of years. I could not wait to show Lucy to Lovejoy. I wished I had more time to think about bipedalism myself, but Aronson was arriving at Hadar, and I would have to spend some time with him. I would also have to continue my own surveying, write a press release on Lucy for the Ethiopians, and then go to Addis to present it.

Aronson arrived in December, and spent two weeks making geological studies and collecting samples. At the end of his visit I returned to the problem of removing the fossils to the United States.

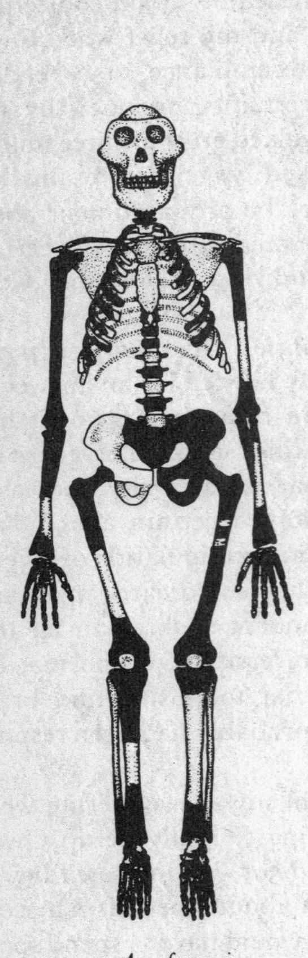

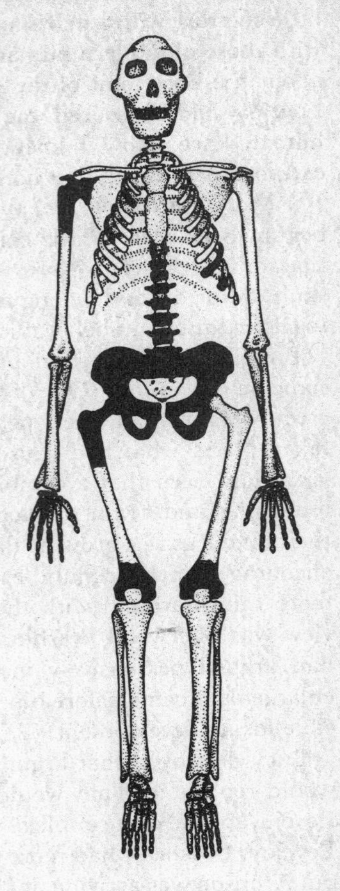

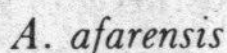

A. afarensis *A. africanus*

If all the Hadar fossils were assembled in one skeleton, they would be represented by the parts shown in black on the figure at left. This makes clear how much better known *afarensis* is than *africanus* or *robustus* (which are similarly marked), even though it is a much newer discovery. The skulls of all three should actually be black also, but that has not been done here so as not to blot out the features. A modern human skeleton is at right for comparison.

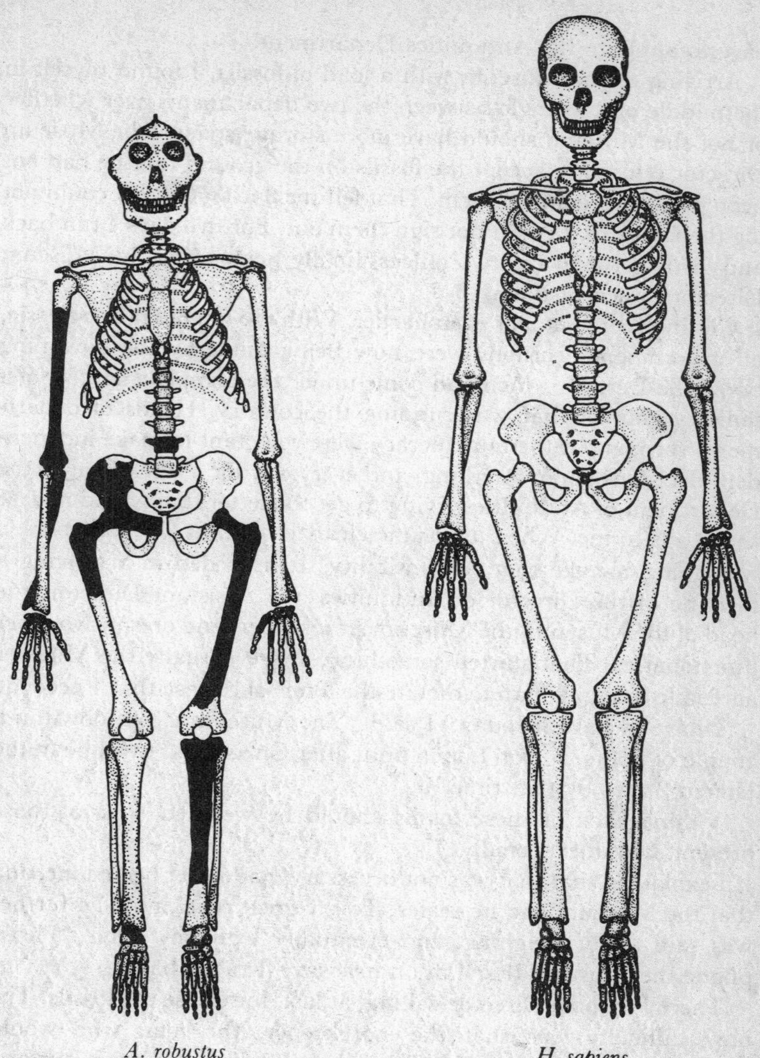

A. robustus *H. sapiens*

To get Lucy out of Ethiopia I had to deal with two bureaucracies: the National Museum and the Antiquities Department. A procedure had been worked out between Taieb and the Ministry of Culture (the very top of the bureaucratic pyramid) whereby all fossils would be numbered, described, logged into the Museum on an entry sheet, and then stored there. But to get them out again, Taieb or I would have to apply to the Antiquities Department.

Arriving at the Museum with a load of fossils, I found myself in the middle of a wrangle between the two departments over whether or not the Museum should have more storage space. The Museum Director refused to accept the fossils on the ground that he had not been provided room for them. That left me flat-footed. If I could not log the fossils in, I could not sign them out. For five days I ran back and forth between the two offices, finally persuading the Museum Director to accept them.

Getting them out was even harder. With the fall of Haile Selassie, all government decisions were now being made by the governing body, the Dergue, which had come under the complete control of a military council that was running the country. Heads of departments throughout the bureaucracy were reluctant to make independent decisions about anything, and every day for a week I sat in the Department of Antiquities trying to get the letter that would release the fossils to me. When it became clear to me that no single person was going to take that responsibility, I succeeded in arranging a meeting of the Director of Antiquities, the Assistant Director, the head of the Museum, the Museum storekeeper, and one or two other functionaries. That started something, and eventually the Minister agreed to assign a man to dictate the letter of release that I needed.

'I have to have it today,' I said. 'The Museum closes down in a couple of hours. It won't open until after Christmas. I will be in the United States by that time.'

'I think each of these fossils should be weighed,' one of those present said unexpectedly.

I explained that that was not necessary, had never been done, and that the Museum had no scales. To my great relief, nothing further was said about weighing, and eventually I got my letter. 'Please phone the Museum that I'm on my way.' I raced back.

There I found a Director willing, at last, to release the fossils. The only difficulty was that the storekeeper, the man who would physically check them off and hand them over, had gone out and hadn't come back yet. It was nearly closing time. I was getting desperate.

'When he comes back he'll have to work late,' I said.

'I can't tell him that. He is a very difficult man. He will go home on the dot of six, no matter what I say.' It was then five thirty.

'Oh, my God.'

'We will do it ourselves,' said the Director. 'But we will have to do it very quickly – before the storekeeper gets back.'

I had a sudden suspicion that I was being set up, that I would be detained at the airport the next morning, the fossils confiscated, and I thrown in jail for the improper removal of a treasure of the country. But I misread the Director. He was going out of his way to be helpful, and was actually cutting red tape. I thanked him profusely, and was going out the door with the fossils under my arm when I spotted the storekeeper at the other end of the hall. I kept going. The next morning, I was braced for the worst at the airport, but at the Customs office I encountered only two uniformed girls who waved me through with a smile. Lucy and I were out.

On a stopover in Paris I had to go through Customs again. An official insisted on seeing what was wrapped up in all those funny little parcels in my suitcase. I explained that they were fossils from Ethiopia. 'You mean Lucy?' said the Customs man. He was an anthropology buff and had read about her in the newspaper. A large crowd gathered and watched as Lucy's bones were displayed, one by one, on the Customs counter. I got my first inkling of the enormous pull that Lucy would generate from then on, everywhere she went. It also dawned on me that I was no longer an unknown anthropology graduate, but a promising young field worker with fossils dazzling enough to match those of paleonanthropology's certified supernova, Richard Leakey.

9 How Old Is Lucy?

The fossil record is a series of local cross-sections of moments of time. But those cross-sections don't spread very wide. You get a good date in one place but you don't in another. The only place you find wide continuous sedimentation is in the deep sea, and that isn't going to do you a damn bit of good.

F. CLARK HOWELL

Maurice Taieb spent two years trying to put together a stratigraphic column of the deposits at Hadar. It was a tough job. He did find a lava flow – a basalt layer – that could be dated. He placed it fairly high in the stratigraphic section, above all the fossils.

TOM GRAY

The trouble with the basalt layer was that it didn't extend to where the fossils were. We had to find correlations. And when we did, we learned that the basalt was actually very low in the section. Many of the fossils were above it.

JAMES ARONSON

With the 1974 haul of hominid fossils out of the country, along with a useful sample of animal bones, there remained the formidable job of cleaning them, organising them according to their local geological history, and finally describing them. Physical description takes a long time. It requires endless measurements with calipers, and then the writing up of all those figures in notebooks. I had done some of this with Alemayehu's jaws, but still had a great deal of it ahead of me. Now came the avalanche of Lucy's bones and a new mountain of measuring. All of this would have to be done before I could address myself to working out a firm scientific hypothesis about what the Hadar hominids were. Paramount in that decision was their age.

James Aronson heats up a sample of volcanic ash in a piece of laboratory equipment that he designed for calculating the ages of volcanic material. For an explanation of how this machine works, see pages 198 and 199.

They needed ironclad, scientifically impregnable dates, and I was uncomfortably aware that the dating at Hadar could not be called impregnable. I was counting on Aronson's recent visit to help deal with the problem.

Taieb, it was true, had succeeded in finding a lava flow – the basalt layer – and had also located some other thin layers of volcanic ash higher up in the section. Since both lava and ash are potential sources of dates, samples of each had been collected the previous year and sent to Aronson's potassium-argon laboratory in Cleveland, where Aronson himself had examined them. He had decided that the ash samples were so badly weathered that they were useless, but that the basalt, although it too showed signs of weathering, was in good enough condition to be run through a machine he had constructed for extracting dates from volcanic samples. He had run several tests and had come up with an age of 3.0 million years for the age of the basalt, with a possible error of 200,000 years either way.

The 3.0-million-year figure was a mean, which he derived by averaging the results of the test runs. The 200,000-year possible-error figure reflected differences in the runs, and it was too large to make Aronson entirely happy. If the lava sample had been purer, the runs would have given figures that were more nearly alike and would have resulted in a possible error of 100,000 years or perhaps only 50,000 years. He was concerned not so much with pinpointing the age of the basalt (although that would have been nice) as with the fact that large swings during a test series throw doubt on the reliability of the sample itself. Obviously, such a sample has suffered alteration; the question is how much. Since he had not collected the samples himself, he could not answer that question. All he could say was that if the sample had been altered it would probably yield too young a date, because some of its argon would have leaked out and been lost.

Rock samples, even of the same kind, are not all alike. The surface of a boulder – as a result of exposure to air and the sun or to running water that contains chemicals in solution – may be significantly different from its centre. Lava surfaces do erode and crumble as a result of weathering. Ash, in addition to weathering, is prone also to unnoticed contamination by blown-in particles of dust that may be ten or a hundred times as old as it is. These niceties are of no concern whatever to the ordinary geologist, who does not have to look with such care at the samples he is collecting. To the dater they are everything. The condition of his samples is as important as the samples themselves. He examines them with the eye of a connoisseur. Over the years he learns to match up field impressions with laboratory results until he has an innate sensitivity to the condition of rocks. He knows them in the way that certain other men know the condition of horses or dogs just by looking at them. He comes to that knowledge very slowly.

'You have to do a lot of looking,' Aronson says, 'and you have to remember a lot. You also have to walk around a lot.'

Jim Aronson is an engaging man with reddish hair, a bright red beard and even brighter blue eyes. He has a gentle, courtly manner that mixes oddly with the hardness of his hands, whose palms are like shoe leather. He is generous with his time, and when I confessed to him once that although I knew in principle what potassium-argon dating was, I had not a clue as to how he went about getting dates, he immediately asked me over for a demonstration.

'The best way is with the machine, on the spot,' he said.

'I'm like a first-year student. You'll have to assume I know nothing.'

'The best way,' he repeated.

I met him in his office at Case Western Reserve University, and he took me across the hall to a room containing a large and complicated piece of equipment.

'This is the machine I designed,' he said. 'Actually, it's rather a small one. I didn't have much money and I had to make it small. Luckily, I was able to pick up some good little pumps and get hold of a mass spectrometer [an extremely delicate measuring device]. It too is small. It's best at measuring very young things.'

'Young things?'

'Things that are only two or three million years old.'

'That's young?'

'For a geologist, yes. There are a lot of people who want to know the ages of things that are really old – petroleum geologists, dinosaur experts, people like that. They want to know the age of samples that may be hundreds of millions of years old. But I got interested in the Plio-Pleistocene, and I built this machine specially for that. When you're looking at it, be careful not to touch that large wire or walk under it. It carries a lot of current and it might kill you.'

The wire was braided. It came out of an electrical panel about six feet high, covered with switches and dials. It was strung across the ceiling and terminated in coils around several sealed glass tubes about the size of milk bottles. Hanging inside the tubes were small metal containers, each with a sample of volcanic ash in it. Pipes came out of the bottoms of the tubes, went through joints, past control points, pumps and meters, and disappeared into another piece of equipment that was connected by wires to a computer.

'The mass spectrometer's at this end. I won't bother to explain it to you. It's just a way of measuring tiny amounts of things. What we're going to do is turn on the electricity to heat up the volcanic samples in these bottles. When the samples melt they will release all their argon gas. We get very little gas – just a few atoms. That's why we need the mass spectrometer; it enlarges the results enough so that we can read them. Until the mass spectrometer was invented, this work was impossible to do. The quantities we get are much too small to be measured in any other way.'

Aronson threw a switch and turned a dial on the panel, gradually increasing the electrical field around one of the glass tubes. In a few minutes the sample inside it began to glow.

'We'll wait until it's melted down. Then we'll move the gas along to the mass spectrometer.'

He explained that the massiveness of the pipe joints was to make the system absolutely airtight. The pumps, though small, were extremely powerful; they had exhausted 99.999,999,999,999 percent of the air inside the system, leaving an almost perfect vacuum.

'The reason for the vacuum is this: there is a lot of argon floating around in the air, far more than in the volcanic sample. If we let any air into the machine, we'd contaminate our sample so severely that we'd be completely swamped by it. We've got to keep out air.'

The argon that Aronson was measuring was the decay product of radioactive potassium, a rare isotope of a common mineral.

'Your body has about a pound of ordinary potassium in it right now. Only twenty milligrams of that is radioactive potassium – or potassium-40, to give it its proper chemical name. That's only one two-hundred-thousandth of a pound. Not much. But it's still a lot of atoms. The potassium-40, being radioactive, is steadily ticking away, gradually changing itself into a stable element: argon, an inert gas. Right now the potassium-40 in your body is ticking over into argon gas at the rate of about 500 atoms a second. You might think that was fast, and that the potassium-40 would get all used up. Well, it isn't fast. Compared with the number of atoms there are, it's very slow. It's been going on since the earth was formed, and it hasn't stopped yet. There are trillions of atoms of potassium-40 in your body. That stuff has been cycling and recycling itself around in rocks and other bodies for millions and millions of years, always decaying at a steady rate.'

He corrected himself. 'I say a steady rate. Actually it's not. The decay rate is proportional to the amount of potassium-40 that's left in the regular potassium. As the amount of potassium-40 decreases, the decay rate slows down. For example, if you were an australopithecine living three million years ago, you would be losing about 501 atoms a second instead of the 500 that you're losing now. So, for our purposes, it's a steady rate – although it would not be for dinosaurs; for them the difference would begin to build up to something meaningful. At the time the earth was created, the decay rate was about 4,000 a second. But in the billions of years that have passed since then, there has been such a huge loss of potassium-40 that the rate has slowed way down. It has had to because the rate of decay, *proportional to what's left,* has to stay fixed.'

Potassium-40, Aronson explained, like all radioactive elements, has what is known as a half-life: the time that it takes for half of it to decay into another element. The half-life of potassium-40 is 1.3 billion years. Therefore, of the amount of potassium-40 remaining on earth today, half will be gone 1.3 billion years from now, half of the remainder in another 1.3 billion years, and so on. As the total sample shrinks, the rate of decay will slow to a crawl. At some unbelievably distant time in the future, if there should be only a hundred atoms of potassium-40 left on earth – and if there still is an earth – it will take another 1.3 billion years to reduce them to fifty.

'You mean, all the potassium-40 in the world is gradually decaying and turning into argon gas?' I asked.

'That's right. There's less and less of it. And more and more argon every day.'

'Where does all that argon go?'

'Ultimately, into the atmosphere. When the earth was created and the atmosphere first formed, there was almost no argon in it. But it has been gradually building up. As you and I stand here talking, we are leaking atoms of argon through our skins and into the air. It is now about one percent argon. That's why I have to take special care to exclude all the air I can from the samples of trapped argon that I measure.'

'You mean that after some billions of years the argon in the air is only one percent?'

'Yes.'

'That doesn't seem like much.'

'It's a lot for argon. It's a lot more than in these volcanic samples. Argon is very rare stuff.'

It occurred to me that if human bodies contained potassium-40 that was steadily turning into argon, why wasn't it possible to measure directly the decay of potassium-40 in fossils? Why did he have to use volcanic samples?

There were several reasons, he explained, two of which were overwhelmingly important. First, there was so little potassium-40 in a fossil bone that it could not be measured accurately. Even more important, the bone had been leaking argon gas for a million or two years; therefore there was no sense in trying to measure what was left. To get an accurate date on anything, it was necessary to use a sample that did not leak. The great virtue of volcanic material was that it was essentially leakproof; it trapped the gas in small crystals. The material that comes shooting out of the earth during an eruption, Aronson explained, under enormous pressure and at a very high temperature, begins to cool as soon as it hits the atmosphere. As it does so, it coalesces, sometimes into something like glass and sometimes into small crystals. This takes place very quickly, in a matter of hours. Therefore the crystals that are formed represent a single moment in time. They come into being absolutely pure. They are clean slates, totally uncontaminated by any older argon. And if they are able to trap all the argon that is released in them from that moment on by potassium decay, it is clear that they hold an accurate key to their own age.

'I just cook the argon off and measure it,' said Aronson. 'Since I know how much potassium I started with, and since I know the decay rate, I can calculate the age of the sample by simple arithmetic.'

It is not quite that simple. For one thing, there is always a tiny amount of air tightly adhering to the surfaces of the sample itself. No matter how good Aronson's pumps are, they never succeed in getting rid of all of it. That atmospheric contamination has to be measured. He does it by taking advantage of a peculiarity of argon.

Like potassium, argon comes in several forms. The form that potassium-40 turns into is argon-40; that is the kind that is so common in the atmosphere. However, already in the atmosphere is a far rarer variety, argon-36. There being no argon-36 in the volcanic sample, any of it that is detected by the mass spectrometer must be from the very small amount of air that is left in the machine after the vacuum pumps have done their work. Since the ratio of argon-40 to argon-36 is constant in the atmosphere – 295.5 atoms of the former for every atom of the latter – Aronson simply measures the amount of argon-36 in his test sample, multiplies that by 295.5 and deducts it from the total. What is left over is the argon-40 that has been released by radioactive decay within the crystal. 'And that amount is very, very small,' said Aronson, 'barely measurable.'

'Why is it so small?'

'Because the sample is so young. I told you that we're dealing with things that are only two or three million years old. There just hasn't been time to produce a respectable amount of argon.'

'How much argon might there be in a three-million-year-old sample of volcanic crystals?'

'Very little. A few trillion atoms.'

'That's a little?'

'Atoms are awfully small,' he repeated.

Aside from the inherent difficulty of handling young samples in his laboratory, Aronson has the equally vexing problem of getting suitably pure samples.

'They have to be clean. They have to be uncontaminated by other things. And they must be free of any damage that might have caused them to lose some of their argon. That was the problem with the basalt sample you sent me from Hadar. From its appearance I suspected that it had become weathered and had leaked a little argon; I couldn't tell how much. Even so, I decided to use it, despite the additional problem of contamination by air adhering to its surface.

'I would have liked to use those first volcanic-ash samples I was sent, but they were far too altered to be useful. Ordinarily I prefer ash; it is apt to contain crystals of feldspar or mica that are rich in potassium. With such young samples you're working on the very edge of your ability to measure; therefore you have to find something that's fairly rich in potassium. If you don't, you'll get such a tiny

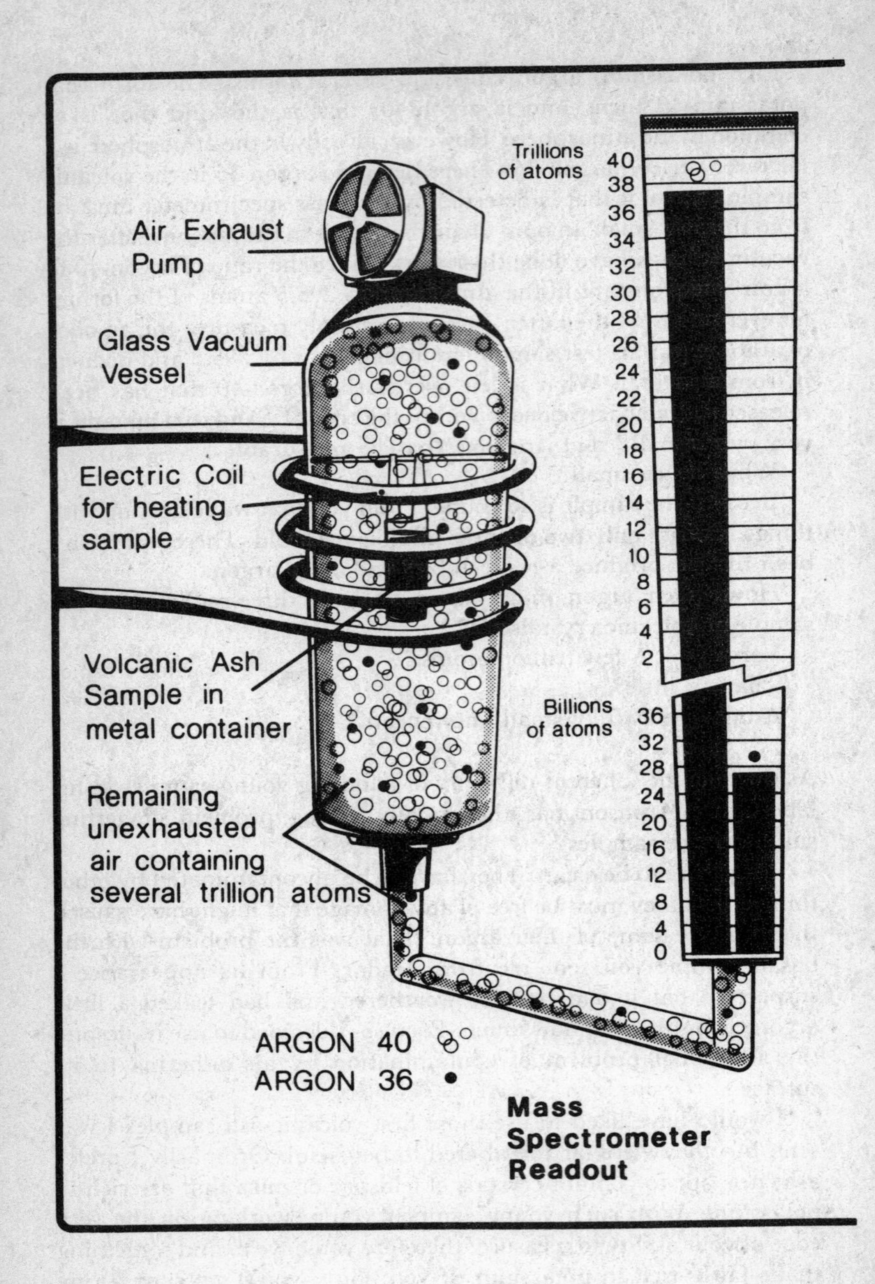
Air Exhaust Pump
Glass Vacuum Vessel
Electric Coil for heating sample
Volcanic Ash Sample in metal container
Remaining unexhausted air containing several trillion atoms
Trillions of atoms
40
38
36
34
32
30
28
26
24
22
20
18
16
14
12
10
8
6
4
2
Billions of atoms
36
32
28
24
20
16
12
8
4
0
ARGON 40
ARGON 36
Mass Spectrometer Readout

HOW TO CALCULATE A POTASSIUM–ARGON DATE

1. *Measure the size of the sample and its richness in potassium.*
   This is easily done by standard laboratory procedures, and shows that the sample contains $^1/_{10}$ gram of potassium.

2. *Calculate the annual decay rate for a sample that size.*
   It is known that the potassium-40 in one gram of ordinary potassium decays to argon at a rate of about 3.5 atoms per second. Therefore:

   $$\begin{aligned} 3.5 \times 60 &= 210 \text{ per minute} \\ \times 60 &= 12{,}600 \text{ per hour} \\ \times 24 &= 302{,}400 \text{ per day} \\ \times 365 &= 110{,}376{,}000 \text{ per year} \end{aligned}$$

   So, $^1/_{10}$ gram of potassium yields 11,037,600 atoms of argon per year.

3. *Boil off the sample,* and send it to the mass spectrometer (along with any contaminating air that may be in the bottle).

4. *Obtain mass-spectrometer reading.*
   This particular sample gives a reading of:

   36,765,875,000,000 atoms of argon-40 (from the air and from the sample)
   27,070,000,000 atoms of argon-36 (from the air only)

5. *Eliminate the atmospheric contaminant.*
   Since there are 295.5 argon-40 atoms to every argon-36 atom in the atmosphere, multiply the argon-36 total by 295.5:

   $$\begin{array}{r} 27{,}070{,}000{,}000 \\ \times\ 295.5 \\ \hline 7{,}991{,}850{,}000{,}000 \end{array}$$

   This is the number of contaminating atoms of argon-40 still in the sample.

   So, from the total reading in the mass spectrometer: 36,765,875,000,000
   deduct the atmospheric contaminant: 7,991,850,000,000
   28,774,025,000,000

   This is the number of argon atoms released by the sample.

6. *Calculate the age of the sample.*
   Since the sample decays at a rate of 11,037,600 atoms a year, it is necessary to divide the number of argon atoms in the sample by that number:

   $$\frac{28{,}774{,}025{,}000{,}000}{11{,}037{,}600} = 2{,}606{,}909.5$$

Answer: The sample is 2.6 million years old.

amount of argon that you may not be able to measure it if the atmospheric-air contamination is large.'

'Does all volcanic ash contain potassium?' I asked.

'Most does. But it isn't always rich enough for you to work with. And ash has other problems, too. You have to make sure that the crystals are not damaged. If they are, either from weathering or from exposure to high temperatures, they may have leaked argon. More serious, they are apt to be contaminated by much older crystals that may have gotten mixed in with them in one way or another. The older stuff may have been lying around when the ash landed, or it may have been moved in much later. Potassium feldspar's pretty common stuff, you know. Suppose I collected a sample from a volcanic layer at Hadar. Mixed in with it might be some feldspar that had washed down from the mountains. That feldspar might be a couple of hundred million years old and absolutely loaded with argon. Mix just a few crystals of that in with your sample and they would overwhelm it. That's why I felt I had to go to Hadar myself, to make sure I got as reliable samples as possible. Your fossils are too important to have bad dates attached to them.'

On his arrival at Hadar in December 1974, Aronson's first task had been to familiarise himself with Taieb's geological reconstruction, get samples from the basalt layer, locate any other datable horizons he could, and procure good samples from them. He and Taieb had taken to each other immediately. Taieb had shown him a diagram of the stratigraphic column he had constructed, then given him a tour of the deposits. From then on, Aronson was on his own. A very hard worker, he was up each morning before daylight, and would pad silently out of camp with his rock hammer and a pack over his shoulder – just another foolish foreigner to the early-rising Afar women he passed in the predawn dusk.

He walked everywhere. 'You can't learn geology from the seat of a car,' he said. He started at the extreme western edge of the deposits and painstakingly trudged through them, making notes and collecting samples. He walked ten or fifteen miles a day. He picked up several marker layers that Taieb had identified, and followed them into and out of gullies. He did this with Taieb's so-called triple tuff: three narrow bands of volcanic ash that had been deposited in rapid succession, with a distinctive layer between them. This was the ostracod layer, named for a dense concentration of the shells of minute aquatic organisms that had had a great blooming at a time when the lake was rich enough in chemicals to encourage their growth. The ostracod layer, with the triple tuff surrounding it, was one of the most easily recognisable marker horizons in Taieb's stratigraphic column. It was almost like a street sign in the gullies.

Aronson slowly followed the triple tuff and the ostracod layer eastward. He identified another horizon that Taieb had located: the gastropod layer, made up of the shells of small aquatic snails. He worked his way along that and along a distinct layer of clay: the 'CC' (confetti clay) horizon – so named because, when eroded, it looked like confetti. All in all, he was highly impressed by Taieb's stratigraphic reconstruction.

Eventually he came to the basalt layer, the datable lava flow that covers the easternmost deposits at Hadar, and whose first somewhat dubious samples had given him an error of plus-or-minus 200,000 years on a provisional date of three million. Aronson was determined to strengthen his confidence in that date and shrink the error. He collected a number of samples, taking great care to get ones that appeared not to have been altered in any way.

The basalt layer had presented a geological problem from the beginning, because its position in the stratigraphic column was equivocal. There would have been no problem if it had extended throughout the whole area, but it did not. To connect it up with the main deposits where the fossils had been found required a matching of layers either above it or below it to similar layers elsewhere. It was impossible to do this with layers above it, because there weren't any. All the material above the basalt layer had been eroded away; it stood exposed to the air, and the geologists could walk about on its surface. Underneath it were some layers of such indistinguished material that finding their counterpart elsewhere in the deposits had proved very difficult. Taieb had done the best he could. He thought he had a match, and had placed the basalt layer about ninety meters above the bottom of the stratigraphic column he had constructed.

One day Aronson walked into camp and announced that the basalt was wrongly placed. It should be older, much lower down in the column.

'What is your evidence for that?' asked Taieb.

'I think I have some.'

All Aronson's trudging had revealed to his sharp eye some things that none of the other geologists who had visited Hadar had paid particular attention to. First, he had located the lip of the lava flow, the point at which it had stopped spreading and had begun to solidify. Others had located that but had failed to examine it carefully. Aronson did. Just in front of the lip he found a sand-filled channel that had been cut by erosion sometime later. He knew it was later because he found blocks of the basalt that had broken off from the lip and fallen into it. If the channel had been cut earlier and had been filled with sand and gravel before the basalt had come flowing over the surface, there would have been no way that hunks of it could have fallen into the bottom of the channel.

Having satisfied himself on that point, Aronson then turned to an examination of the contents of the channel. Those were disappointingly uninformative; they resembled much of the other sandy material that was so prevalent throughout the Hadar deposits. But in the layer just above the channel he found a distinctive pattern of ancient root casts. With rising excitement, he traced the root casts out into the main section of the deposits and was able to anchor them twenty meters below the ostracod layer: a solid linkup at last.

'I'm afraid you're going to have to drop the basalt down to here,' he said.

That could have been a sticky moment. But it was not with Taieb. He is a scientist, anxious to get at the truth. He does not get hung up on pride. I know anthropologists and geologists who will cling to a point long after it has become overwhelmingly clear that they are wrong. Not Maurice. When Aronson came up with that really brilliant piece of geological detective work, Maurice was delighted.

Camp was breaking up. Taieb had returned to France. Those of us who were left – Aronson, myself and a couple of others – were reduced to eating up the odds and ends that Kabete, the cook, was doing his best to stretch out. He had been serving mostly goat meat: roast goat, fried goat, fricasseed goat, goat kebab, goatburgers. Now we were down to our last goat, a small kid that Aronson had become very much attached to.

Sitting down at the table for his last supper, he said, 'What's this?'

'Goat.'

'What goat? There aren't any more goats.'

'That's *your* little goat.'

He left the table, walked out into the desert and refused to eat any supper at all. The other campers didn't see him until morning.

Kabete, the camp cook, did all his work out of doors, with his cloths, pots and pans, and food hanging in a nearby acacia tree. Here he makes bread. Two goats hang ready to be cooked.

'We have to do those things,' I said.

'I know. I know.' But he was obviously hurt by the slaughter of his little friend, the only small, carefree, innocent thing he had seen during his entire visit to that bleak land. He left the same day, with his samples, for the United States.

Those samples told Aronson a number of things. First, his test runs came up with a series of dates that clustered closely around 3.0 million. That was most reassuring. Not only did it confirm his earlier test, but the near-uniformity of the new figures made it possible for him to reduce his possible margin of error from plus-or-minus 200,000 years to plus-or-minus 50,000 – a swing that was only a fourth as large as his previous one.

Still, he was not happy. He had examined all the basalt samples under the microscope, and had detected in even the best of them what he took to be minute signs of weathering, which meant that they probably had leaked a little argon. If that was true, then the basalt was older than 3.0 million – by how much he could not say, because he had no way of telling how much argon had been lost.

He decided to check his suspicion. He had deliberately collected a number of samples that had been more severely altered by the elements. Now he tested these, assuming that the greater the alteration, the greater the argon loss would be, and the younger the sample would be. That assumption turned out to be correct. A sample that had been weathered to a point where it was grayish in color, instead of being nearly black, gave a date 400,000 years younger. A still grayer, more crumbled one gave a date 500,000 years younger; it had lost considerable argon. Inasmuch as the 'bad' samples showed (to the naked eye, and to a more extreme extent) the same kind of weathering that the best samples showed to a lesser extent under the microscope, Aronson reluctantly concluded that they too were slightly altered and that there was indeed a small but, at the moment, indeterminable error in the 3.0-million-year date.

'You can be sure that the basalt is at least three million years old. My own hunch is that it's older,' he said to me.

'How much older?'

'I can't say yet. But we should get some help from paleomagnetism in figuring that out.'

As it turned out, paleomagnetism (the magnetic properties of the earth itself) could help. The earth is a magnet. Like all magnets, it has a positive and a negative pole. Today, which is considered 'normal', the positive end of the magnet is at the North Pole and the negative end at the South Pole. But for reasons that apparently have to do with the ebb and flow of molten magnetic material deep below the crust, the poles flip over from time to time, and there are considerable periods of 'abnormality', or reverse polarity, during which the North Pole is negative and the South Pole positive.

By studying the lineup of magnetic crystals in rocks, geologists have been able to chart these flip-flops of polarity, going back for some millions of years. The flip-flops say nothing themselves about when they took place, but they do give a very clear indication of the order of flip-flopping. In 1972, taking advantage of the torrent of absolute dating that was pouring out of laboratories as a result of radioactive measurements, three scientists, working separately, produced near-identical timetables of paleomagnetic shifts that were tied to actual dates. They were good the world over, and could be applied directly to Hadar as soon as some magnetic crystals from rocks there had been collected.

A continuous series of four hundred rock samples of increasing age was assembled by a geologist, Tom Schmitt. When they were subjected to magnetic analysis, it quickly was established that the basalt had been laid down during a period of reverse polarity. But which one? Was it the so-called Mammoth reversal, now known to have lasted from 3.1 to 3.0 million years ago? Or was it the slightly older Gilbert reversal, which went from 3.6 to 3.4 million? Nothing between would have been possible, because there was a 300,000-year period of normality during that entire time.

'So,' I said to Aronson, 'by applying paleomagnetism, you're telling me that the basalt has to be either 3.1 to 3.0 million years old, or else it has to be more than 3.4 million years old.'

'That's right,' he said. 'My hunch is that the older date may turn out to be the better one, although I can't prove it. If the basalt is altered at all – and I suspect it is – that will mean that it is somewhat older than the tests show. If that is so, we will have to move it out of the Mammoth reversal at 3.1 million. It can't be 3.2 or 3.3; that would put it in a normal period. So, if it's pushed out of the Mammoth, it will have to go all the way back to the Gilbert and to a date of 3.4 to 3.6.'

'Okay, so which one do I pick?'

'That's up to you. The safest thing would be to pick three million. That's what the potassium-argon dates give. Until they're proved wrong I think we should go with them. But I would hedge and say *at least* three million. That's what I'm going to recommend we say when we publish our dating paper.'

That lack of precision in the date of the basalt was extremely bothersome to Aronson. He made up his mind that he would try to improve it by taking one of his assistants, a young volcano expert named Bob Walter, to Hadar the following year. He had learned during his long trudges through the deposits that the volcanic history of Hadar was extremely complex, and he hoped that Walter might be able to make some progress in unraveling it and, as a possible bonus, come across some volcanic tuffs that might lend themselves to dating. With the basalt as the only useful dating horizon found so far, he lacked the kind of cross-checking that is so important in this work.

Walter went to Hadar in the fall of 1975, and rather quickly spotted something of tremendous importance: three shallow layers

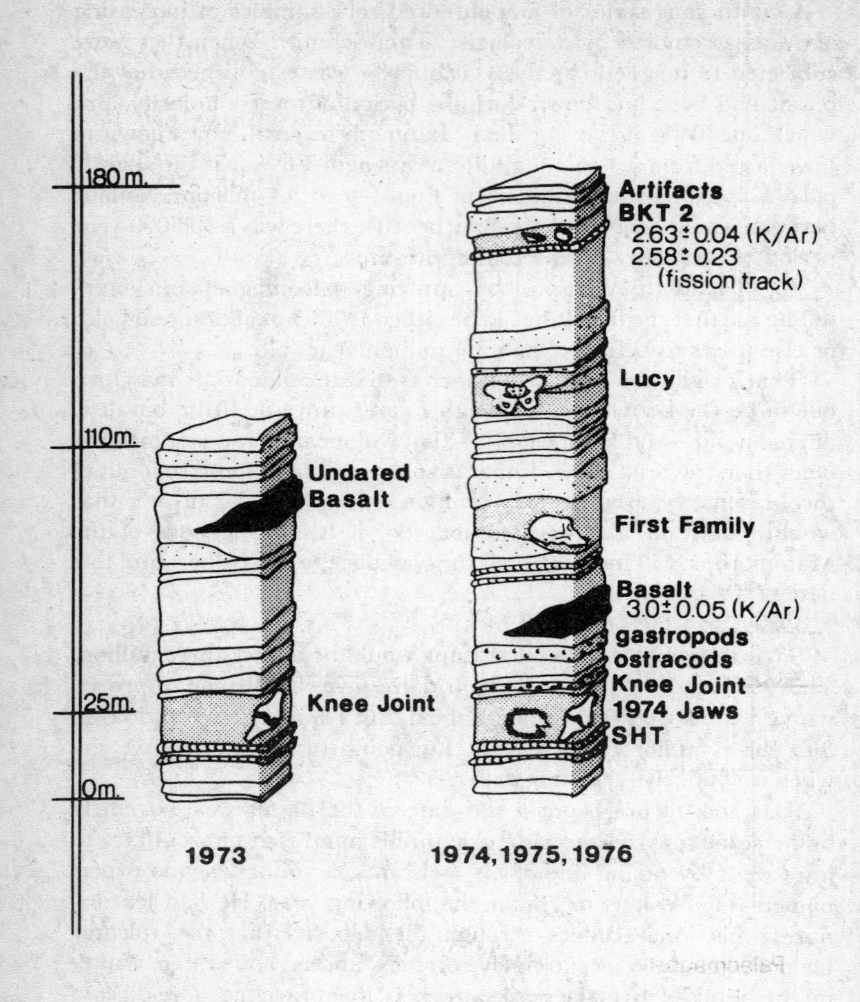

Dating the Hadar fossils took seven years and required the synchronisation of five techniques: geology, potassium-argon dating, fission-track dating, paleomagnetism, and biostratigraphy. All were necessary for the gradual development of a detailed and accurate stratigraphic column. The one labeled 1973 shows what Taieb was able to learn solely through geology on his first attempt. The next three years produced a flood of fossils and dates,

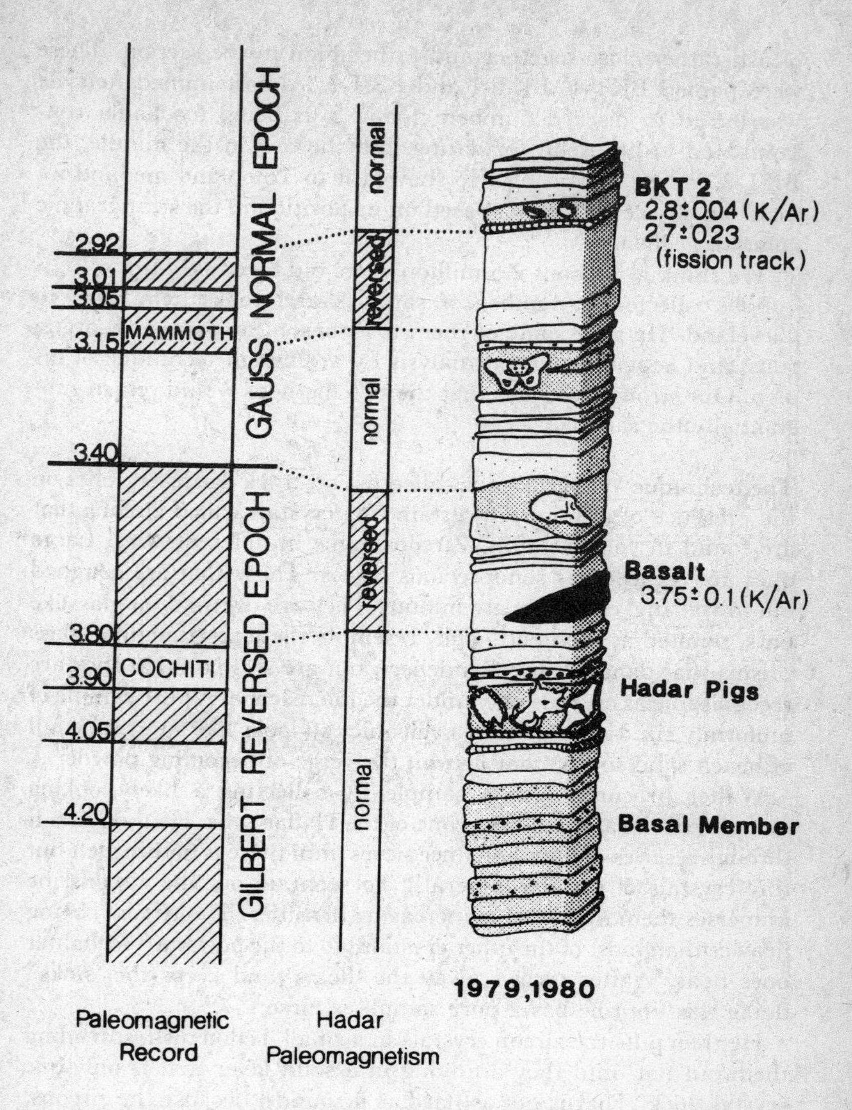

which were corrected in 1979 and 1980 by the paleomagnetic record and by the evidence of pig fossils (biostratigraphy). From all these calculations Aronson is now reasonably sure that the basalt is 3.75 million years old. That would make Lucy and the First Family close to 3.5 million years old, the jaws and knee joint close to 4.0 million years old.

of ash rather close together and rather high in the section. These were named BKT-1, BKT-2 and BKT-3. Almost immediately he was forced to discard numbers 1 and 3 as being too badly contaminated to be useful for dating, but the one in the middle, the BKT-2, looked promising. He showed it to Taieb and me, and we made an off-the-cuff guess, based on its position in the stratigraphic column, at its age.

'We think it is about 2.5 million years old,' we told him.

Bob collected a number of samples and took them back to Cleveland. He gave some of them to Aronson for potassium-argon runs, and kept the rest for analysis by a different technique of his own. Our strong hope was that the two methods would return substantially the same answers.

The technique Walter used is called fission-track dating. It relies on the presence of uranium in certain tiny crystals called zircons that are found in volcanic tuffs. Zircons come in different sizes. Large ones are regarded as semiprecious stones. Those that are coughed out in volcanic eruptions are minute. They are transparent glasslike bars, pointed at each end. They resemble the long triangular glass prisms that dangle from chandeliers, but are so small that they are recognisable as crystals only under the microscope. A pure sample of uniformly sized zircons from a volcanic tuff looks like a test tube full of beach sand so fine that it is on the verge of becoming powder.

Walter procures such a sample by collecting a likely-looking scoopful of volcanic ash from one of the Hadar tuffs. He then sifts it through a series of finer and finer sieves until there is nothing left but tiny crystals of various minerals. To separate out the zircons, he immerses them in liquids of increasing densities. The zircons, being heavier than most of the other crystals, go to the bottom. The lighter ones float. Walter throws away the 'floats' and keeps the 'sinks', doing this until he has a pure sample of zircon.

He then puts the zircon crystals in a small Teflon dish, spreading them out flat until they are lying in a solid layer that is only one crystal thick. That is not as hard as it sounds, because the zircons, being several times as long as they are wide, naturally lie on their sides. Glued in place, they can then be polished to remove any weathering or scratches on their upper surfaces. This step is necessary because Walter is looking for other tiny scratchlike lines formed by uranium explosions within the crystals. He will count

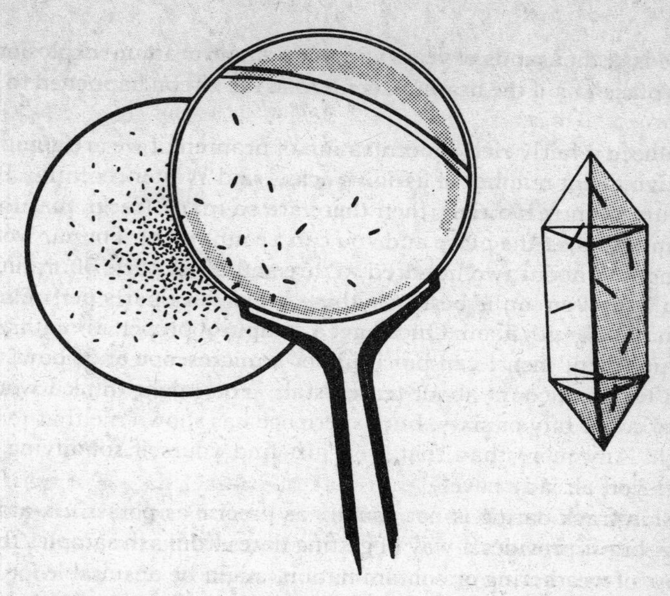

Fission-track dates are obtained from tiny zircon crystals, which are first screened to yield a pure sample, then polished so that all surface scratches are removed. Any marks that remain in the crystals are the work of uranium atoms exploding one by one over a long period of time. The number of marks – fission tracks – in a crystal indicates its age.

them to determine the age of the crystals, and therefore must start with a surface that is unblemished by the slightest mark, however small.

Uranium-238 is a radioactive element whose atoms decay at a slow but steady rate into lead, just as potassium-40 ticks over into argon. The difference is that the potassium-argon transformation is a quiet one, whereas the uranium-lead one is accompanied by a minute release of energy. This fires off simultaneously from both ends of the atom with sufficient power to dislodge some of the other atoms in the zircon. The effect is like that of a man walking through a wheat field; it leaves a faint line of damage to mark that single tiny movement of energy. The line that is left in the zircon is called a fission track, and if it is delicately etched with a chemical to enlarge it, it can be seen under a microscope.

Uranium-238 fission decay is even slower than the decay of potassium-40. It has a half-life of many billions of years. At that lethargic rate of change, a zircon crystal might lie around for hundreds of thousands of years without a single uranium explosion's taking place in it if the uranium content in the zircon happened to be low.

'We need a fairly rich concentration of uranium if we are going to get a significant number of fission tracks,' said Walter recently. 'But if the uranium is too rich, then there are so many tracks that they crisscross all over the place and you can't count them. For our work, a strength of about two hundred to three hundred parts of uranium per million of zircon is best. Anything under fifty parts per million we can't work with at all. Once I get a sample of proper strength and get it polished, then I can put it under a microscope and count the fission tracks. I count about ten crystals. You might think I would have to count fifty or sixty, but experience has shown me that ten is enough. Any more than that, you just find yourself solidifying an answer you already have.'

Fission-track dating is not usually as precise as potassium-argon dating, but it provides a way of getting dates from ash samples that, because of weathering or contamination, might be unsuitable for the latter method. More important from Aronson's point of view, it relies on a totally different technique. It is as if one were to estimate the age of a tree by counting its annual growth rings; then check the result against a series of old family photographs with the tree in the background, noting its age from the known ages of people in the photographs; and finally interview a couple of old men who claimed to remember when the tree was planted. If each test substantiated the others, one could be pretty sure that he had the age of that tree nailed down.

Similarly with Walter's fission-track test. When it was done, the confidence in the Hadar dating soared. It indicated that the BKT-2 tuff was 2.58 million years old – a figure nearly identical with Aronson's differently derived potassium-argon date of 2.63 million years.

When Aronson and Walter had completed their work, the Hadar stratigraphic column looked like this: It showed a volcanic-ash layer near the top at about 2.6 million years – a firm date that checked in two different ways. Deeper in the column was the basalt layer, dated at 3.0 million years, and now checked by half a dozen potassium-argon runs that duplicated one another with reassuring consistency. Lying between them was Lucy, at a level that suggested her age was 2.9 million years.

'Except that we still don't *know* that,' I said to Aronson. 'If you really think the basalt is older, then that would make Lucy older. What do I do about that – check the animal fossils again?'

'At present, that would be your best bet.'

That meant turning to Basil Cooke and his pig sequences. These had already straightened out a dating puzzle at Lake Turkana and shoved Richard Leakey's 1470 *H. habilis* skull forward from 2.9 million to less than 2.0 million. Perhaps they could do it for Lucy too. But in this case they would be stretching her age, not shrinking it. Looking at the extreme primitiveness of our collection, I began to get a stronger and stronger impression that that was what would happen.

My suspicion, however, could not be converted to belief. Cooke, it turned out, did have pig sequences running all the way back to four million, but he was not ready to talk about them. In fact, he said nothing until 1978. In that year he published a paper that compared the Hadar pigs with the Omo pigs, saying, 'An age of 3.0–3.4 would give a better fit than the 2.9-million-year age for Lucy presently obtained from the basalt.'

That apparently innocent statement packed a wallop. It meant taking the basalt out of the Mammoth reversal and putting it in the Gilbert. It meant that Lucy could be as much as 3.3 or 3.4 million years old, and that some of the jaws that came from lower down in the section, beneath the basalt, might be as much as 3.7 million years old.

I wish I had known that in 1975. It would have simplified things. Perhaps I should have waited to hear from Cooke. But I was under increasing pressure to publish about the Hadar fossils. You cannot announce them and then just sit on them. People begin pressing you about that. So I published, and I made a mistake. In fact, in my early papers I made several.

In the first paper that Taieb and I put out, we located the basalt layer too high in the section. I then published a second paper about Lucy and the jaws, and I made a second mistake: I said that Lucy and the jaws probably represented different species. Then on the matter of dates I went with three million for the basalt; now that seems to have been a third mistake. That is about par for the course. A mistake in geology, a mistake in paleontology, and a mistake in dating. But you correct as you go. It all comes out in the wash.

10 The Third Hadar Field Season: The First Family

When Don showed me the first knee joint, I told him to go back and find me a whole animal. He obliged with Lucy. So I told him to go back again and get me some variety. The next year he found Mom and Pop and the kids.

C. OWEN LOVEJOY

Back in the United States in 1975, I found myself in quite a different situation than I had been in the year before. That time I had overrun my grant budget, had very little to show for it, and felt myself to be in imminent danger of being labeled an unreliable field leader. Now I had what was probably the most electrifying fossil in the world, and people everywhere were eager to work with me. Money ceased to be a serious problem. The *National Geographic*, ever generous after the mine begins producing nuggets, asked for an article, and agreed to send a photographer, David Brill, to Ethiopia the next fall for a major illustrated piece on the Hadar work. At the same time, a French film team was planning a visit. The National Science Foundation renewed its grant.

More important to me was that Aronson's dating, plus the paleontological work done by Gray and myself, had convinced most scientists that Hadar was a meticulously worked site with high professional standards; they could risk their own time and reputations by becoming part of its activity.

But Lucy had also brought increasing responsibilities. In 1974 I was appointed Curator of Physical Anthropology at the Cleveland Museum of Natural History. I had to set up that program. I had to begin phasing in the best of my students who were graduating at Case and wanted to do postgraduate work with me at the museum. Finally, I was in demand for lectures. All in all, I scarcely had time for the paper I wrote in collaboration with Taieb and Coppens on the 1974 season. The gist of it was that two kinds of hominid had been found. One kind (Alemayehu's jaws) represented some sort of extremely primitive *Homo*. Another, smaller kind (Lucy) represented something else.

I would withdraw that paper today if I could. It stands as an object lesson to me not to be too hasty in the future. I am not ashamed of the descriptions I wrote. I tried to model them on those which J. T. Robinson made of the South African australopithecines, and I believe they are as good as his. But the interpretations – well, I have learned more since, and would make them differently today. Despite the encouragement Mary and Richard Leakey gave me to think so, I no longer regard Alemayehu's jaws as belonging to human beings. Nor do I think that Lucy is different from the jaws. She is just a small version of the same thing.

My greatest concern in 1975 was getting back to Hadar. After the end of the 1975 field season the Ethiopian political situation had begun to change radically. The military was now in firm control. Nothing stirred without its permission; no decision was made that it did not make. The various ministries were becoming close to impotent, afraid to function on their own.

To add to the internal turmoil, war with Eritrea was building up. When I arrived in Addis Ababa in September 1975, I was advised by the U.S. Embassy that a trip to the Afar was out of the question: 'We don't want you to go out there. We'll just have to rescue you.' I learned that the Eritrean threat was a real one from the Ethiopian point of view, but not so much from mine. Raider bands would do their harassing along the border. To get to Hadar they would have to traverse a wide stretch of desolate territory inhabited only by Afar tribesmen who would certainly attack any invaders encountered. The chance that an Eritrean armed party could penetrate to Hadar and get out alive was considered remote. If the invaders were encumbered by hostages, their escape was thought to be virtually hopeless.

I also got some encouragement from my Addis supply contact, the archeologist and historian Richard Wilding. Wilding is a perfect British colonial type. Addis Ababa has been his home for years, and he probably will never leave it. Cool, unflappable, he cheerfully endures the inconveniences of life in a politically unstable environment. If a bomb goes off down the street, he ignores it. If the water supply fails or if the local market runs out of supplies, those are the prices he must pay for living in a country that he loves. He hoards and he scrounges. He does without.

Braced by Wilding and by his assessment of the Eritrean danger, and convinced that the Dergue would support our work – it might make operations difficult for us by tying us up in red tape, but we were certainly in no danger from it – Taieb and I called a meeting of all the expedition members the day before they were scheduled to truck out to Hadar from Addis. I said that any foray into the desert in a country as combustible as Ethiopia carried a certain degree of risk. I had assessed that risk as well as I could, and had decided, for myself, that it was worth taking. Any member of the expedition who felt differently was free to leave without prejudice; the air fare out and back would be taken care of by the expedition.

Nobody said anything. I put the question again. All said they wanted to go. I wasn't surprised. They were an extraordinary group. It is really uncomfortable at Hadar, but they love it, just the way I do. There is something about the place – its stillness, remoteness, and beauty – that holds you. And the work – there is always the feeling that something momentous is about to happen. If you are hooked on the past – and all of us are – where else in the world would you want to be?

Furthermore, not one of us thought there was any real danger. We had spent two seasons with the Afar and had gotten along famously with them. They all carried weapons, of course. We didn't. I had made it a rule that nobody in the expedition would carry a weapon of any kind. Consequently, we began to feel more or less as if the Afar were protecting us. I say, *we* felt that way. The Amhara workmen who came down from the highlands with us did not. They were afraid of the Afar. They sensed their dislike and stuck very close to camp. I remember one Amhara workman being told when he got into some kind of argument with an Afar, 'When the trouble starts, you're the first one I'm going to kill.' That was no idle threat. Police patrols were a long way off. They seldom got down into this part of the country. Even though it was ostensibly under the control of the central government, it was actually under the control of tribal leaders.

We did take the precaution of working out a danger-signal system with Wilding. Having no radio (that was forbidden by the government), we relied on Wilding to make a weekly air flight not only to deliver supplies and remove fossils, but also to make sure that nothing had gone wrong. The signal would be in the way the vehicles were arranged in camp. If they were parked in a row, then the pilot who buzzed the camp before going on to land at the airstrip some miles away would know that everything was going smoothly. If the vehicles were disarranged, that meant trouble.

As it happened, one of Wilding's first flights came on the day a large elephant skull had been found, and a truck and trailer had been dispatched to get it back to camp. When Wilding arrived by plane, the trailer was in front of the paleontology tent and the truck was somewhere else. Wilding did not even land. He flew straight back to Addis, alerted the U.S. Marines, started making arrangements for evacuation by helicopter, and even notified the U.S. Navy in the Red Sea. Then, as a precaution, he flew back to Hadar – and found all the vehicles neatly lined up again. The Marine alert was cancelled.

Wilding's biggest problem was with his pilots. It took all his persuasiveness to get them to make the trip. The airstrip they would land at was a forty-five-minute drive from camp. The pilots did not want to leave their plane unguarded, but at the same time, they were afraid to stay alone with it after Wilding had been met and driven off to camp. They too were anxious about what the Afar might do to them. None of them was molested, however, and the flights continued without incident throughout the season.

One person who was unperturbed by the Afar was David Brill, the photographer sent out by the *National Geographic*. He turned out to be seven feet of legs with a little chest perched on top, and a baseball cap on top of that. When the plane buzzed the camp to notify us that it was landing at the airstrip, Gray drove over there, and at first didn't see anything. The plane was gone. Then down at the far end of the strip, he spotted a small piece of cloth tied up in a bush. Dozing under its shade was Brill. When he got to camp he said to me, 'You must be Doctor Don.' He might have been tempted to say 'Dr Johanson, I presume,' but he managed not to. We had warned the *Geographic* to send somebody who could rough it, and they really came through. Not only was Brill a fantastic photographer, but he was one of the most adaptable and uncomplaining men I have ever met.

I had begun to suffer from a craving for sweets. I found myself dreaming about them at night, and had asked that Brill bring some with him. 'Did you get any chocolate?' was one of the first things I said to him.

'Hold out your hand and stand back,' said Brill.

I did. Brill rummaged around in his pockets and managed to produce a driblet of three or four sticky half-melted M&Ms.

'Is that all?' I asked.

'Well, I was in a great hurry at the airport. Also I got hungry and I ate most of them myself.'

That night at supper Brill went on about the chocolates and what a shame it was that he hadn't managed to do better. He brought it up again at bedtime in the tent he and I shared, saying that as I fell asleep I would probably be dreaming about the chocolates I might have been eating at that very moment if he hadn't been in such a hurry.

'Why don't you just shut up about the chocolates?' I said.

'Okay,' said Brill. 'But meanwhile. . .' He opened his knapsack. In it were eighteen bars of delicious Swiss chocolate.

Brill was less flaky about his work. He walked the deposits with the same kind of determination that Aronson did, but with a different purpose. He wanted to see how they – and the escarpments – appeared under different light conditions. When a fine saber-toothed cat skull was found, he complained, 'Why do you always find things in the middle of the day? Why don't you find them late in the afternoon when the light is good?'

I didn't answer.

'When are you going to find another hominid?'

'Tomorrow,' I said in exasperation.

Later, Brill would say that all one had to do to find hominids in Ethiopia was ask for them, because 'tomorrow' turned out to be one of the most extraordinary days in the history of paleoanthropology. I was out surveying with Mike Bush, a young medical doctor with an interest in archeology. Mike is a very quiet man. He is a plodder, who never says anything but keeps his eyes open. In fact, the French became irritated with him because he never spoke. 'Why doesn't he talk?' they wanted to know. 'What's wrong with him?'

Well, there was nothing wrong with him. We had been out an hour or so, and he called to me from over a rise, very quietly, 'I think maybe I've got something.' I went over. Sure enough, there was a block of stone with two hominid premolars sticking out of it. They were extremely inconspicuous. How he ever saw them I'll never know. I'm not sure even Alemayehu would have spotted them.

Brill wandered up, and I said, 'Here's the hominid I promised you.'

'You've done it again,' said Brill. 'Right smack in the middle of the morning you find one – an impossible time to take photographs.'

'Very well, Your Majesty, exactly when would you like us to continue to find hominid fossils?'

Brill checked the sun. 'Tomorrow morning at eight o'clock.'

We suspended work, and gathered the next morning to begin removal of Bush's fossil. Brill set up his equipment to photograph the operation. Also present was a French husband-and-wife movie team. The wife's name was Michèle. She was always enthusiastic about finding something, but she couldn't tell a fossil from the end of a Coca-Cola bottle. While we were working down at the bottom of a rather steep slope, she climbed halfway up to sit in the shade under a small acacia bush, and she practically sat on a couple of bones. 'What are these?' she called out, holding them up. I climbed up to look at them and was bowled over. One was a hominid heel bone. The other was a femur exactly like the femur I had found two years before, except that it was nearly twice as large. All of this was photographed by Brill and by the French movie people – the first time ever that professional photographers have taken pictures of fossils actually being found. All the others that you see in books – *all* of them – are re-enactments.

Finding teeth at the bottom of the slope and two other bones halfway up encouraged a closer scrutiny of it. Almost immediately other fragments were spotted. Looking on the other side of the acacia bush, I had the unnerving experience of picking up, almost side by side, two fibulas – the smaller of the two shin bones in the leg. Another Lucy? No, these were both right legs, indicating the presence of more than one individual. Meanwhile, others were shouting over finds of their own, all of them hominid. Fossils seemed to be cascading, almost as from a fountain, down the hillside. A near-frenzy seized us as we scrambled madly to pick them up.

For a little while I scarcely knew what I was doing. I had never seen anything like it. I had never *heard* of anything like it. We were like crazy people. Finally the heat got to us, and we settled down.

The bone-strewn slope was duly entered on the master map as site 333. The rest of the field season – and most of the next year's as well – we devoted exclusively to working it. First the entire hillside was combed systematically – not helter-skelter, as in that first insanely euphoric hour. Then the gravel surface was carried down the slope – tons of it – to be sifted, load by load, through coarse sieves. Ultimately the hillside would yield about two hundred teeth or pieces of bone. Duplication of specific parts made it clear that at least thirteen individuals were represented: men, women and at least four children. It was not possible to fit many of the bones together into partially complete skeletons, as had been done with Lucy, because of the way they were jumbled and scattered down the slope. The possibility also cannot be ignored that considerably more than thirteen individuals may be represented.

Just as there is a residual faint hum or static buzz when a radio is turned on, so in Hadar there is a ubiquitous faint background 'noise' of animal fossils. Collect anywhere, and fragments of animal fossil show up, more prominently in some places than in others, but nowhere are bits of stray animal bone entirely absent, attesting to the opportunities for deposition presented by the sheer passage of time, to the slowness of geological activity, to the incessant wandering and dying of unnumbered animals – back and forth, endlessly, while hundreds of feet of deposits are laid down over hundreds of thousands of years. What was odd about the material sifted from the 333 hillside was that animal fossils were virtually absent. The background scatter that normally could be expected to result from an intensive collecting effort in any fossil-bearing stratum was somehow missing; everything that came from that hillside was hominid.

How was this to be explained? Had an entire band suddenly been wiped out, so suddenly that there had been no time for the expected deposition of animal bones? It would seem so. The hominid fossils appeared to come from a common source in a stratigraphic horizon near the top of the slope. What was found lying on the slope was what had presumably washed out during the last few rainy seasons. This line of reasoning was strengthened when the horizon itself was found, and some preliminary excavation into the hillside produced nearly twenty more fossil pieces. The inference was clear that before erosion had begun to scatter the bones, they had all lain in close proximity, and therefore had presumably died together.

But what had killed them? An epidemic? If so, what would have prevented their being found and gnawed by scavengers? Also, what would have prevented the addition of the normal 'background noise' of animal bones to their own as they lay open to the elements,

Maurice Taieb examines the cutaway face of the 333 site for geological clues as to what might have happened to explain the sudden deaths of so many australopithecines at one time.

gradually being covered by silt?

The longer I thought about these two anomalies, the less likely it seemed that there had been an epidemic. There was no way that the natural deaths, over a period of days or weeks, of an entire band of hominids could have resulted in their preservation in such a pure state.

What else, then? A fight? Was this the first recorded example of the kind of ferocious aggression that Raymond Dart claimed to be the natural heritage of hominids: one australopithecine band wiping out another? That too seemed wildly unlikely. There was simply no evidence, other than speculation, for it. Also, the absence of corpse molestation by scavengers, and the mystifying scarcity of other animal bones, were not to be explained by mass murder. Indeed, the Dart theory called for the eating of australopithecines by their brethren; that was one of their principal sources of protein. If this had been a deliberate slaughter, the bones would have been cracked open for their marrow.

Drowning? Had a band been trapped in a narrow ravine by a sudden flash flood? This seemed more plausible. Desert people the world over shun wadis or defiles as campsites. A journalist traveling recently in the Sinai desert found that the Bedouin there have a definite phobia about being caught in ravines. They have learned through bitter experience that in a stony landscape with no vegetation to act as a sponge to soak up sudden rain, a ravine can be a death trap, with a flash flood hitting hours after and miles away from the original downpour. In such floods the water can rise ten or twenty feet in a few minutes and drown everything in its way.

I discussed this problem with Taieb, who did his best by studying the stratigraphy of the hillside, to see if there was any evidence of sudden flooding. He determined that all the bones came from a single stratum that consisted almost entirely of fine clay. The thinness of the layer suggested a single event like a flood. The clay did not; rushing water would have carried a mixture of larger particles. What could have happened, said Taieb, was that everything – bones, mud, sand – went down in a rush to a lake edge or some other flatter, more open area where the bones could have settled slowly and quietly, covered by finer material. Taieb stressed that while the foregoing may be logical, it is unprovable. Geology is wonderfully informative, thanks to sedimentation, volcanism, faulting, and so on, but it has not yet reached a stage of refinement that can interpret local events that took place overnight three million years ago. The strongest evidence for drowning, he continued, would have to be the absence of animal fossils. Animals would not have frequented a ravine inhabited by hominids, and thus would not have been caught with them in a catastrophe. By the time they began wandering about again in the vicinity, the hominids would have been safely buried in their layer of clay. That animals did eventually return is evidenced by the presence of their bones – background noise – in strata above the 333 clay layer.

That was as far as Taieb could go. The unique concentration of a hominid band remains a mystery. Its exact size will have to wait final analysis of all the bones recovered so far, and the full excavation of the hillside, for I am confident of finding within it other bones in a far less disturbed state. I hope for another Lucy, perhaps a more nearly complete one. I hope for a better skull than Lucy provided. Most of all, I simply hope for more bones to build up my population sample. For there lies the true value of site 333. It provides a representative mix of sexes and ages that is in its way even more valuable than the priceless Lucy. Lucy is extraordinary, there is no question about that. She is like something seen during a lightning flash, her details lit by an unearthly clarity for just an instant. During that instant we take in a lot about her. We see her almost whole. But with nothing to compare her with, how do we figure out what she is? Is she representative of her kind? If so, her kind was extremely small. It had an odd lower jaw that came together in a V-shape at the front. Alemayehu's jaws were not V-shaped. Did they represent a different species? Were they like the new teeth and jaws from the 333 site, now dubbed the First Family? Between the anthropologists and answers to those questions lay the bones themselves, a vast jumble of them which had to be cleaned, sorted, described, and finally interpreted – a job that would take several years.

In December 1975 the third Hadar field season wound down. Taieb and his associates returned to France to spend the winter in mammal sorting, pollen analysis and geological studies. Aronson and Walter returned to Cleveland loaded down with volcanic samples. I took the First Family to Addis Ababa and braced myself for the task of getting an exit visa for them.

Having run that gauntlet once before, I felt more confident this time. Taieb and I held the inevitable press conference to establish the importance of the fossils, then went the rounds of the ministries to secure the necessary papers. All went smoothly. I sailed through Customs and flew with Tom Gray to Nairobi to show the fossils to my friends Mary and Richard Leakey and to a number of other scientists who had been working as part of the Koobi Fora field team, several of whom I already knew. I also met for the first time some people who had been working for Mary Leakey at Laetoli, a site in Tanzania a few miles south of Olduvai.

When I spread out the haul of new bones, they were an instant sensation. Nothing that combined their extreme antiquity, their remarkable quality and their profusion had ever been encountered before. There was an animated handing-round of fossils and a babble of discussion about them. Mary and Richard Leakey studied them with absorption and voiced the opinion that the First Family tended to confirm their conclusions reached during their visit to Hadar the year before: these were *Homo* fossils, albeit of a very primitive nature. Lucy, whom they had had a chance to examine by this time, was, they felt, something different. I was inclined to agree with them.

In the general rush to handle the fossils there was one person who hung back: an owl-eyed young man with thick glasses, lank blond hair and a white lab coat, who stood off to one side. I attributed that to shyness. I found out that the young man was an American who had worked at Koobi Fora for a couple of seasons as a paleontologist for Richard Leakey and who was now working for Richard's mother. His name was Tim White.

Not long ago I reminded Tim of that day. 'You were lurking in a corner as if you were too timid to come out. Do you remember? I went over and introduced myself and we got talking.'

'I wasn't timid, for God's sake. I was just being sensibly cautious. Here you were, the smooth young hotshot, shooting off your mouth about all your great fossils. I'd never met you before. I didn't know if you could tell a hippo rib from a rhino tail. I was just waiting for you to fall on your face, say something really dumb.'

That brought a cackle from Owen Lovejoy, who was listening to the conversation. 'Tim's a show-me man,' I explained. 'The original prickly, stubborn, I-won't-believe-it-until-you-can-prove-it-with-fossils type. He'll argue with anybody about anything.'

'Don's a nail-polish salesman, a real operator. I felt I had to watch him. Also, he was buying all that guff about *Homo* at three million that the Leakeys were dishing out.' He turned to me. 'You were, you know.'

'I wasn't buying it,' I said. 'I was *thinking* it. In the context of what we knew at the time, it made sense to think it.'

'It made sense if you swallowed a lot of junk about dates, and didn't bother to look at primitive dental features. It made sense if you were careless and naive.'

'Me careless? Naive?'

'Yeah, both.'

For all White's suspicions about me as a fly-boy at that first meeting in Nairobi, and for all my misinterpretation as shyness of what was actually an uncompromisingly cold approach to science, coupled with an idealism about it that made it hard for Tim truly to admire more than one or two people in the field, we got on well from the beginning. Tim discussed some of the fossils Mary Leakey was beginning to find at Laetoli. Finally he stepped forward to examine the Hadar specimens. After looking at them he said something that I will never forget: 'I think your fossils from Hadar and Mary's fossils from Laetoli may be the same.' He showed me a couple of them. They seemed nearly identical.

That was a stunner. The two places were a thousand miles apart. Provisional dating of the Laetoli specimens gave them an age of 3.7 million years. That was three-quarters of a million years older than the 3.0 million assigned to the Hadar basalt by Aronson. Granted, he had cautioned that the basalt might be older. But Basil Cooke's pig evidence that might confirm an older date was three years in the future.

Were the Hadar and Laetoli fossils one species? How on earth was that to be dealt with?

Tim and I agreed to keep in touch.

11 The Fourth Hadar Field Season: Cleaning Up

Knowledge rests not upon truth alone, but upon error also.
C. C. JUNG

You know . . . everybody is ignorant, only on different subjects.
WILL ROGERS

When I was a boy of fourteen, my father was so ignorant I could hardly stand to have the old man around. But when I got to be twenty-one, I was astonished at how much the old man had learned in seven years.
MARK TWAIN

After returning to the United States in early 1976, I found myself frantically busy. The new haul of site-333 fossils, the First Family, stood in front of me like an Everest of organisation and description that I would have to scale. I began scouting around among my graduate students for bright young people to whom bits of this work might be delegated. Ultimately I found two: Bill Kimbel, a husky young giant with a mop of curly black hair and a Fu Manchu mustache; and Bruce Latimer, a tawny blond who so closely resembled the ideal of All-American young manhood that he has been nicknamed 'The Surfer'. Both have proved to be exceptionally conscientious workers, and are now fixtures in my laboratory at the Cleveland Museum. Kimbel is deep in a study of a juvenile skull from the 333 site, the first of its kind complete enough to justify com-

The Hadar fossil collection, together at last, is spread out in the new Cleveland laboratory to give an idea of its size and diversity. The knee joint, some jaws and other miscellaneous pieces are in the foreground. Next is Lucy. Back of her is the entire 333-site First Family, arranged according to skeletal part. In the background is Tim White, standing next to some chimpanzee skull from the Museum's reference collection. The Laetoli hominids are at left, foreground.

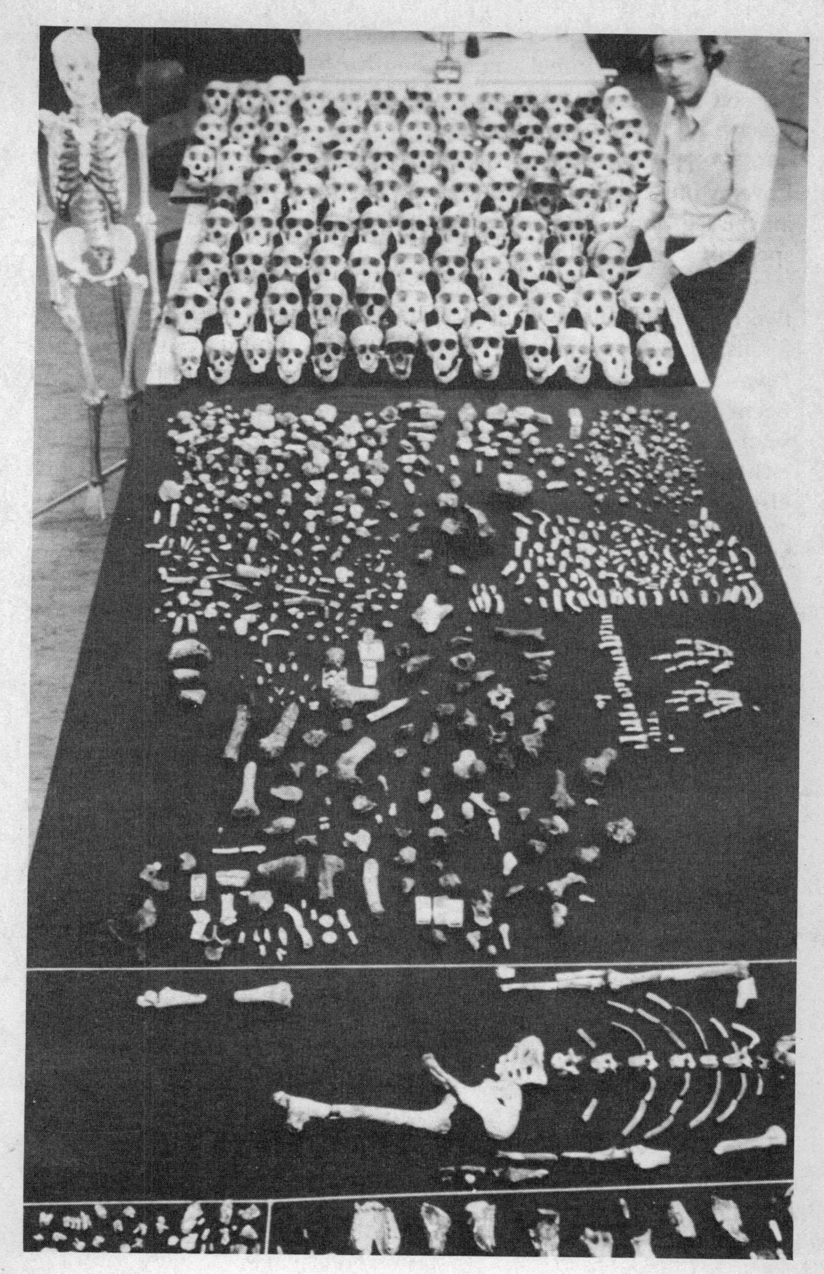

parison with that other famous juvenile, Raymond Dart's Taung Baby. (In 1979 Kimbel went to South Africa to make such a comparison.) Latimer has been entrusted with the responsibility of describing and analyzing the Hadar foot bones. Mike Bush, who had spotted the first fossil at the 333 site, got the hands. To owen Lovejoy of Kent State, the locomotion expert, went responsibility for all the leg bones, the pelvises and the vertebrae. I stuck to my own speciality: teeth and jaws.

These plans had to be worked out against the distractions of two tearing problems. The first had to do with space. With all the fossils that were pouring in, and with the graduate-student force I was assembling to work on them, the small quarters that we had been assigned at the Cleveland Museum began to burst at the seams. I finally issued a despairing ultimatum to the Museum: either I would have to be given more space or I would have to look elsewhere.

The Museum agreed to open up a large storage area, but insisted that I raise some of the money needed to turn it into a proper laboratory. I threw myself into a fund-raising drive, and within a few months $200,000 had been raised, enough to outfit an excellent laboratory. It was officially opened in December 1976, scarcely in time to cope with problem number two.

In my negotiations for removal of the 333-site fossils from Ethiopia, the authorities had given me permission to keep them for only a year. Somehow or other, during that short time, I would have to get them cleaned – a very picky and laborious task – and then have casts made of them. The latter step depended on the setting up of a casting room in the new laboratory. I had already found an expert caster, Bill McIntosh. Under the lash of that tight schedule McIntosh was installed and hard at work well before the rest of the laboratory conversion had been completed.

During that year I also wrote several scientific papers and a popular article for *National Geographic*. The first paper was the one containing a diagram of the preliminary stratigraphic column made by Taieb and showing the basalt in the wrong place. That error was not serious. The paper contained a footnote to the effect that its geological conclusions were preliminary and subject to correction.

With the torrent of new fossils came a flood of information describing the geological findings and dating of the recently completed field season. This required the publication of another paper, correcting the first and greatly enlarging it. This second paper was the one that relocated the basalt layer in accordance with Aronson's findings; it was also the one in which it was agreed to date the basalt at 3.0 million years, but with the cautionary remark that the figure 'should be regarded as a minimum age'. The paper was a formidable one and bore a formidable array of names as cosigners: James Aronson (potassium–argon dater), T. J. Schmitt (the American geologist who had assembled the four hundred samples), Bob Walter (fission-track dating), Maurice Taieb (geologist), J.-J. Tiercelin (a French geologist associated with Taieb), C. W. Naeser (an American fission-track expert), A. E. M. Nairn (a British paleomagnetic expert), and myself. Taieb and I now had the active support of outstanding specialists, and we made the most of it.

Another of my 1976 papers – one that I was under considerable pressure to publish – was an assessment of the fossils found. The anthropological world had heard about the first knee joint, about the jaws, about Lucy. How did all that material fit together? People wanted to know.

Indeed, they wanted to know before I was really ready to talk. I had found so many fossils so fast, and was still so deeply embroiled in the task of sorting them out and describing them, that I had not yet had time for the slower exercise of fitting them into an overall scheme. I was also under strong pressure to show the fossils to other scientists. I did not want to keep them locked up; I felt I *should* make them available to others for inspection. And yet I was aware of the damage that might be done if faulty descriptions or erroneous statements about them, made by others, worked their way into the scientific literature. I got around that problem by agreeing to make the fossils available to qualified people, provided that nothing was written down about them until I had had time to publish myself. This, of course, put additional pressure on me to do so. I published a joint paper with Taieb and Coppens in French, and another with Taieb in the British publication *Nature*, in whose august pages all the past salvos over the Taung Baby and *Homo habilis* had been fired. In both papers I supported the conclusion that the Leakeys and I had first reached at Hadar in 1974, and again in Nairobi in 1976: the large individuals represented by Alemayehu's jaws were *Homo*, as were the newly found representatives of the First Family. Lucy probably was not.

Whereas the first, slightly flawed geology paper was soon

corrected and gave me no trouble, the *Nature* paper made problems for me. To correct it, I first had to overhaul my own thinking. From there I was led on to larger conclusions about hominid evolution which, if accepted, would affect the way all paleoanthropologists would have to look thereafter at the pattern of human ancestry.

None of these possibilities was apparent to me when I signed the paper and sent it in for publication.

Also in 1976 I met Timothy White again, this time at a conference in Nice. Tim had with him some casts of Mary Leakey's Laetoli fossils that he had been working on. He and I sat down to compare them with the Hadar specimens. He had already expressed in Nairobi his belief that the two collections might represent the same kind of hominid. When he repeated it now, I checked him instantly: 'Hominid? Singular?'

'Singular. One kind.'

'But we have two kinds from Hadar. Little Lucy and the big ones.'

'Maybe you don't. We'll have to study that.'

On that tantalising note we parted: Tim to Michigan to finish his doctoral dissertation, I to the multitude of chores that were piling up on me – in particular, the problem of getting the 333 fossils cast and described. I decided that I would not be able to go back to Hadar that fall; I had too much to do at home. I put Tom Gray in charge of the American team, then called my graduate student Bill Kimbel into my office and said, 'Your work has been fine. You've done your internship, made your commitment to anthropology. How would you like to go to Hadar as Tom Gray's assistant?'

Kimbel was speechless. By this time Hadar had become the most coveted spot on earth for Plio-Pleistocene fieldwork. There was not a graduate student in the country who would not have envied Kimbel this chance.

I notified Taieb of the changes, and learned to my dismay that the Ethiopian political situation had changed even further. Taieb told me that no matter how busy I was in the United States, I would have to stop what I was doing and join him in an effort to launch the next field season; otherwise there would not be one.

'We may not get one anyway,' said Taieb. 'We certainly won't unless we both are there to argue for it.'

We met in Addis in September 1976. I realized immediately that Taieb had been right. Ministerial affairs were at a near-standstill because of paralysis at the civil service level. Stories about disappearances and deaths were common.

We checked in at the French and American embassies, both of which advised strongly against any kind of expedition. According to the American Embassy, 'major developments' were expected any day. I could not get anybody to explain what that meant, but I was told, 'If you are caught out in the field you will be stranded.'

'Can you stop us from going?' I asked.

'No. But we want to put it to you as strongly as we can that we don't want you to go. You will be just another nuisance to us, somebody else to worry about and keep an eye on – and eventually rescue.'

Even Wilding, the imperturbable Britisher, advised against going; the expedition's proverbial buffer, the Afar tribesmen, were themselves in turmoil and might no longer be dependable.

Maurice and I decided to check that out ourselves. We located the headquarters of an Afar chieftain at a cotton plantation about fifty miles north of Hadar. We got word to him that we would meet him at the plantation, and flew out there in a rented light plane. We were met by armed men at the plantation airstrip and driven to a house with guards at the door. This had all the flavor of a military headquarters. Inside, sitting at a table, was a small granite-faced man named Habib, with more armed guards behind him. He was the son of the previous sultan of the area who had been exiled some years before to the Middle East by another sultan, Ali Mira.

Ali Mira had been the power in the region through all the previous field seasons. He had collected his own taxes, administered his own justice, had no contact whatsoever with Haile Selassie other than an apparent unspoken understanding that he could do what he wanted as long as he did it in the Afar and nowhere else. With Selassie gone, the political storm at Addis had spread ripples as far as the Afar. Ali Mira had been chased out by local enemies loyal to their former sultan.

Now that sultan's son was getting his innings. He was an impassive little man who sat very still behind his dusty desk. But he had a stern hawk's eye and radiated great resolve. He asked us what we wanted.

We replied that all we asked for was some assurance that local disturbances would not disrupt the expedition. We were afraid that if any serious fighting broke out among the Afar themselves, we might be caught in the middle.

We were told not to worry. There were a few roving bands of Ali Mira supporters still at large, but they were being picked up rapidly. We could consider the area secure. Because of our past good relationship with the tribesmen and our total noninvolvement in politics, we were welcome.

The new chief designated a subchief to watch out for us. 'His name is Muhammed Goffra. He will live with you. His men will guard you. I can assure you of his loyalty to me – which means that you will be safe.'

The sense of being in an armed camp came back to me as I listened to this speech. And yet I felt reassured. There was something about raw power, directly applied, that was appropriate to this dry frontier setting. It was simple and trust-inspiring. It had nothing of the cobwebby complexity of the capital. On the plane going back, I said to Maurice that I felt better.

'So do I.'

'What do you say we stick to our plan?'

'I am for that.'

The 1976 expedition went off as scheduled, with Gray in charge of the American effort. I returned to the United States to try to clean up some of the backlog of work there. In December, at the very end of the field season, I went back to Hadar with the fossils from the previous year, which had been earmarked for return to the Ethiopians under the terms of our agreement with them.

I wanted to ask for a year's extension on their return, but was reluctant to risk it, given the political situation. I had to maintain our credibility with the Dergue at all costs; the ongoing stakes were too high. I had learned that the surface collecting at the 333 site was about completed, that it had yielded a great deal of material, and that I would therefore be faced with the problem of getting the new haul out of the country in a few weeks. Furthermore, a small incision had been made into the 333 hillside, revealing the presence of more fossils within it; I felt confident that a large-scale digging effort into the hill the next year would produce dramatic finds – possibly another mother lode of undisturbed individuals. In my mind's eye I could see more and better Lucys. Finally, a magnificent jaw and teeth had been found just across the river from camp in deposits that had not been surveyed at all. What else lay over there I could only guess at.

Returning fossils that had not been cast or thoroughly described, in exchange for new ones and the chance of finding still newer ones, was a revolving-door operation. What was the sense of borrowing an interesting fossil, keeping it in my pocket for a little while, then taking it out and returning it in order to make room for another?

At that supremely frustrating moment, I decided that the best I could do was take my caster, Bill McIntosh, with me, set up a temporary casting operation in Ethiopia, and hope that McIntosh could finish with at least some of the uncast fossils before they had to go back to the National Museum.

Gray, meanwhile, had been having his own problems. That year the big difficulty was transport. The expedition had a new van and had planned to use it for a weekly delivery of meat, fruit and vegetables from Addis. Last year's driver, frightened by the turmoil in the country, would not sign on. Gray scouted around for a new man and took him on a test drive in Addis traffic. Within three blocks he had smashed up the van by running it into a truck. While it was being repaired, Gray went back to the old driver and in a burst of eloquence persuaded him to change his mind. He made two trips and quit again. His mother, he said, was sure he would be ambushed and murdered when he slowed down at a river crossing. Gray dug up a third man whose eye lit up when he saw the new van. When tested, he drove skillfully, but he was a racer at heart. On his own he went flat-out as if he were at Le Mans and quickly landed

upside down in a fifteen-foot ditch, totally wrecking the van and injuring his head.

Efforts to supply the expedition from Addis by vehicle were abandoned. A few plane runs were made, but they too stopped when the Ethiopians decided that the risk of being shot down had gotten unacceptably high. Kabete had to make do again with local goats.

Despite these inconveniences, I found Gray and Kimbel in high spirits when I arrived in camp. Kimbel, in particular, was bubbling. He told me about the jaw find across the river. It had been brought into camp by a small Afar boy who, by keeping his eyes open and by listening to his elders talk about which among all the bones strewn over the landscape were the ones that the addled foreigners prized most highly, had learned all on his own how to recognize hominid jaws.

The one the boy produced was the right half of a lower jaw with all its teeth in place – a truly marvelous specimen. He said that the other half was across the river; he knew where it was, and would take Kimbel to it if he could be hired as a kitchen helper. That afternoon Kimbel and Nicole Page waded across the river and followed the boy for about half an hour as he unerringly guided them to the other half of the jaw. Put together, the two pieces constitute the finest lower jaw in the entire Hadar collection.

Another significant achievement of the 1976–1977 field season was the first finding of stone tools at Hadar. This had been accomplished by one of Taieb's associates, a French archeologist named Hélène Roche, who had picked up several from the surface of the ground in a gully about three miles from camp. Unfortunately, she had had to return to France before being able to investigate this extremely important new development more thoroughly. That left the camp with no archeologist – with nobody trained to study the artifacts or the presumed habits of early hominids – so I asked Maurice if it would be all right with him if I invited one to join the team for a couple of weeks. I had in mind a young New Zealander named Jack Harris, an extremely able archeologist with some experience in stone tools, having worked on them at Koobi Fora under Glynn Isaac and Richard Leakey. I needed Maurice's consent to this because the French were strong in archeology and felt that they 'owned' it at Hadar. Maurice knew Harris and thought highly of him. He considered the possible repercussions that might result among his colleagues from the addition of a non-French archeologist to the team, and decided that it was justified.

Stone tools have played a mixed role in the elucidation of Early African prehistory. They have turned up from time to time in South Africa, and repeated efforts have been made to link them with australopithecine fossils there. If such linkups were successful, that would mean that the tools would probably be not less than a million years old and possibly as much as two million years old. However, those link-ups have never been established; no one has been able to prove that South African australopithecines were tool users. Indeed, evidence is accumulating to indicate that no australopithecine *anywhere* was a regular maker or user of stone tools. Where very old ones have been found, there is usually associated fossil evidence of *Homo*. The South African tools, for example, appear to have been the work of *Homo erectus*. In Olduvai Gorge, where the tools are nearly two million years old, *Homo habilis* is present to account for them.

Calculating the age of a stone tool is not easy. The stone does not lend itself to radiometric dating even though it may have been made of volcanic rock. The reason is that the hominid who made it may have picked up a piece that was fifty or a hundred million years older than he was. He may even have made the tool in one place and dropped it in another. Furthermore, tools that have been chipped and banged about may have suffered so much alteration to their surfaces that any dates derived from them will contain an unacceptably large degree of error. Therefore one must look at the tool itself and at its geological and fossil associations.

The tool itself often will say almost nothing. The trouble with it is that it is about the most durable object on earth. Once made, it defies destruction. It can be made, discarded, buried by deposits, and then surface a couple of million years later, and show little effects of its journey through time or geology. Therefore, if it is found on the surface of the ground and has no peculiarities of workmanship that might date it, what is there to say how old it is? Put another way: if men were making similar tools a thousand years ago and leaving them lying about, how would an archeologist distinguish between one of those and a tool like it made a million years earlier?

There are differences in workmanship that do say something about a tool's age. There is the extremely primitive Oldowan industry, identified and named by the Leakeys at Olduvai Gorge and now known to date back at least 1.8 million years – probably back to two million, on the evidence of similar artifacts found at Omo and Lake Turkana. There is also the more advanced Acheulean industry, generally believed to be associated with the appearance of *Homo erectus*. Tool type and human type both seem to have appeared rather suddenly at about 1.5 million, and to have evolved little thereafter. Both were stubbornly resistant to change for at least a million years. It would seem that the needs of an Early African hunter-gatherer were adequately served by the Acheulean toolkit and that there was little or no reason to improve it.

Harris touched on that point when talking to me about tools. He said that it really was impossible to date a surface-found tool at Lake Turkana because modern humans who needed rough blades to chop animals were making similar implements in profusion as recently as a thousand years ago, and that there were even a few people who were making them there today. Thanks to their indestructibility, Harris observed, there was a background surface 'noise' of stone artifacts scattered about Africa like the background noise of animal fossils, but less useful by far because the tools showed so little evolutionary change. The only way to date a tool was to dig, to find it undisturbed in the horizon where it presumably had been dropped by the individual who made it.

Digging meant encroaching on the preserve of the French archeologists. Maurice thought it would be better if Harris kept away from the spot where Hélène Roche had been working. Since neither of them knew exactly where on the slope of a particular gully this was, I instructed Harris to make sure to work only the other side of the gully.

Harris set out, and I settled down to a few days of fossil cleaning. I had some of the previous year's site-333 items with me, still encrusted, still undescribed. I was able to work on them in the field because I had brought with me a small pump which produced enough compressed air to operate the kind of instrument that is used to clean fossils in the laboratory: a pneumatic device that works like a miniature jackhammer to blast away the encrustations of rock that cover a fossil. The device must be used with great care on teeth, because their surfaces must be bared with no scratching or scarring. If it is not, it will blur the traces of wear that are revealed in the facets that are formed on all teeth by years of chewing. To avoid damage, the final work on teeth and jaws is always done with dental picks.

I had scarcely gotten settled in with the fossils when Harris was back; he had what he thought was a tool-bearing location, and wanted some help with it. I had had experience in field archeology on Indian digs in the States when I was a student. I didn't have to be told how to hold a surveying rod or use a transit or divide the area into squares. We went right to work and did it. We excavated an area twenty-five feet on a side, going down a couple of feet. We found

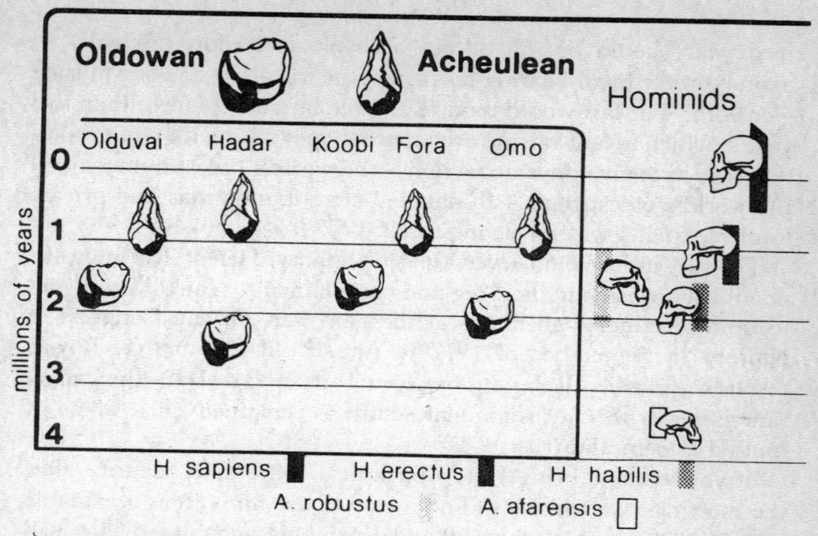

An arrangement of Oldowan and Acheulean tools according to site and to age reveals that the Oldowan are probably the work of *Homo habilis* and that the Acheulean are the work of *Homo erectus*. Oldowan tools at 2.5 million from Hadar suggest that *Homo habilis* may ultimately turn out to be that old also, since there is no good evidence from anywhere to indicate that australopithecines made tools of either type. Apparently stone tools are an invention of *Homo*.

about two dozen artifacts in there, along with part of an elephant molar and some other bone fragments. Looking around at the geology, I had a pretty good idea of where we were in the stratigraphic column because I thought I could see a layer of BKT ash, the material that Aronson and Walter had been using for dating in another part of the deposits. But to be sure, I had Walter take samples back to Cleveland. He has since said that as a first-shot guess the tools do come from between the BKT-2 and the BKT-3 tuffs. That would make them two and a half million years old – the oldest in the world.

The tools that we found were of basalt, deliberately and consistently made, actually of somewhat better workmanship than those found at Olduvai. They were a stunning surprise. And they tended to strengthen my published opinion that the large hominids at Hadar were *Homo*. Could australopithecines have made such things? I thought not. I began turning over in my mind what the work priorities for the next year should be: an all-out archeological effort on tools, an assault on the fossil-bearing horizon at the top of the 333-site hillside – or both? The last would require an even larger field force than this year's, which meant raising even more money. But with two properties of such tremendous attraction – the world's oldest humans and the world's oldest tools – to dangle before foundations and private funders, I felt I was on strong turf.

The turf in Paris, however, turned slippery. Hélène Roche heard about the invasion of 'her' dig and complained to Taieb. There had to be a meeting of all hands at the next Pan African Congress in Nairobi in September of 1977 to thrash out the matter. It was resolved amicably. Relations between Taieb and myself – the important issue in this jurisdictional scuffle – remained on a plane of mutual esteem and trust.

'If you want to bring Harris back next year,' said Maurice after the meeting, 'we will try to find a way to use him. He is a valuable man. I am sure that when he and Roche get acquainted they will work well together.'

'I hope so. If we are really going to date tools at 2.5 million, we'd better have some terrific archeology to back up that claim, because it's going to put everybody else right smack on their tails in astonishment.'

'That is surely so,' said Taieb.

The tool meeting actually took place some months after the end of the field season. Camp had already been closed and everyone was gone. I had already wrapped up the hominid fossils and departed for Addis, leaving the final pack-up to Gray and a couple of others. As usual, food had run short during the last few days. On the final morning Kabete had cooked an enormous omelet made from a single ostrich egg. It had served about eleven people and tasted delicious.

Gray had a big cleanup job that year. There was a large collection of animal fossils that had to be crated for storage, since they were too bulky and numerous to be shipped to the United States. Tents, bedding, cooking equipment, water filters, tables, camp chairs and portable lamps all were loaded into trucks and trailers. All the tin cans, cardboard boxes, lengths of string and wire, sheets of plastic and other throwaway things from the season's activities were claimed by the Afar, whose women moved in and began picking them over even before the caravan began groaning its way up out of the deposits to the tableland for the long jolting trek to the capital.

Meanwhile I had gone up to Addis with all the new site-333 fossils, plus the small selection from the previous year that I had managed to give a last-minute cleaning in camp.

The bulk of the 1975 333-site items had been left in the United States so that they too could continue to be worked on as long as possible. I had arranged that Bobbie Brown, a member of my team, would fly them to Addis at the last minute and report on how much remained to be done with them. I would make up my mind then whether or not to ask the Museum if we could hold them longer. The fact that I had taken the trouble to bring them all back would, I hoped, be held as a sign of my reliability and good intentions.

Bobbie arrived on schedule and handed over her fossils, pointing out that some had not yet been cast. I took them all to the authorities. Inasmuch as they had been signed out the previous year under the eye of Alemayehu, the Ethiopian representative to the expedition, he had to be present again to supervise their signing in and to verify that the manifest was complete. In all the confusion I was not sure that he would show up. But to my intense relief, he did. He checked the fossils over and declared the list to be complete. This all went so smoothly that I decided to risk asking if I could have certain ones back to finish work on them. To my surprise, the request was granted.

With the 1975 fossils so felicitously disposed of, I then turned to the processing of the 1976–1977 collection. This required the presence of another Ethiopian, Getachew, who had replaced Alemayehu in the field that year. Another miracle: Getachew showed up also. The next day he accompanied me to the Ministry of Culture to secure the papers authorising the signing-in and the signing-out of the new collection. This was handled by the Permanent Secretary to the Ministry, a dynamic young man who signed the papers with a smile and a flourish. That was at five-thirty in the afternoon. Half an hour later the Secretary drove home and was shot and killed on his own doorstep.

I knew nothing of this. I found out about it the next morning when Getachew and I returned to the Ministry to have the papers processed. The Ministry was in a state of shock. In this atmosphere of panicky paralysis, I moved from desk to desk murmuring my sympathies to people who obviously were very much upset, getting them to check photographs, look at bones, initial documents. I reminded balky clerks that the Secretary was to be buried that afternoon and that it would be a good idea to get this last bit of business out of the way in time for everyone to attend the funeral. By the end of the morning I had gone the rounds and secured every signature. The papers were given to Getachew to keep overnight. The next morning they would be presented at the Museum and the fossils withdrawn.

'What's wrong with those people?' I said to Getachew as we left the Ministry. 'They seemed badly frightened.'

'I would also be frightened,' he said.

'It seemed more than that. A couple of them looked as if they felt somebody might be gunning for *them.*'

'Yes.'

'What do you mean, "yes"? Why would anybody want to gun down some civil service people?'

'I don't know. There are rumors. Something is happening.'

'Are you going to go to the Secretary's funeral?'

'No,' said Getachew. 'I think it would be better not. For you too. Better not.'

I skipped the funeral and decided to call the cultural attaché to the French Embassy, a man with whom I had become friendly through Taieb. The attaché urged me to come to dinner. 'Pack your bag and bring it with you. Spend the night here. There is a curfew tonight and you may not be able to get back to your hotel.'

I accepted his offer. At eleven, the attaché received a phone call from the French Ambassador and was jolted to learn that there had been a coup in Ethiopia that morning and that the country was now being governed by a new set of military leaders. This had happened while I had been making my rounds at the Ministry. There had been no public announcement of it. Still, I wondered what connection, if any, the murder of the Secretary the day before had with the takeover. 'A coup?' I said to the attaché. 'A real coup?'

'That is what the Ambassador said.'

During the evening, the Ambassador said more. I learned that sometime during the morning there had been a meeting of the Dergue called by its presiding officer and de facto head of the country, a colonel named Tefarra Benti. There had been some disagreement at the top and some sudden executions. A new leader, Mengistu, had emerged, and he has ruled the country ever since.

I spent a sleepless night, regretting my decision to let Getachew hold the papers; I was sure I would never see him or Alemayehu again. In a country collapsing into anarchy, ruled by weapons, who would there be to listen to the plea of an obscure foreigner to take some bones away with him? I could see the last two weeks of paper chasing as thrown away, the last two years of fieldwork as wasted. With only part of the First Family to work with, what use would it be? I got up the next morning, said goodbye to the attaché, checked out of my hotel and dragged myself to the Museum, profoundly dispirited. To my amazement, Alemayehu and Getachew were both there, ready to begin the business of checking out the fossils, all of them now neatly logged in. First the uncast ones from 1975 were checked out to me, then the entire 1976 collection. Each piece was carefully wrapped and stowed in a large box. Dazed, I accepted it.

There was still the airport. I arrived there with my heart in my mouth, half-expecting it to be closed down and under military control. But everything seemed normal. A few weeks before, when coming through Customs, I had passed out copies of my *National Geographic* article, thinking that a story about Ethiopian fossils would interest the women who examined my luggage. They remembered me: the fossil man.

'More fossils in here?'

'Yes. Very interesting and important ones.'

They waved me through. I literally ran to the airplane, holding tight to the box.

12 Koobi Fora and Laetoli: Arguments Over Dates and Footprints

I have steadily endeavoured to keep my mind free so as to give up any hypothesis however much beloved (and I cannot resist forming one on every subject), as soon as facts are shown to be opposed to it. Indeed, I have had no choice but to act in this manner, for with the exception of the Coral Reefs, I cannot remember a single first-formed hypothesis which had not after a time to be given up or greatly modified.

CHARLES DARWIN

Arguments are to be avoided. They are always vulgar and often convincing.

OSCAR WILDE

In February 1977 I was back in Cleveland. With me, assembled in one place for the first time, was the entire Hadar collection: the knee joint, the jaws, Lucy, and all of the 333-site First Family members. That ingathering of bones was, in retrospect, a near-miracle; a mixture of good timing, good luck and persistence – laced with a dash of brashness. Together, they added up to a sizable collection. It was not nearly as large as the combined South African collections, but far surpassed them in quality. It was much more representative of the complete skeleton, it was far better preserved, and of course, it had the kind of precise dates that the South African material lacked.

Here were more than three hundred and fifty separate fossil pieces making up a group of males, females and juveniles, with enough variability between individuals to offer strong assurance that they could be evaluated as a population. From that evaluation it should be possible to proceed to a determination of what they were: *Australopithecus africanus*, *Homo habilis*, or what. Finally, whatever they did turn out to be, that species would be better known and more clearly perceived than ever before.

It would be up to me to make that analysis. Even though it would come at the virtual start of my career, I was well aware that this could be the most important contribution I might ever make to paleoanthropology. I was the proprietor of an unparalleled collection. It threw a spotlight on a short period of time – around three million B.C. – that was virtually unrepresented elsewhere on earth. The Omo fossils of comparable age were hard to evaluate because of their fragmentary nature and their wear. Richard Leakey's skull 1470, although he still stoutly claimed that it was 2.9 million years old, had begun to slip, in most scientific circles, down to an age of less than two million years. For all practical purposes, that left the Hadar hominids standing by themselves.

Alone in my office one night in the basement of the Cleveland Museum, I got out all the jaws and lined them up on the table. It was utterly quiet down there. The laboratory was like a bomb shelter: concrete-walled, belowground, windowless. In that silence I stared at the jaws, at the rows of pearly gray teeth, the rough brown jawbones. Sitting there, unlabeled, unidentified, they seemed to mock me. 'What are we?' they whispered. 'We are three million years old.' Against every chance of remorseless old geology and crazy new politics, I had found them. Now what was I going to do with them?

I thought again of Laetoli and the fossils that Tim White and I had looked at in Nairobi. The existence of that other population, small as it was, intrigued me tremendously. No matter that Laetoli was a thousand miles from Hadar; no matter that its deposits might be 800,000 years older (on the other hand, they might not); the fossils in the two places were so remarkably alike that logic compelled that their stories be intertwined. I decided that the time had come to get hold of Tim again for another and more detailed study of the two collections.

My two previous encounters with Tim had gone very well. I liked his directness, his willingness to introduce an idea just to get an argument started, his willingness to abandon that argument when it seemed wrong. Most of all I liked his uncompromising skepticism, his refusal to be satisfied with speculation, his insistence that something was worth believing only if the fossil evidence for it was clear, the dating unassaliable, and the laboratory analysis detailed and accurate.

What bothered me about Tim was his lack of diplomacy. The same uncompromising quality that made him a good scientist also made him a blunt and prickly man. Although Tim had originally admired Richard Leakey, and had done valuable work for him at Lake Turkana, they had gotten into arguments and their friendship had cooled.

Tim's disagreement with Leakey came about over the dating at Lake Turkana. Remember that when the Leakeys and John Harris visited Hadar, they had exhibited enormous interest in the mammal fossils there. Harris, in particular, had made a strong effort to find three-million-year-old *Equus* teeth in the Hadar collection to bolster a similar age claim for *Equus* teeth found at Lake Turkana. Although I had convinced him that there were no *Equus* teeth at Hadar – that all the horses there were of an older genus, *Hipparion* – Harris went away dissatisfied. He believed strongly in the dating of the KBS tuff at Lake Turkana. The biostratigraphic evidence that was coming in from other places to contradict it disturbed him profoundly.

Questions about the KBS tuff had arisen as early as 1972, when Vincent Maglio, an elephant specialist working at Lake Turkana, came to the conclusion that the elephant fossils he was studying there were not as old as the KBS tuff said they were. He called Leakey's attention to this without much success, and eventually drifted away from anthropology and into practicing medicine.

The next assault on the KBS tuff was a more serious one. It was launched by Basil Cooke, the pig expert. He had assembled a detailed sequence of several separate pig lineages over a period of a couple of million years. His basic work was done at Omo; therefore his dates were reliable. He then proceeded to make comparable studies from other places in the East African rift system – pigs from Olduvai, pigs from Hadar, every pig that could be dated. All told the same story. The evolution of pigs was consistent through time and over a wide expanse of geography – everywhere but at Lake Turkana, where there was an 800,000-year (more or less) discrepancy. Either the Turkana pigs were wrongly dated, or all the others were.

Central to Cooke's argument was a genus of pig named *Mesochoerus*. This animal was common throughout eastern Africa at two million. But at Lake Turkana the KBS dating placed it at nearly three million.

Cooke's paper was widely read and made a considerable stir. However, it did not appear to shake Richard Leakey's confidence in the potassium-argon date given to the KBS tuff. He went to a conference in London in 1975 and found it to be the major topic of conversation. By this time a good many other scientists were beginning to be drawn into the debate because the date of skull 1470 was tied to the date of the KBS tuff. If that was changed, there would have to be a major revision in how science looked at human evolution. *Homo* would probably not go back to three million, as Richard Leakey claimed. Instead, there might be an australopithecine back there that could qualify as an ancestor – again counter to what Richard believed and what his father before him had believed.

The London meeting was known as the Bishop Conference, because it had been organised by William Bishop, a much-admired British geologist who would die unexpectedly and tragically shortly afterward. It turned out to be a highly charged affair. Glynn Isaac, a South African-born archeologist who had done a great deal of work at Lake Turkana as a codirector with Richard Leakey, arrived at the conference wearing what he called his pigproof helmet; that would protect him against the pig men. Cooke showed up wearing a necktie with the letters MCP woven into it. Such ties were being widely sold in the United States that year, the letters meaning 'male chauvinist pig'.

Those frivolities were forgotten when the meeting got under way. Cooke made a biostratigraphic presentation that was overwhelming in its logic. He told the British volcanic daters Fitch and Miller that their estimate of the KBS tuff was wrong, and challenged them to dispute him on the basis of the fossil correlations he was getting. He cited *Mesochoerus* as a two-million-year-old pig everywhere but at Koobi Fora, where it was claimed to be nearly three million years old. When the meetings adjourned for lunch, they broke into vociferous groups that continued the arguments of the morning. I joined the Lake Turkana group. Richard Leakey said nothing, but Glynn Isaac spoke up vigorously in behalf of the Lake Turkana dating. He maintained that the lack of correlation between the Omo and Turkana fossils could well be explained by ecological differences in the two places. I remembered that argument well; it was the same one that John Harris had tried to advance earlier at Hadar.

Then Cooke, who had been silent up to then, pointed to his necktie and said, 'You may think you know what MCP stands for, but you don't. It really stands for "*Mesochoerus* correlates properly".'

A roar of laughter ended the argument, but it could not conceal the fact that Cooke's statement had been a devastating summary of the proceedings of the morning. Nearly everyone but the Lake Turkana team went away convinced that the KBS tuff and the skull-1470 dates would have to be corrected. John Harris, still a holdout

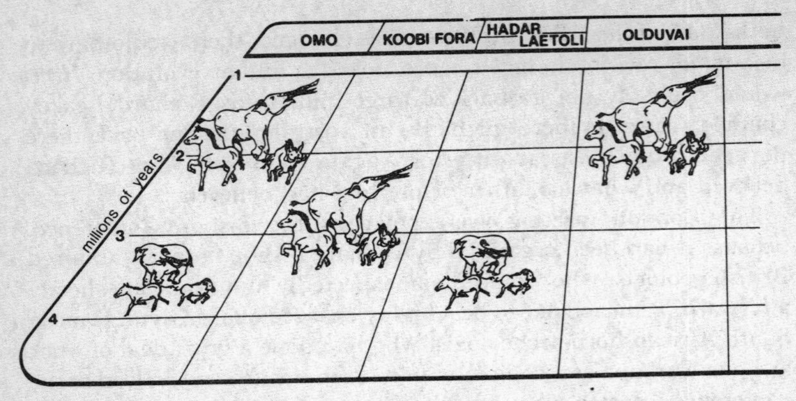

This shows why skull 1470 cannot be 2.9 million years old. Biostratigraphic evidence (comparison of mammal fossils at different sites) makes clear that the mammals living at three million – a primitive horse, an early elephant and a pig – were entirely different from the types that occurred at two million. This evidence is consistent at all sites except Koobi Fora, suggesting that the Koobi Fora dating is wrong, and that the animals found there should be redated from three million to two million. That would also require the redating of skull 1470.

and still believing the potassium-argon figures of Fitch and Miller, was unhappier even than he had been at Hadar when he had been hopefully combing over the *Hipparion* teeth. He resolved to make his own biostratigraphic analysis to prove the validity of the KBS date, and began looking around at the other members of the Turkana team for someone to help him. His eye lit on Tim White. Tim agreed to help. The two men made an exhaustive study of the Turkana pigs. When it was done, Harris was appalled. It indicated that Cooke was right and that the Lake Turkana dating was wrong.

'We can't just sit on this,' said Tim. 'We'll have to talk to Richard about it.'

They went to Leakey, but he did not want to hear about discrepancies in the dating. When they said they thought the matter was important enough for them to write a paper about it, he advised them not to. He felt that the flow of papers coming out of the work at Koobi Fora should be controlled by him. If everyone started writing papers, there would be no way of coordinating them or of avoiding contradictions in the Turkana research; disagreements about the dating would only cause confusion. Leakey, as it happened, had just sent a paper to *Nature* about a pelvis that had been designated as coming from below a three-million-year-old tuff. Tim asserted that the pelvis was a case in point; not only was that date suspect, but the pelvis did not come from below the tuff. Leakey hastily withdrew the paper.

By the summer of 1976 discussion of the Turkana geochronology had become intense. Leakey, realising that both the geology and the dating on which he had been depending to bolster his ideas about the antiquity of *Homo* were now thoroughly suspect, convened a meeting at Koobi Fora to discuss revisions. Glynn Isaac and Kay Behrensmeyer were there. Ian Findlater, Leakey's principal geologist' flew in from England. All listened intently to what Harris and White had to say about their pig findings. Tim then pointed to what he considered to be discrepancies in the stratigraphic column. Findlater defended the column, insisted that it was correct. Tim, not being a professional geologist, could say no more. After a great deal of discussion it was decided to adopt a new method of labeling and numbering everything at Lake Turkana. This was the so-called Collection Unit scheme. It provided a new keying system and different numbers, based on stratigraphy, for fossil sites. Now, if a fossil was described as coming from something called Area 108, Tuff T-2, Collection Unit A, it would be difficult to extract from that labelling the information that what was really being said was that – relative to the KBS tuff – the fossil was less than two million years old.

Tim was infuriated by this. He considered it a smoke screen designed to cover up mistakes in geology. The Lake Turkana people stood up for it on the eminently defensible ground that as collections and knowledge grow, revisions in a keying system become necessary; otherwise the system itself becomes tangled and hard to use.

Whatever the merits of the new scheme, Harris and White agreed to abide by it in a paper they finally decided they had to write. They showed it to Leakey, who objected strenuously to statements about the local hominids that Tim had inserted. He pointed out that this was a paper on pig evolution and should be confined to that. Tim countered that it had a logical and inescapable bearing on how hominid relationships would have to be interpreted.

'The trouble was,' said Tim, 'that Richard had already published papers about those hominids. They were "his" and other people weren't supposed to talk about them without his permission – at least, people working for him weren't. He couldn't control people who were not working for him, of course.'

Unable to get Leakey's approval of the paper as first written, Harris, more diplomatic than Tim and with a longer relationship with Richard Leakey, decided to tone it down by removing some of the hominid references. Whereupon Tim said he wished his name removed as coauthor. Harris ignored that request and sent the paper to *Nature*. It was rejected. Tim suspected pressure. He recalled how Robert Broom's spirited defense of Dart's Taung Baby had been suppressed forty years earlier by the *Nature* editors, and felt that he and Harris were the victims of a similar act of suppression. However, he has not pursued that matter, nor has anybody else. Harris and he got their paper back, rewrote it and submitted it to *Science*, the American equivalent to *Nature*. It was published in January 1977. Tim was not asked back to Lake Turkana.

Richard Leakey did not stand quietly by while all the pot shots were being taken at 1470 and the KBS tuff. He was disturbed enough by the rising tide of mammal-fossil evidence to ask Fitch and Miller to do another potassium-argon run. They did, and this time came up with a date of 2.4 million years. To the pig-fossil men this was a step in the right direction, but far too small a one; they felt that another dater should be tried. They were pleased when Thure Cerling, a University of California graduate student, turned up at the Berkeley campus with some samples of the KBS tuff that he had brought back with him from Lake Turkana. He gave the samples to Garniss Curtis, the acknowledged dean of potassium-argon dating and a pioneer in its application to Plio-Pleistocene fossils.

Curtis ran tests on Cerling's samples, one of which returned a date of 1.8 million, the other a date of 1.6 million. Those were almost exactly what the pig-fossil men would have predicted, and they were delighted with them. These results gave pause to Leakey. In a thorough review of the Lake Turkana hominids that he wrote for *Scientific American* in 1978, he noted the discrepancy between the Fitch–Miller findings and those of Curtis, but did not indicate which he favoured. For me, the Curtis dates clinched it. There was now no way that 1470 could be more than two million years old.

I asked Jim Aronson not long ago what he could say about the 700,000-year swing between Fitch–Miller and Curtis.

'I think it has to do with the purity of the samples,' he said. 'One of the first things Curtis and his group at Berkeley documented was that the KBS sample they had been given was contaminated by a few grains of much older material. Since we are always operating on the outer fringe of the capability of the potassium-argon technique, accidental intrusions like that can loom very large – only a few grains can throw you way off.'

The Berkeley group, he went on, had spotted some grains of very old feldspar under the microscope mixed in with the much younger feldspar crystals that represented the true KBS volcanic event. They had been able to handpick out the older contaminant before making their test runs. The results were reassuringly consistent; they were all bunched around 1.8 million years. Even more reassuring was that Curtis had tested two kinds of material in the sample: feldspar crystals and glasslike pumice, both of which had usable amounts of potassium. Both returned the same age estimates.

'It is rather ironical,' said Aronson, 'that the KBS tuff is now probably the most reliably dated in all of East Africa. The problem ahead is to match it to the fossils. The geology at Lake Turkana is extremely confused, and it is going to be tricky to make those link-ups. That will be a major priority for the Kenya team working at that rich site.'

Having left Lake Turkana, Tim White joined Mary Leakey to work as a paleontologist at Laetoli, which lies about thirty miles south of Olduvai. Of all the fossil sites in the Rift system, Laetoli has always been the odd one. While Omo, East Turkana and Afar have all changed climatically since the Plio-Pleistocene, Laetoli has not. Once they were lush lake regions swarming with game, laced with winding rivers and thick stands of tropical forest. Now they are near-deserts. Stable Laetoli was drier than they were then, and is greener now. Today there are several small lakes in its vicinity and a good deal of vegetation. The great wildebeest migration is apt to pass through each year on its way to the Serengeti Plain in a moving mass of flesh that must be seen to be believed. To the tourist, Laetoli is far more benign than any of the desert spots, but to fossil collectors it is much too thickety. All that vegetation gets in the way of surface collecting; they can see nothing.

For all that, Laetoli has attracted paleontologists for more than forty years because its deposits were believed to be very old, extending well back into the Pliocene. Recent potassium-argon tests have confirmed that belief. Two tuffs have been dated: one at 3.59 million years, another at 3.77 million. The hominid fossils that White was engaged to study fell neatly between them at about 3.7 million years.

For several decades Laetoli had just missed as a hominid fossil site. Louis Leakey had a try there in 1935, but came up empty-handed. He did not know that a tooth he had sent to the British Museum labeled as a baboon's was a hominid canine. Not only was it the first adult australopithecine tooth ever found, but it was the first of any kind since the discovery of the Taung Baby. Nevertheless, it lay unnoticed in the Museum collection until 1979, when it was spotted and properly identified by White.

Leakey, meanwhile, not realizing that he had had in his hand the oldest hominid fossil then known, packed up and moved to Olduvai. He was followed at Laetoli in 1938–1939 by a German named Kohl-Larsen, who recovered a bit of an upper jawbone with a couple of premolars in it, and a well-preserved alveolus – or socket – for a canine tooth.

The trouble with those early Laetoli finds was that they were far too old and far too primitive for anyone then to dream that they were not apes or monkeys; the imagination of the 1930s was simply not elastic enough to accommodate them, even though that same imagination was saying to itself, 'Look deeper into time for older ancestors.' This is an odd, schizophrenic view that still persists today.

By 1974, when Mary Leakey decided to have a go at Laetoli, her mind at least was ready to recognize and accept very old specimens of *Homo*. When one of the Leakey-trained Kenyan field experts, Kamoya Kimeu, took it upon himself to cut a road in to the deposits through the thicket and came out with a hominid, Mary Leakey moved in with a team of her own. In the next couple of years she or her workers found forty-two teeth, some of them associated with bits of jawbone. One in particular, LH-4 (Laetoli Hominid 4), was a fine specimen, a mandible with nine teeth in place.

But what sets Laetoli apart from every other site in the world is some footprints that have been found there, certainly one of the most extraordinary cases of preservation and discovery in all of paleoanthropology.

Laetoli has a nearby volcano, Sadiman, that is extinct today. Not quite four million years ago it was active. One day it spat out a cloud of carbonitite ash. This stuff has a consistency not unlike that of very fine beach sand, and it powdered down over the surrounding landscape in a layer that reached a thickness of about half an inch before the eruption stopped. This fall of superfine cinders must have been extremely unpleasant for the local animals and birds while it was coming down, but there is no evidence that it did more than make them uncomfortable, because they stayed in the area. That first puff of ash – probably not lasting more than a day – was followed by a rain. The ash became wet and, almost like a newly laid cement sidewalk, began taking clear impressions of everything that walked across it: elephants, giraffes, antelopes, hares, rhinos, pigs. There were also terrestrial birds like guinea fowl and ostriches, and even the small tracks of millipedes.

In the hot sun of Laetoli the wet ash layer quickly dried and hardened, preserving the footprints that crisscrossed it. Then, before it could rain again, Sadiman spoke a second time. Another cloud of ash drifted down, covering the first and sealing in the footprints. This happened a number of times over a period estimated to have been no longer than a month, producing a single volcanic tuff about eight inches thick. But because of the periodic puffing of Sadiman and the periodic hardening of the ash that fell, the tuff is actually composed of between a dozen and two dozen distinct thin layers. Some of these layers have been exposed recently by erosion, and are visible here and there at Laetoli in the form of a gray substrate wherever the mat of coarse turf above them has been carried off.

One afternoon in 1976, some of the more boisterous members of Mary Leakey's field team were amusing themselves by throwing hunks of dried elephant dung at each other. This may seem a peculiar pastime, but recreational resources are limited on paleontological digs, and there are times when young spirits need to blow off steam. One who felt this urge was Andrew Hill, a paleontologist from the National Museum of Kenya, who, while ducking flying dung and looking for ammunition to fire back, found himself standing in a dry stream bed on some exposed ash layers. One of these had some unusual dents in it. When Hill paused to examine them, he concluded they probably were animal footprints. That diagnosis was confirmed when a larger area was surveyed and other prints found. But no serious effort was made to follow up this extraordinary discovery until the following year, 1977, when a number of large elephant tracks were found by Mary Leakey's son Philip and a co-worker, Peter Jones, and alongside them some tracks that looked suspiciously like human footprints.

The world heard about the footprints later that year when Mary

Leakey came to the United States to report on them in a series of press conferences and interviews. To many it seemed almost inconceivable that anything so ephemeral as a footprint should have been preserved for so long. But Mary was positive about the hominid ones. She went on to describe the latter as having been made by a creature that was an imperfect walker; the prints indicated that it had shuffled. She also reported the probable presence at Laetoli of knuckle-walking apes and the existence of a water hole around which the animals and birds appeared to have clustered. She even saw some evidence of panic in the tracks, suggesting that the animals had been fleeing the eruption.

Those revelations by Mary Leakey electrified everybody who heard them. She resolved to devote much of the next season's effort at Laetoli to footprints, and asked the American footprint expert Louise Robbins to join her team. White went to the Laetoli site for the first time that year, and found three other young scientists there: Peter Jones, Paul Abell and Richard Hay. These men had some doubts about Mary's interpretation of the footprints. White questioned the presence of knuckle-walking apes; he had examined those prints and said that they had been made by large extinct baboons that walked flat-footed. Jones said there had been no panicky exodus from the area, because birds, which could have flown away easily and quickly, continued to walk about in the ash – it was crisscrossed with their tracks. Hay could find no evidence of a water hole.

These disagreements made for a good many nights of heated argument in camp, during which the supposedly human footprints had their ups and downs. No one could agree on them. Then Paul Abell, prospecting alone one day, found a broken impression – but a much clearer one – that he said he was quite confident was a hominid print. White and Jones made some Polaroid shots of it and came back with a strong impression that Abell was right. They recommended that excavation in the area be started immediately. But Louise Robbins, the footprint expert, examined it and declared that it was the print of a bovid (a hoofed animal). She told Mary Leakey that further investigation would be a waste of time. The men objected.

By then Mary Leakey had become thoroughly exasperated by all the arguing that had been going on. She announced that there would be no excavation. Jones, now convinced that it was a hominid, continued to plead with her for permission to make an excavation. A very small one, he said, was all he asked. Mary was adamant. Louise Robbins, the authority, had spoken; there was too much incomplete excavation at the site already. If there was going to be any digging, let it be done by somebody who had nothing better to do. She pointed to Ndibo, the maintenance man, the man in camp with the least archeological training.

Ndibo, however, proved equal to the task. He returned to camp the next day and reported not one, but two footprints. One was very large. He held his hands up, about a foot apart.

'Those Africans are always exaggerating,' said Mary. But she did go out to have a look, and there they were. White was permitted to start an excavation.

The direction of the prints indicated that their maker had been walking north under some sections of turf that had not yet been eroded. Because of the dense tangle of roots at the bottom of the turf, the task of exposing a clear ash surface without destroying it – not to mention the exact ash out of a dozen or more thin layers of it – turned out to be extraordinarily difficult. But Tim is an extraordinarily patient and determined man. He found another print, and then another. He proceeded to protect the prints by hardening them with a preservative, which he poured into them in very small amounts, letting the material dry and then strengthening it by adding more. Working with agonising slowness, he inched his way farther and farther into the turf and discovered that the trail consisted of the tracks of two hominids.

Now he had the riveted attention of the entire camp. Others joined the work and ultimately were able to reveal more than fifty prints covering a distance of seventy-seven feet. Louise Robbins, her interest in the footprints suddenly rekindled, issued another opinion: indeed there were two hominids; they were probably walking together; one (with slightly larger prints) was a male; the other, possibly pregnant she said, was a female; on the evidence of the prints, this type of hominid had been an erect walker for at least a million years.

These are entertaining speculations. There is no way of telling what sex the makers of the footprints were, if one was pregnant, or how long their ancestors had walked erect. The hard truth is that 3.7 million years ago erect hominids of indeterminate sex did walk through fresh-fallen ash at Laetoli and leave an imperishable record

(Opposite) This is the trail of footprints uncovered at Laetoli. They show a large and a small hominid walking in the same direction, but give no clue as to whether they walked together – or despite some assertions, as to their sex. They lie just below the present surface of the ground in some shallow layers of ash, and are extremely fragile. The picture above shows White at the moment of having successfully removed a mold of one of the prints without destroying it.

of their passing. After seventy-seven feet their trails disappeared under the overlying ash; the particular layer that marked it has been washed away. Tim's work on the footprints stopped at that point, which also marked the end of the season. But he felt strongly that the trail could be picked up again a little farther along and that it would yield more prints if proper excavation were carried out. Work in that direction was done in 1979 by Ron Clarke, and the trail picked up again.

Tim was not a party to this further work. His arguments with Louise Robbins over interpretation of the footprints have made him as unwelcome now at Laetoli as he is at Lake Turkana – a pity, because in each instance he was only trying to help the proprietors.

Tim's concern today is that as more prints are found, they be handled with the utmost care. They are supremely fragile, and the slightest mistake in excavating them can destroy them completely. Some have already been damaged. They are not like fossils, those rocklike models of durability. They are only spaces, mere shapes in a relatively soft and frangible matrix. If that matrix is nudged incorrectly, it will crumble – and the footprints will be gone.

But, by a wildly improbable linkage of random events, they are there. Sadiman had to blow out a particular kind of ash. Rain had to fall on it almost immediately. Hominids had to follow on the heels of the rain. The sun had to come out promptly and harden their footprints. Then another blast from Sadiman had to cover and preserve them before another obliterating shower came along.

All this had to happen over a period of only a few days. And the volcano had to synchronise its activity with that of the seasons. If its bursts had not come just when they did – at the beginning of the rains – the footprints would not have been preserved. A month or two earlier, during the dry season, the ash would not have had the consistency to take a sharp imprint. It would have been a hopelessly blurry one, a mere dent, like the one a passerby today makes in the dry sand on the upper margin of a beach. If it had come later, at the height of the rainy season, it is overwhelmingly likely that there would have been too much rain; the footprints would have been washed away before they could have been baked hard by the sun. Indeed, there had to be just what the beginning of a rainy season produces: sporadic showers interspersed with intervals of hot sun.

All things considered, the preservation and recovery of the Laetoli footprints are nothing short of a miracle. They confirm without a shadow of a doubt what Lucy confirmed at Hadar: that hominids were fully erect walkers at three million B.C. and earlier. At Hadar the evidence is in the fossils, in the shape of leg and foot bones. But at Laetoli, where the fossil remains – some extremely scrappy and enigmatic postcranial bits, jaw parts, and some teeth – are of very poor quality, there is no way without the footprints of deducing how those hominids got around.

'Make no mistake about it,' says Tim. 'They are like modern

footprints. If one were left in the sand of a California beach today, and a four-year-old were asked what it was, he would instantly say that somebody had walked there. He wouldn't be able to tell it from a hundred other prints on the beach, nor would you. The external morphology is the same. There is a well-shaped modern heel with a strong arch and a good ball of the foot in front of it. The big toe is straight in line. It doesn't stick out to the side like an ape toe, or like the big toe in so many drawings you see of australopithecines in books.

'I don't mean to say that there may not have been some slight differences in the foot bones; that's to be expected. But to all intents and purposes, those Laetoli hominids walked like you and me, and not in a shuffling run, as so many people have claimed for so long. Owen Lovejoy deduced all that from studying the Hadar bones. Now the footprints prove him right. I think they rank with the most wonderful and illuminating discoveries in decades. Although it didn't end too happily for me, I'm still grateful that I was lucky enough to have participated in the work on them.'

I worried about Tim and his arguments with the Leakeys, because the Leakeys were friends of mine. They had always gone out of their way to be nice to me when I went to Nairobi. Sometimes I stayed with them. We exchanged visits at our sites and discussed our hominids. Richard invited me to join the board of a research organisation he had formed: FROM, the Foundation for Research into the Origins of Man.

I was even friendlier with Mary. I kidded her a lot; we got along very well. I remember going to see her at Olduvai one time when she was expecting a visit from Louis' sister and her husband, the Archbishop of East Africa. Mary didn't know him well, and as a contributor to the collapse of Louis' first marriage, she wasn't particularly looking forward to that evening. She is not a sociable woman. She prefers to be by herself or with a few friends, and was glad to have me there to break the ice.

I had brought some live lobsters with me down from the coast. Her African cook had never seen a lobster and did not like the look of them, so I offered to cook them when the guests came. But they didn't come.

It got darker and darker as we sat having drinks, looking up at the huge shape of the Ngorongoro Crater above us in the sky. The road from the crater rim down to the Serengeti Plain runs past Olduvai and occasionally we saw the lights of a car creeping down. But it always went past and out to the plain. So we had another drink. Mary got to talking about hunting fossils, and said, 'You know, you're like a Leakey; you can find fossils. You know where to look for them and where to find them. That's a Leakey trait.'

I was flattered by this, and said so. She went on: 'You're like us in another way, too. You're lucky. Don't undervalue luck. Look at poor Clark Howell. If there was ever anyone who was unlucky, it's Clark. He's been looking for hominids for years and hasn't really found much of anything.'

I agreed that I was lucky, and we had a couple more drinks. When the Archbishop and his wife finally arrived, we were feeling no pain at all. I cooked the lobsters and more or less jollied everyone through the evening. The following morning, after they left, Mary thanked me and said she could not have gotten through that evening alone.

Those were good relationships. I prized them not only for personal reasons but for the scientific value they had for me. By that time Richard and his mother were the two best-known anthropologists in Africa – although neither of them, strictly speaking was one. Richard was self-taught, he had no degree; and Mary was an archeologist. Still, they both had priceless backgrounds in paleoanthropology, and each was in charge of a very important fossil site. It was becoming clear to me that for any sense-making analysis of hominid fossils anywhere in East Africa, the data from Laetoli, Hadar, Lake Turkana and Olduvai would have to be conformed. They linked together to make a nice time-chain: Laetoli at about 3.7 million, Hadar at about three million (maybe), Lake Turkana from two to one million. Delicacy and cooperation in puzzling out the hominid story would be more useful than arguments. With Tim White's relations with the Leakeys deteriorating, I had to wonder how they would regard my close association with him.

On the other hand, Tim was *the* man on the Laetoli fossils. He had written all the descriptions of them. He knew them and understood them better than anyone else. He is a supersharp paleontologist. I finally decided that I could not proceed without him, that the risk of getting mixed up on the fossils was greater than the risk of annoying the Leakeys by associating with Tim.

PART THREE

What Is Lucy?

13 The Analysis Begins

What happens when a new work of art is created is something that happens simultaneously to all the works of art which preceded it. The existing monuments form an ideal order among themselves, which is modified by the introduction of the new (the really new) work of art among them. The existing order is complete before the new work arrives; for order to persist after the supervention of novelty, the whole existing order must be . . . altered.

T. S. ELIOT

I feel confident that one day we will be able to follow man's fossil trail at East Rudolf back as far as four million years. There, perhaps, we will find evidence of a common ancestor for Australopithecus *– near man – and the genus* Homo, *true man.*

RICHARD LEAKEY

That's what Don found at Hadar – a common ancestor – but when Richard heard about it he wouldn't buy it.

TIMOTHY WHITE

By the summer of 1977 I was feeling less pressed than I had been for several years. I could not go back to Ethiopia, and was spared the responsibility of organising another field season. I had also managed to get most of the major Hadar fossils cleaned, and felt it was time to do some hard analytical work on them. Tim, I learned, also had some time on his hands. He had finished his Ph.D. dissertation. The paper that he and Harris had written on pigs was published. Knowing that he had casts of the Laetoli fossils that he had been describing for Mary Leakey, I asked him if he would bring them to Cleveland for a careful comparison with mine to see if our first impression of them had been right: did the two collections represent the same kind of hominid?

Up to that time I had not had a good look at the Laetoli specimens. My private suspicion was that when I did, I would find enough differences between them and mine to be able to distinguish between them. The most I hoped to do was write a paper with Tim that was purely descriptive, highlighting those differences but coming to no hard conclusions about possible relationships.

Then Tim began putting the Laetoli casts out on the table. Compared with the Hadar material they were very skimpy. There was one mandible in fairly good condition, plus thirteen other fragments, some of them single teeth. Skimpy or not, one fact came bursting from the surface of the table: the two sets of fossils were startingly alike. Wherever direct comparisons could be made, they were virtually identical. Over a period of days and with great care we checked the specimens against each other, tooth by tooth, cusp by cusp. It was an uncanny experience. Finally I said, 'My big ones are the same as Mary's big ones.'

'That's what I told you.'

'But not Lucy. She's different.'

'Yes, she's different because she's female. We may have sexual dimorphism here. Also allometry.'

'Now, wait a minute.'

'Think about allometry,' said Tim.

Allometry is a phenomenon well known to anatomists. It touches on the fact that among different-sized individuals in a population the relative size of certain bones or teeth may vary. This is particularly true among the primates – true in two ways. Males are not only larger than females; they are also proportioned differently. A male baboon, for example, is not merely a blown-up replica of a female baboon. It has an extremely large canine tooth, far larger in proportion to the rest of its teeth than the canine in any female. A paleontologist who knew baboons only through a collection of female jaws and teeth, and who suddenly came across a single male canine, could be excused if he assigned it to a different species. 'How could such an enormous tooth,' he would say to himself, 'be accommodated in any of the jaws I know? The jaws themselves would have been differently shaped.'

This discussion is relevant to Lucy because she has notably small front teeth. That is one reason her jaw has a distinctive V-shape: her four incisors are not big enough to require that the jaw be wide in front. In striking contrast are the jaws and teeth found by Alemayehu: not only are they absolutely bigger than Lucy's, but the front teeth are *bigger* yet. Keeping all the other congruences of cranium and jaw – that is, maintaining the same overall proportions, despite differences in absolute size – the result would have to be a widening and rounding of the front of a bigger jaw to accommodate its still bigger teeth.

'You claim allometry,' I remember saying as I held up Lucy and looked at her intently. 'But why should I accept it?'

'I don't claim it,' Tim said. 'I'm just suggesting it.'

'Well, it's a dumb question.'

'All right. But if this is the only significant difference between Lucy and the large ones, then it's worth thinking about.'

'Okay, I'll think about it.'

I did. I looked at Lucy again. I wasn't buying allometry. I was convinced that there were two species at Hadar.

My thinking was complicated by a great many things. Listing some of them may give an idea of how hard it is in paleoanthropology to keep your head clear. Not only do you have to keep up with what is going on everywhere, but you have to be aware of your own prejudices and of what effect your association with various other people has on your thinking. There is no such thing as a total lack of bias. I have it; everybody has it. The fossil hunter in the field has it. If he is interested in hippo teeth, that is what he is going to find, and that will bias his collection because he will walk right by other fossils without noticing them.

I think the most important of my biases when I started my analysis of the Hadar fossils was in favour of a multidisciplinary approach to site development and interpretation. In other words, try and make all disciplines work for you. Start them working at the beginning of your field effort and carry them right through. I learned that at Omo from Clark Howell. Overall, I think that is a healthy bias, and I am glad to have it.

Some other biases were not so healthy. In everybody who is looking for hominids there is a strong urge to learn more about where the human line started. If you are working back at around three million, as I was, that is very seductive, because you begin to get an idea that that is where *Homo* did start. You begin straining your eyes to find *Homo* traits in fossils of that age.

I was also biased – and this is a bit tricky – in the direction of thinking that anything gracile from the Plio-Pleistocene that did not look like a South African gracile was probably a *Homo*. Howell and I had made that exhaustive study of the South African fossils, and I had their characteristics burned in my brain. Their teeth had already evolved away from an ape condition and were clearly more like those of humans. But they also had traits that made them different from human teeth. An australopithecine has very big back teeth. Some of them are truly massive, twice the size of a modern human molar. Second, the enamel on those back teeth is extremely thick, much thicker than in human teeth. Third, the front teeth are remarkably small in comparison with the back teeth. When you pick up an australopithecine jaw, those are the things you notice.

When you pick up one of the Hadar jaws you don't see those things. You see, on the average, smaller molars, and you see larger front teeth. Those are human traits. That is what Richard and Mary Leakey responded to when they first saw the Hadar jaws. I don't blame them. Richard, in fact, had already made up his mind about how to classify the fossils he had been finding at Lake Turkana. He put them all in two groups. The ones with massive jaws and molars he was calling *Australopithecus*. The ones with small jaws and molars he was calling *Homo*. When he got to Hadar and found specimens there that were small-toothed like the Turkana specimens, it was natural for him to regard them as *Homo* too.

I won't pretend that anything I said to the Leakeys influenced their judgment of the Hadar fossils. If anything, it was the other way around. I was the new boy on the block. They were the ones who had been in the business for years. Mary, in fact, had been acquainted with Plio-Pleistocene fossils for forty years by that time. It was their judgment that strengthened mine to a point where I tended to brush aside some other peculiarities of the jaws that I would find myself paying more attention to later on. There, my bias was at work.

With Lucy I had no problem. She was so odd that there was no question about her not being human. She simply wasn't. She was too little. Her brain was too small. Her jaw was the wrong shape. With those seemingly 'primitive' traits staring me in the face, I interpreted other things in her dentition as primitive also, as pointing away from the human condition and back in the direction of apes. That the larger jaws had some of those same primitive features did not seem so significant to me. (Another bias.)

Tools were also on my mind. Don't forget that we had found them at Hadar at a tentative date of 2.5 million. That was a real eye-bulger, because there were no other tools in the world that were that old. What was I to make of that when I knew that people had been trying for years to associate stone tools with australopithecines and had never succeeded? What was I to make of it when I thought about Louis Leakey at Olduvai, searching around among the tools there, and finally coming up with *Homo habilis* as their maker – a human? That put both tools and *Homo* back close to two million years. If *only* human beings could make tools – and we had found tools at 2.5 million – then the logic was very strong to push *Homo* at least that far back.

Logical, maybe, but also biased. I was trying to jam the evidence of dates into a pattern that would support conclusions about fossils which, on closer inspection, the fossils themselves would not sustain. At any rate, that was my general line of thought when Tim and I began our first thorough review of the Laetoli and Hadar collections. I was strongly biased in considering that the larger jaws represented some form of very early *Homo* – the earliest *Homo* yet known. I was fairly sure that Lucy was something else. Tim was not so sure of the uniqueness of Lucy, but he too felt that there were strong *Homo* affinities in the rest of the collection. It was only after we had started looking at the fossils carefully that we began to have second thoughts. What were we to make of features that were not like those of either *Homo* or *Australopithecus* but seemed markedly more apelike?

Privately we were worrying about that, but neither of us said anything to the other until one day, when we had one of the site-333 First Family specimens on the mat, an adult partial cranium numbered AL 333-45. It was a little thing, with a strong apelike arrangement of muscle markings on the back of its skull. Tim took a careful look at it and said, 'That's a really weird fossil. Tell me, honestly, are you going to try to call that *Homo*?'

I said, 'Are you going to try to call it *Australopithecus*?'

That stumped him; if he took the South African fossils as models he couldn't.

I think it was at that moment that it dawned on us that Richard Leakey's dichotomy would not serve us. Up to then we had found his categories useful. Quite suddenly it was clear that 333-45 would fit neither of them.

'A third thing?' Tim said.

I nodded. 'Maybe.' I didn't want to say that there also might be a fourth: Lucy. Not right then. Discussion of Lucy with Tim more often than not led to arguments which could not seem to resolve themselves.

That was a very confusing moment. When big ideas begin rushing around in your head and you have not had time to pin any of them down, they seem almost too big to handle. You look for something smaller and simpler to get hold of. I remember wishing that there were some orderly way that we could proceed, a safe retreat into the step-by-step procedures that all scientists find so useful – and so reassuring; something we could work on as technicians to relieve ourselves of the hugeness and the cloudiness of the ideas that these fossils kept pushing into our minds.

Which of us suggested that we organise our thoughts by going back and reviewing the famous old Le Gros Clark paper, I don't now remember. But the instant it was suggested, we both jumped at it as a useful way of arranging our fossils, a way of seeing where they stood against a tried-and-true measuring stick.

What Le Gros Clark had done back in the 1950s was address himself to the ongoing confusion about the identity of australopithecines. Despite the growth of knowledge about them, there were many scientists who still clung to the belief that they did not qualify as hominids. Le Gros Clark decided to set that doubt at rest. He listed eleven clear and consistent differences in the teeth of apes and humans. He then laid australopithecines alongside for comparison, and found that in every respect the australopithecines were humanlike and not apelike. His paper describing that comparison was a landmark in paleoanthropology; it demolished forever any lingering doubt that the Taung Baby and all the other South African fossils were hominids and not some peculiar kind of erect-walking apes.

Why not, we thought, subject our fossils to the same kind of review to see where they fell? Privately I suspected the answer would be *Homo*, which would mean that I would not have to spend the rest of the year arguing with Tim about that. Privately he had begun to suspect the opposite and that he would no longer have to argue with me. Each of us had his bias. Tim was looking at overall primitive characteristics. I was looking at small manlike back teeth. One of those biases would have to give.

By the middle of the summer of 1977 we were ready to go. We had chimpanzee and gorilla specimens. We had a fine collection of South African australopithecine casts. We had the Laetoli and Hadar material organised. To analyse it properly we would have to consider three questions:

1) Did we have something new, or was it too much like something already known to deserve the label 'new'?
2) If it was new, how did it relate to other known material? In other words, where might it fit on a family tree?
3) What should it be named?

The Le Gros Clark analysis, we felt, should supply us with an answer to the first question, and also help with the second. For the third, we would be on our own.

There is a long, high counter running down the centre of the lab in Cleveland. Late every day Tim and I would settle down there after the other workers had gone home. We could hear doors slamming as they left, and could picture them walking out of the building through a cavernous concrete storeroom lined to the ceiling with wooden boxes, each box containing a human skeleton labeled adult or juvenile, male or female. These had been collected in years past from the city morgue and constituted one of the largest assemblages of modern *Homo sapiens* skeletons in the world. Beyond the boxes was another room full of cages in which sat a number of live hawks and owls that the naturalists in the museum were studying. When we left, later in the night, we too would pass those stacked skeletons and those cages of birds. The hawks would be asleep by that time, huddled motionless and seemingly headless on their perches. But the owls would be staring out at us with their round yellow eyes which can blaze with such a fierce light when the bird is free and healthy, but which always seemed to me to be dull and remote, almost lifeless, after months or years of captivity in that cellar. There are owls in Ethiopia, many of them with similar habits and similar appearances to those that live in North America. The caged birds in Cleveland would carry me back to nights at Hadar when, lying in my tent, I had heard owls calling down by the river.

Since large raptorial birds appear to evolve slowly, certainly no faster than hominids, it is nearly certain that the lakeshores, the rivers and forests that made up the Afar landscape three million years ago were inhabited by familar ancestors of the owls that are there now. If so, the bones Tim and I were setting out on their soft foam pads had once been animated by blood and nerve endings and by eyes and ears attentive to the calls and to the evening flight of those owls. Paleoanthropology is full of those sudden twitches into the past. Much as they look and feel like rocks, fossils throb with a life of their own. They remind us of emotions and feelings almost inconceivably distant now, of satisfactions, fears, anger and pain – all experienced by our own forebears, but filtered through brains so dimly like our own that we have no way of truly recapturing those long-lost perceptions. What did the world really taste like on the tip of an old hominid tongue? What a hopeless question.

After the last door had slammed I would shake myself and forget about the owls. Tim and I would pull up a couple of tall stools and get to work on our comparisons. We would check one feature against another, and another. I insisted on a remeasurement of all the tooth and jaw dimensions, having learned that the published figures did not always exactly match the fossils or the fossil casts. Once again I noted the accuracy of the work of J. T. Robinson, the South African authority on australopithecines.

For an understanding of the interpretive problem we had to deal with, the reader should have an opportunity to familiarise himself with those differences. Here are eight of them, taken verbatim from Le Gros Clark's paper, along with our comments on each:

APES (Pongidae)	HUMANS (Hominidae)
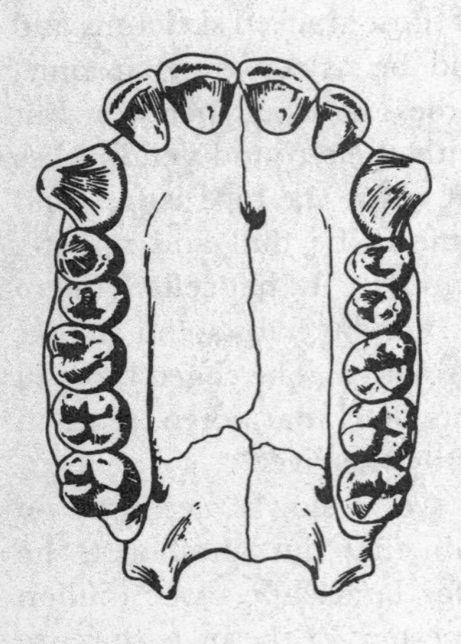	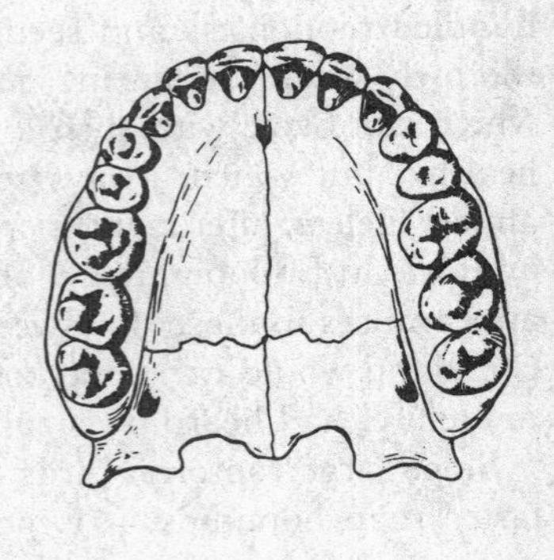
Chimpanzee upper jaw	Human upper jaw
The canine and postcanine teeth form approximately straight parallel rows.	The dental arcade consistently has an evenly curved parabolic contour.

Le Gros Clark's first difference between the two is obvious. The boxlike parallel tooth row is common to all apes. The human jaw is consistently rounded, and has been for a long time. *Homo erectus*, a million years ago, had a curved dental arcade essentially like modern man's.

Chimpanzee canine
Canines relatively large, conical and sharply pointed, with a well-marked internal cingulum commonly prolonged backward (in the lower canine) into a talonid.

Human canine
Canines relatively small, spatulate and bluntly pointed, with the internal cingulum reduced to a basal tubercle. No projecting talonid.

Again an obvious difference. The ape canine is more cone-shaped, and comes to a sharp point at the tip. The human one is not pointed. Instead of being conical, it is broader and more flattened: 'spatulate.' The cingulum is the extension to the rear at the base of the tooth. The talonid is the small bump at the back. Both are lacking in the human canine.

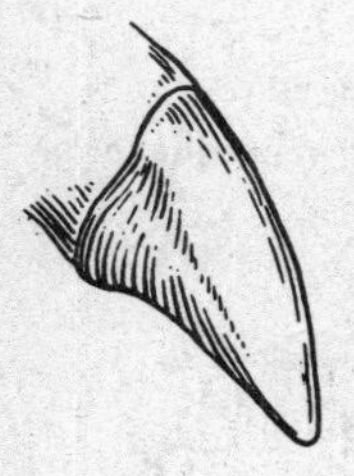

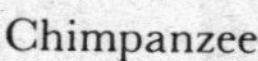

Chimpanzee

male female
Canines show a pronounced sexual dimorphism.

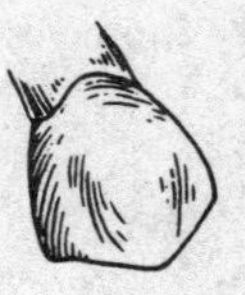

Human

male female
Canines show no pronounced sexual dimorphism.

Once again a clear distinction. Not only are male chimp canines absolutely larger, because males themselves are larger than females, but they are relatively larger as well. Male human canines are larger than female ones only because males themselves are larger. Scale a human male body down to female size, and the difference in their canines would be negligible.

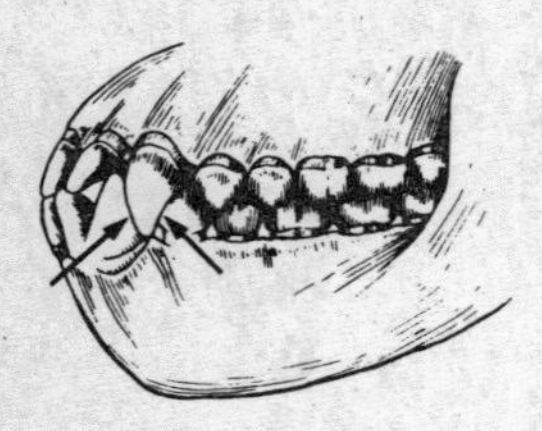

Chimpanzee mandible

At an early stage of attrition, the canines show facets on anterior and posterior aspects of crown and remain projecting well beyond occlusal levels of postcanine teeth.

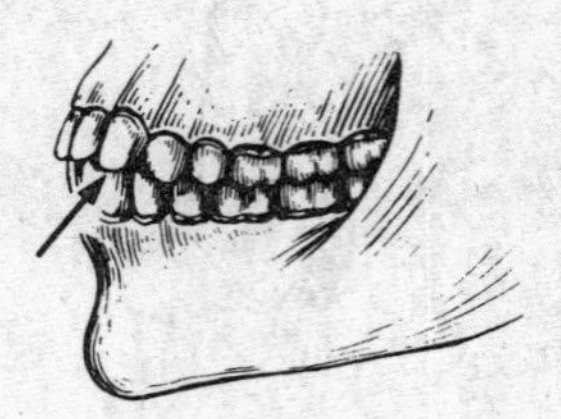

Human mandible

Canines wear down flat from the tip only and at an early stage of attrition do not project beyond occlusal levels of postcanine teeth.

With its oversized lower canine (and with a comparably large one projecting down from the upper jaw), there is a meshing of teeth when the ape's jaw is closed. The canines meet along their sides instead of at their ends. As a result, the canines develop wear patterns at the points shown by arrows. By contrast, human teeth do not mesh; they meet at their occlusal (grinding) surfaces. As a result, the human canine wears flat at the tip, so that it is the same height as the other teeth in the jaw.

Chimpanzee upper jaw

Upper incisors almost always separated from the canine by a well-marked diastema

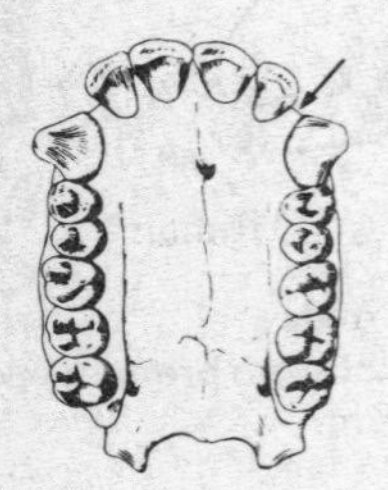

Human upper jaw

Uniform absence of true diastema

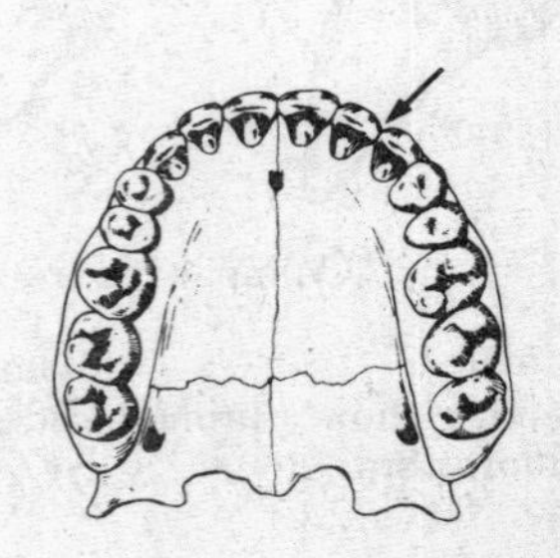

The diastema (at arrow, opposite, bottom) is simply a space in the tooth row. It is necessary in the chimp jaw to make room for the large, upthrusting lower canine when the jaw is closed. Humans, who lack projecting canines, do not need a diastema.

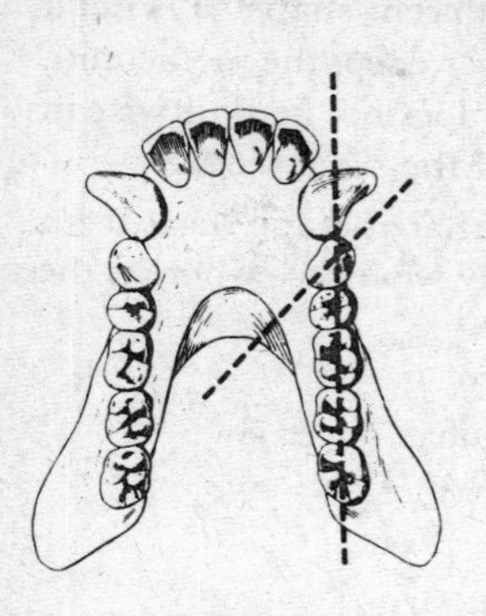

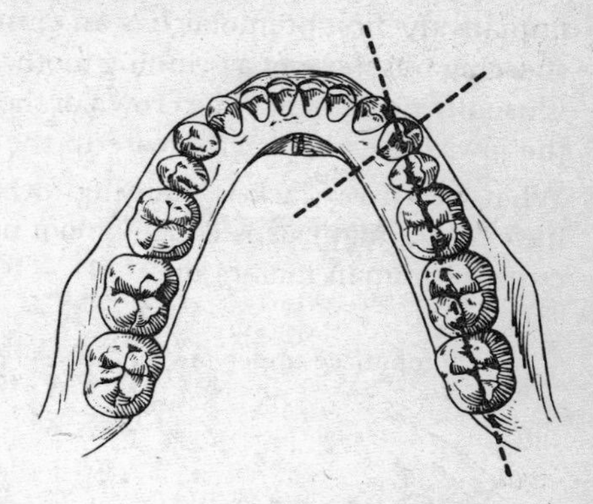

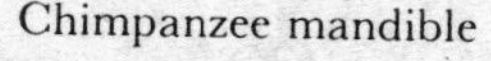

Chimpanzee mandible

Human mandible

The anterior (or forwardmost) premolar is the next one in the jaw after the canine. Sectorial means 'shearing' or 'slicing'. This tooth, in an ape, has only one large cusp (the protoconid) on its biting surface for the shearing action that the tooth provides. The

Protoconid (A)

Metaconid (B)

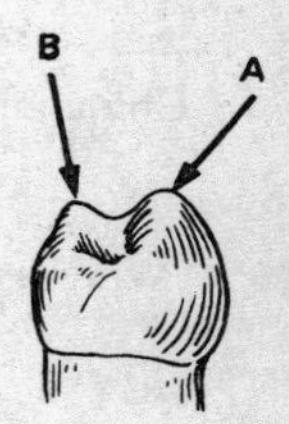

Premolar from top

Premolar from rear

Premolar from top

Premolar from rear

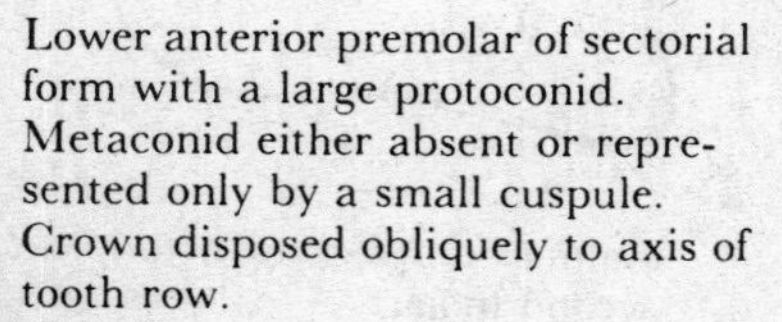

Lower anterior premolar of sectorial form with a large protoconid. Metaconid either absent or represented only by a small cuspule. Crown disposed obliquely to axis of tooth row.

Lower anterior premolar of bicuspid, nonsectorial type, with metaconid often approximating in size to the protoconid, the two cusps lying in a transverse plane. Well-defined anterior and posterior fovea.

metaconid is a smaller cusp on the inside of the tooth. It seldom appears in apes. When it does, it is an insignificant bump (a cuspule). The crown of the ape tooth, the part that extends above the gum, is set at an angle with the others in the tooth row. In humans the first premolar has an entirely different shape. It is not a shearing but more of a grinding tooth. Its inner cusp, the metaconid, is usually prominent. The crown of the tooth does not lie slantwise in the jaw. Instead it is crosswise to the line of the back teeth.

What Le Gros Clark is pointing to here is that the ape back teeth have rather high cusps which do not normally wear flat, whereas the cusps of human molars do..

Chimpanzee upper jaw

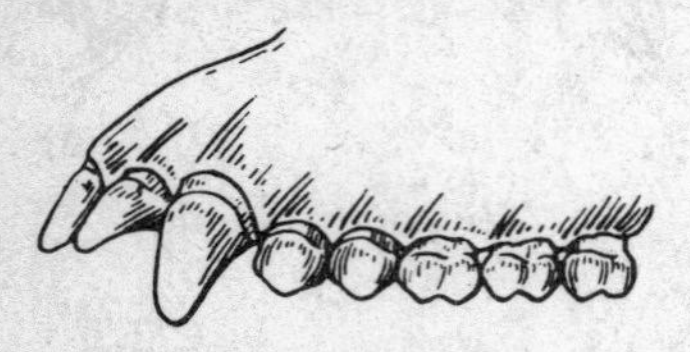

Molar teeth do not wear down to an even flat surface except in extreme cases of severe attrition.

Human upper jaw

Molar teeth commonly wear down to an even flat surface in the earliest stages of attrition.

That is his only weak comparison. It is not always consistent.

Le Gros Clark also noted three differences in the deciduous, or 'baby' teeth of chimps and humans, but they need not be listed here,

Chimpanzee mandible

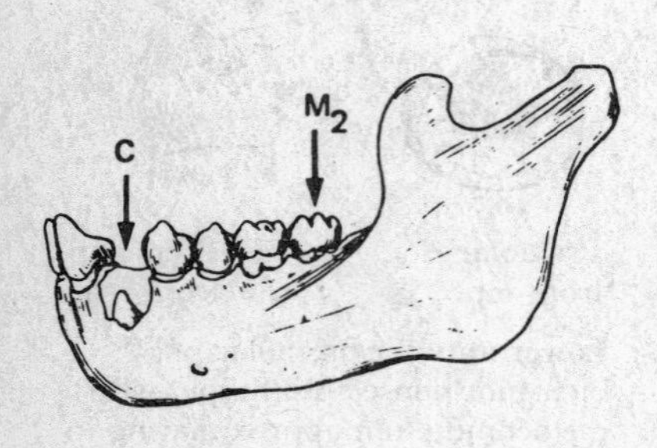

Canines erupt later, after the second molar.

Human mandible

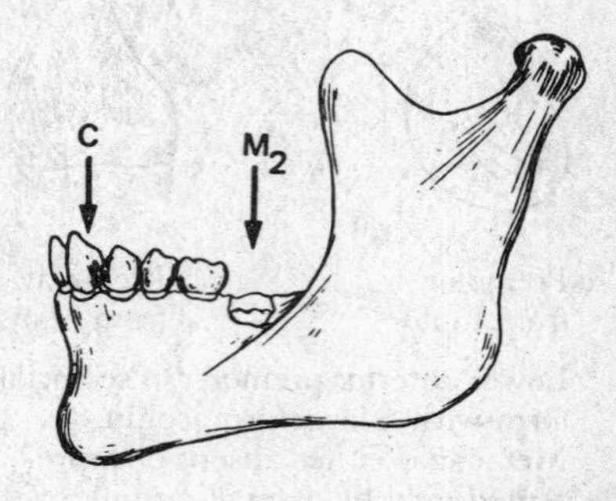

Canines erupt early, before the second molar.

since the examples given for adult teeth make their point clearly enough.

There are, of course, many other differences in the skull, pelvis, and arm and leg bones of chimpanzees and humans, but Le Gros Clark did not concern himself with them. His exercise was designed to position australopithecines vis-à-vis apes and humans, and since his supply of australopithecine fossils consisted mostly of teeth and jaws, he properly concentrated on them in his ape–human comparison. When he placed the australopithecine fossils between them, his case was made. In every instance, the australopithecine fossil was hominid and not pongid.

The clarity of Le Gros Clark's analysis and the economy of his presentation are all the more remarkable when one realises that it was made nearly thirty years ago, when the fossils were not as well known as they are now, and when there were considerably fewer of them. His conclusions are brutally clear today, *after* his analysis was made. But they do not begin to reflect the confusion and the arguments that went on before.

Having satisfied ourselves that Le Gros Clark's comparisons were still sound, Tim and I began putting our own fossils down on the table. I got quite a shock. That was not so much because our jaws

APE

LAETOLI-HADAR
Dental arcade and diastema

HOMINID
(*Australopithecus* and *Homo*)

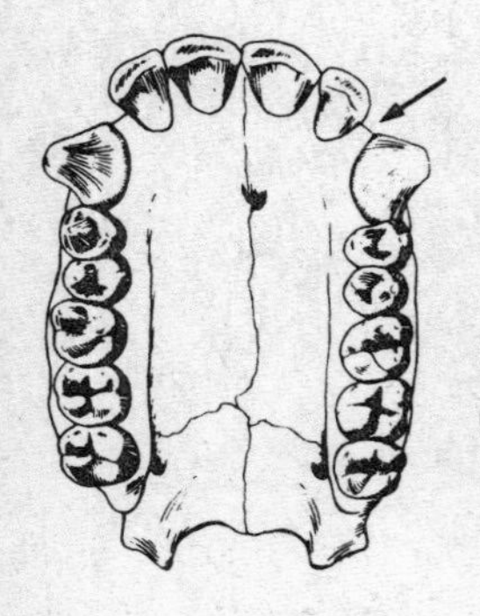

Chimpanzee upper jaw

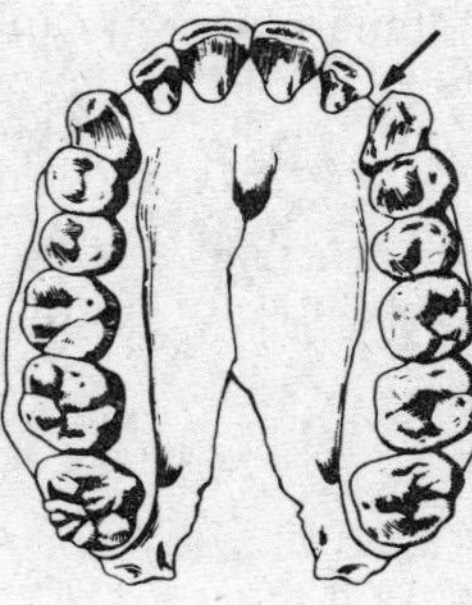

AL-200

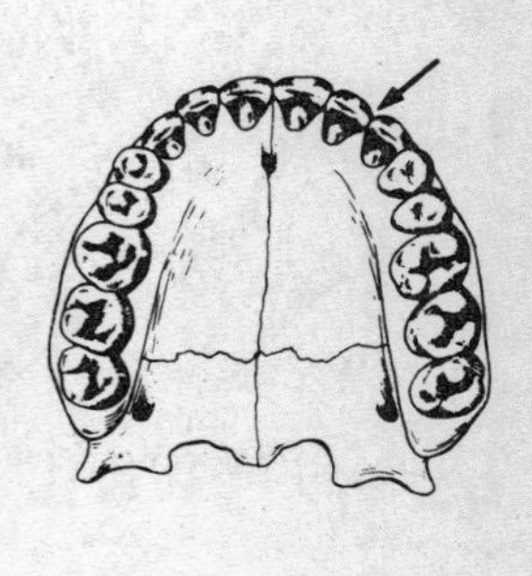

Human upper jaw

looked primitive. I had expected that. It was because they looked *so very* primitive. Instead of being human with apish tendencies, they seemed more apish with human tendencies. What was crystal-clear was that Laetoli and Hadar stood somewhere between apes and humans and appeared to be neither one nor the other.

AL-200 is an upper jaw from Hadar. The back teeth are in pretty much of a straight line (like an ape's) except for the rearmost molar, which kicks in a bit to give a slight curve to the tooth row. The diastema (at arrow, previous page) is marked in the ape, small in the Hadar specimen, absent in the later hominids – which places Laetoli-Hadar between the two others.

Canine shape, canine wear

Chimpanzee

AL-200

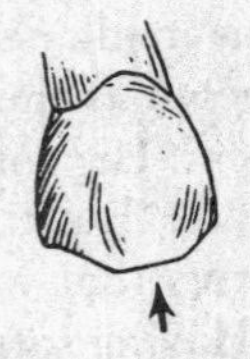

Human

The canine also has in-between characteristics. It is apelike in being conical and not spatulate. It sticks up somewhat above the tooth row but not very far, and comes to a fairly good point. Like the ape canine, it is worn down on the sides. Unlike it, it is also worn down on the tip.

Sexual dimorphism in canines

Chimpanzee

Human

AL-199 333X-3

The size variation in the Laetoli-Hadar canines is more apelike than manlike. Given the analogy of other primates, we take this to be sexual dimorphism. Lucy, an undoubted female, is the smallest adult in the entire collection.

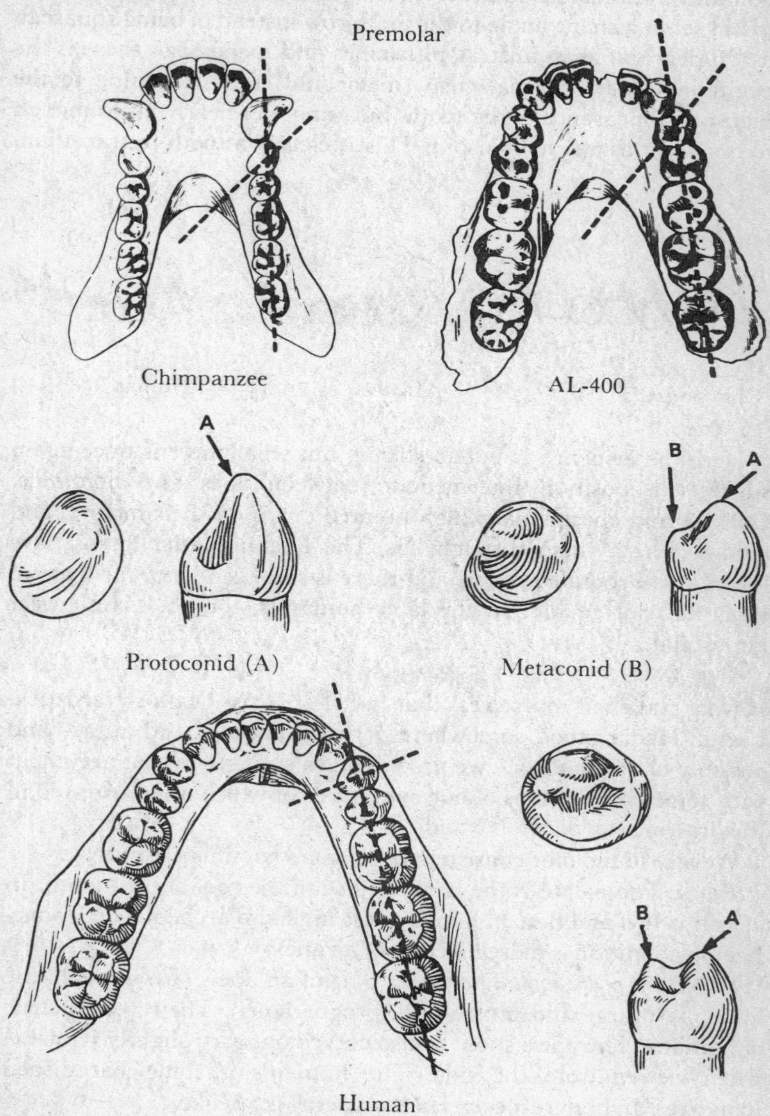

This tooth is one of the most interesting in the Laetoli-Hadar collection. It has already departed from the truly sectorial form found in apes, but has not yet reached the bicuspid form of later hominids. Its shape is rounder – more like a bicuspid – but its crown is still set at a sharp angle to the tooth row instead of being square to it (90 degrees) as in australopithecines and men. Also, there is the beginning of an internal cusp (metaconid) that is adding to the bicuspid appearance of the tooth, but is not yet nearly large enough to qualify it as a true bicuspid. This is clearly a tooth in transition.

Molar wear

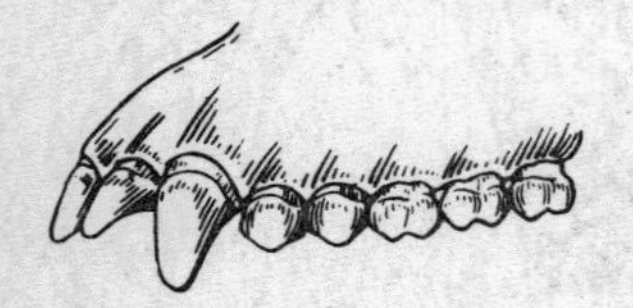

Chimpanzee

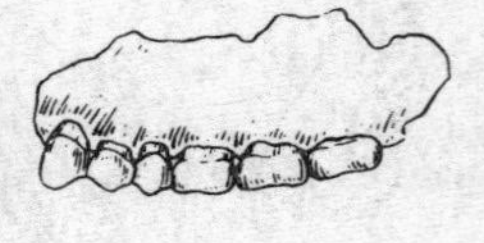

AL-199

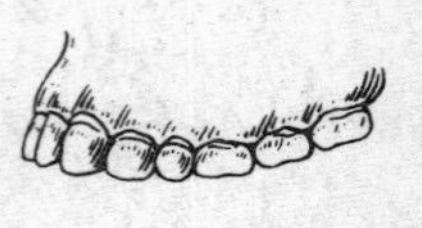

Human

Here the evidence is not so strong, but what there is once again suggests a position intermediate between apes and hominids. Gorillas and chimps have high pointed cusps on their molars, and tend to retain them through life. The Laetoli-Hadar fossils have lower, more rounded cusps, and there is some tendency for them to wear flat during life. In the later hominids, the back teeth wear almost flat.

Having satisfied ourselves that by Le Gros Clark's yardstick, Laetoli-Hadar stood somewhere between humans and apes – and possibly closer to apes – we proceeded to strengthen that argument with some comparisons of our own, three of which are illustrated in the drawings on pages 276 and 277.

We also found four consistent differences not illustrated here:

Palate. The palate is the hard surface in the roof of the mouth. In apes it is low and flat. In humans it is high and arched. The Hadar-Laetoli condition is more like that of an ape.

Proportions of the face. The upper part of an ape's face is small, the lower part large and protruding (prognathous). The reverse is true of humans. The ape's snout is also curved outward slightly (convex) when viewed from the side. The human's is somewhat dished (concave). In both respects, Hadar-Laetoli is apelike.

Canine tooth roots. In apes the canine roots are so large that they cause a swelling of the jawbone in which they are embedded. This swelling takes the form of a vertical bulge, or column, on either side of the nose. It is also true of Hadar-Laetoli, but not true of humans, whose canines are so small that they fit into the jawbone without causing any appreciable bulge.

Cranial capacity. In chimpanzees this runs from 300 to 400 cc. In Hadar-Laetoli it runs from 380 to 450 cc. In humans it goes all the way from 460 (the low end of *habilis*) to more than 2,000 (the high end of *sapiens*). What is noteworthy here is that Hadar-Laetoli overlaps the ape slightly but does not overlap the human.

The preceding analysis was the core of the case we would later build for our fossils. No single one of the many differences we identified, taken by itself, would have made that case. But when a large number of differences is found, and when they are consistent through a large number of specimens, then one can begin drawing conclusions about them with increasing confidence. In paleoanthropology one does not have brilliant flashes of insight. There are no $E=MC^2$ revelations. Recognition comes slowly, almost by hindsight. You catch yourself saying, 'So that's the way it is; why didn't I think of it sooner?' We worked our way through those fossils one trait at a time, only gradually coming to an understanding of what it was we were dealing with. By the end of the summer we had logged a sufficient number of differences in our sample to convince us that the Laetoli-Hadar individuals were distinct from apes and from any of the later hominids.

In short, our point-by-point review had answered our first question: we definitely had something new. The only quibble was: did we have one or two new things? I held out for two; Tim, for one. Bill Kimbel, who is now the deputy in my lab and who had joined us for some of the analysis, agreed with me. He kept saying, 'Lucy is different.' Tim would say, 'Come on, Kimbel, let's get out the chimps and do a little more comparing and a little less hollering.' But the next day Tim would be hollering himself. He would come into the lab yelling, 'One thing, one thing.' We would yell back, 'Two things.'

Although Kimbel and I were convinced by that time that the collection covered such a wide range of sizes that Lucy's smallness was in itself no barrier to her being included with the others, we

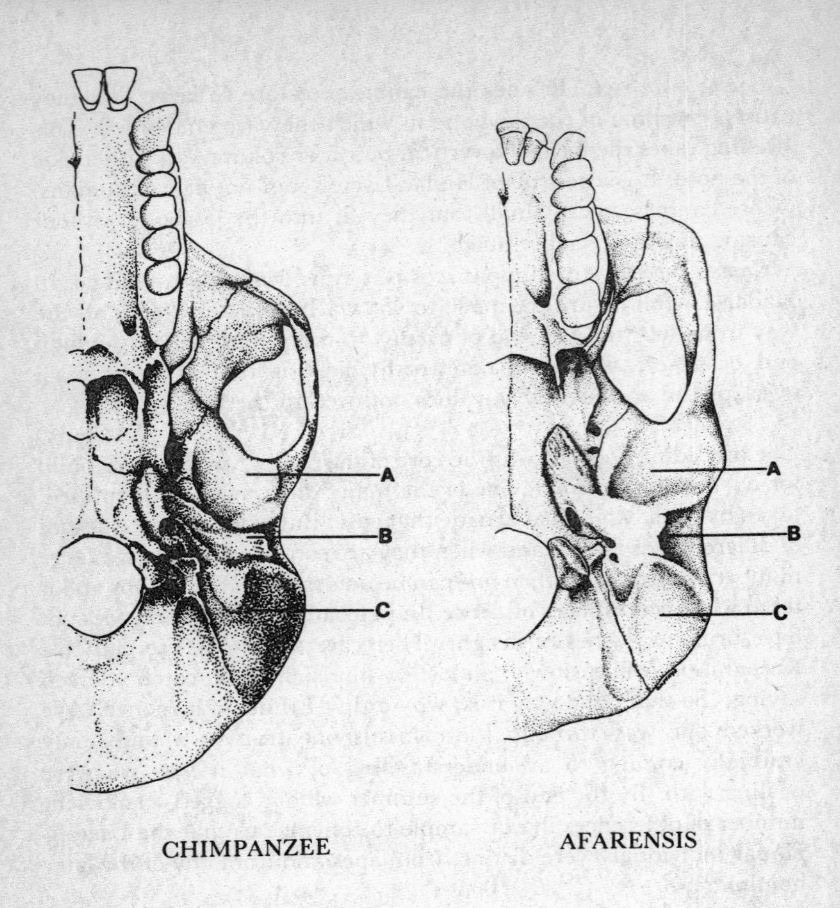

These three differences between a chimpanzee, a Hadar-Laetoli specimen and a human can only be seen by looking at their skulls from the under side. The area marked A in each case is the mandibular fossa, the place where the hinge of the lower jaw is anchored. In apes it is flat. In Hadar-Laetoli it is nearly so. But in humans it is bordered by a distinct ridge along its front (a).

The object marked B in each drawing is the tympanic plate, a small tubelike bone that starts at the ear opening and runs inward. In apes and Hadar-Laetoli it does indeed resemble a tube. But in humans it has been squeezed into more of a ridge than a tube.

The area marked C is the mastoid process. In apes this is triangular in shape and flat. In Hadar-Laetoli it is still triangular but beginning to be curved on its surface. In humans that curve has increased and has a wrinkle in it. The entire process is no longer triangular but has become oval in shape.

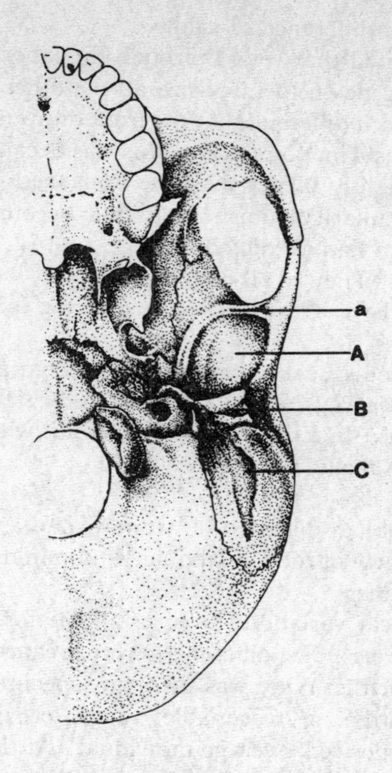

HUMAN

continued to hold out that the V-shape of her jaw was different. Also, there was an understandable reluctance to back down from an earlier position taken publicly and in print. Finally, Tim laid down a series of fossils in a row on the table. He had selected them for size – a graduated series of jaws running from the largest to the smallest in the collection. When Lucy's jaw was placed at the end of the series, it became plain that she belonged there. There was a compelling shrinkage of features down to her. She differed *only* in the narrowness of the front of her jaw. Otherwise she had the same set of primitive teeth that all three of us, after months of mulling over them, could now recognise instantly as being representative of our fossil collection – and no other. But I was stubborn.

'How can you ignore twenty other features – an overwhelming similarity – and get stuck on this difference in the jaw?' asked Tim.

'Because it is a difference,' I said.

Tim then insisted that we go through the laborious arithmetical process of scaling down to Lucy-size all the other jaws. When this was done, the width differential at the front disappeared. Some were a bit wider, but as Tim pointed out, this was because the front teeth of males were slightly bigger than the front teeth of females. After that, Lucy's peculiarity vanished. It had become an exercise in allometry, as Tim had predicted.

'Okay,' I said. 'They're all one kind.'

'You really believe it? You're not just giving in?'

'I believe it.'

'It's not like Louis Leakey hammering and hammering on Phillip Tobias about *Homo habilis* until he had him beaten down?'

'No, no. I believe it. You should learn to quit when you're ahead.'

'I just want to make sure.'

'I'm sure. I'm *sure*.'

The Lucy problem disposed of, we now had a remarkably clear and consistent view of the Hadar-Laetoli hominids. They displayed these characteristics:

1) Despite great variability in size, all the fossils were samples from a single species. Size differences were greater at Hadar than at Laetoli, but in neither place was there any specimen that did not fit comfortably within a reasonable range of variation from a mathematically derived 'average individual'. At the low end among the adult specimens was Lucy, who was little more than three and one half feet tall and probably weighed about sixty pounds. At the high end were individuals five feet tall and weighing up to one hundred and fifty pounds. In addition to size variations among individuals of the same sex, there was also variation between sexes. This was particularly marked in the jaw, and explained the tendency to a V-shape that the smaller females had.

2) Although small, these were extremely powerful creatures. Their bones tended to be thick for their size, and had markings on them that showed them to have been heavily muscled. As Tim said one day: 'Although I'm bigger than a chimp, I'm not nearly as strong. I would not want to go one-on-one unarmed and in a locked room with one; he'd certainly kill me before I could kill him. Our hominids look to have been at least as strong as chimps.'

3) They were fully bipedal. This was demonstrated by fossil evidence at Hadar and confirmed by footprints at Laetoli.

4) Their arms were slightly longer for their size than the arms of humans.

5) Their hands were like human hands, except for a tendency for the fingers to curl a bit more. Certain of their wrist bones were extremely apelike.

6) Their brains were very small, of a size comparable in scale to the brains of chimpanzees.

7) Their overall appearance could be summed up as follows: smallish, essentially human bodies with heads that were more ape-shaped then human-shaped. Their jaws were large and forward-thrusting. They had no chins. The upper parts of their faces were small and chimplike. The crowns of their skulls were very low. Male or female, they probably were hairier than modern humans. How much hairier cannot be determined. The color of their hair is unknown. It could have been black or brown like a gorilla's or a chimpanzee's, or reddish like an orang's. It could even have been silvery like the hair of some monkeys. No one will ever know. The colour of their skin is also unknown, but was probably dark, since the skin of gorillas, chimpanzees and all tropical humans is dark.

8) There is no evidence that they made or used stone tools. This does not mean that they did not, only that no link between their fossils and tools has yet been found. The earliest known tools from anywhere are those found at Hadar by Roche and Harris. They appear to be a million years younger than the youngest fossils in the collection. The actual fabricator of those tools awaits discovery.

9) They flourished from about four million years ago to about three million years ago. During that time they underwent little or no evolutionary change.

So informed, and now secure in our conviction that we were dealing with a new and distinct species of hominid, we found it no longer possible to write a merely descriptive paper, as we had originally planned. Our investigations had carried us beyond that; we had an obligation to report what we had learned. That brought us face to face with the second question: how did our species relate to other hominids already described and named? In other words, what sort of family tree should we draw that fitted what we knew about our species to what we knew about *Homo habilis*, *Australopithecus africanus* and *Australopithecus robustus*? More specifically, was our species to be classified as a human being, or was it too primitive?

Here Tim and I got into a second argument. As I have said before, I had been inclined from the beginning to regard the larger Hadar fossils as showing *Homo* affinities. And they did, in one very important respect: they had small back teeth and large front teeth. That is the modern human condition, but it is not the australopithecine condition. Since I, like nearly everyone else who had been studying paleoanthropology since the 1950s, had been educated to believe that australopithecines were probably ancestral to humans, it seemed logical, on studying their massive back teeth, to conclude that 'large is primitive' on the hominid line. From that, it seemed just as logical to conclude that 'small is manlike'.

With nothing older than the South African and East African collections to study, that bias became very hard to shake. Now I clung to it in the face of an overwhelming mass of evidence from other parts of the jaw and cranium that the Hadar type was primitive.

'It has to be primitive,' said Tim. 'It's a million years older. It can't be *Homo*.'

'It's got small molars like a human. It has to be *Homo*.'

'You're doing just what you did with Lucy. You're building a case on one characteristic and ignoring all the others.'

We were stuck. To resolve the argument, we decided to make measurements of the length and breadth of all the teeth of the various types in our collection. As representative examples we selected the third molar, the first premolar and the canine, and made the kind of biometric analysis that Robert Broom would have been scornful of. We then translated our figures into graphic form. The diagram that resulted (opposite page) shows very clearly that as far as tooth size goes, the Hadar fossils are a great deal more like *Homo* than they are like the australopithecines.

But the diagram shows something else, too. It makes clear that the robust form is the least human in having the largest molars. Seeing that expressed so clearly by the position of the teeth in the diagram, I was reminded of something I had not thought much about since I had examined the South African collections with Clark Howell seven years earlier.

I had spoken to Clark then about John Robinson's brilliantly reasoned conclusion that large, heavily enameled back teeth were a dietary specialisation. I had suggested to Clark that if this *was* a specialisation, it would have developed gradually over time and

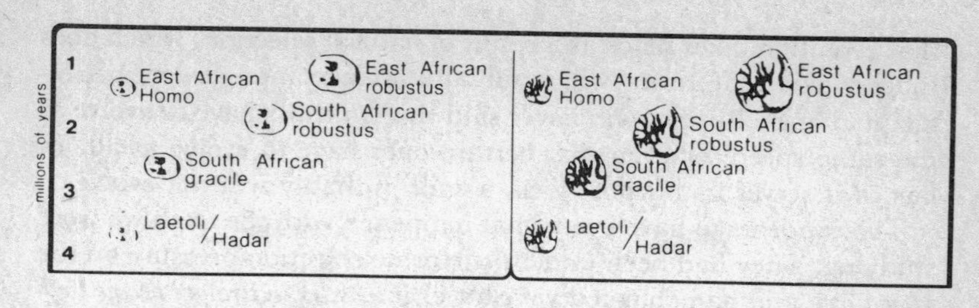

It took a biometric analysis to demonstrate conclusively that the Hadar hominids were not *Homo* but some kind of early australopithecine. This chart, which positions teeth from left to right in increasing order of size, and from top to bottom according to age, makes clear that 'small' is 'primitive', and that large molar size never did occur in humans but appeared later among australopithecines, reaching its maximum in the younger robust types. The examples shown here are premolars (left) and molars (right).

would have been the most pronounced among the youngest specimens in the collections. That fitted the evidence from South Africa very well. The robust type was younger than the gracile type, and its back teeth were larger.

As I reflected on all that, my thoughts prompted by the visual evidence of the diagram, it occurred to me that 'large is primitive' might be incorrect. The diagram was saying, 'large is late australopithecine'. Had I been looking through the wrong end of the telescope around? What if I accepted the Hadar type as the prototype? Then, on the evidence of its teeth, I would have to say, 'small is primitive'. Just the reverse of what I had been thinking.
prototype? Then, on the evidence of its teeth, I would have to say, 'Small is primitive'. Just the reverse of what I had been thinking.

It is hard for me now to admit how tangled in that thicket I was. But the insidious thing about bias is that it does make one deaf to the cries of other evidence. I recognised my bias quickly enough when I did turn the telescope around. Everything fell into focus. On the Aristotelian principle: if A (primitive) equals B (small), and if B (small) equals C (*Homo*), then A (primitive) equals C (*Homo*). Euclid put it even better: things that are equal to the same thing are equal to each other. Either way, I now saw that what I had taken for a late human trait was actually a primitive one. A better word here would be 'old', because primitive suggests something less good, less highly evolved, whereas in truth it may be perfectly good. Darwin had said that evolution takes place as a result of natural selection, which does indeed hint that traits will gradually change into newer, better-adapted ones. But Darwin never said that *all* traits have to evolve at the same speed, or even that certain ones have to evolve at all. In fact, if it serves its purpose well, a trait probably will not evolve.

That appears to have been what happened with the teeth we were studying. They had been under little or no selection pressure on the *Homo* line and had changed very little. *It was the australopithecine teeth that had changed.* They had gone in a direction of their own to satisfy a life-style somewhat different from that being lived by early humans – a life-style that would become increasingly specialised and lead to the development of larger and larger teeth.

Our two arguments, along with our analysis of the fossils, had taken the entire summer. Night after night of work had left us groggy with fatigue, but also in harmony. We now agreed on the fundamentals of a scenario that seemed thoroughly sound to both of us. We argued some more, but like lawyers testing each other, trying to find weak spots in a brief. We found none.

'What next?' I said to Tim.

'Well, if we've got something new and different and older – are we absolutely sure about that?'

'Yes.'

'Then it's our responsibility to report that. We'll have to name a new species.'

'Species?'

'You're not suggesting that we name a new *genus*, are you?' said Tim.

'My God, no. Even a species will raise a big enough stink. I'm just thinking that we can't settle on a name until we've decided where our species fits with respect to the other hominids. I'm thinking we haven't got an answer yet to our second question, the family tree question. What are we going to call it – *Homo something*?'

'A bit awkward since we've just decided it isn't *Homo*.'

'What would you call it?'

'At this point in time, I'm damned if I know.'

At that point in time the summer was over, and we were worn out. Tim had to go to California. I just wanted to go somewhere and sleep. We agreed to think about the name and the family tree, and meet again in December.

14 The Analysis Is Completed

What's in a name? That which we call a rose
By any other name would smell as sweet.

WILLIAM SHAKESPEARE

Classification is not intended to be an adequate expression of phylogeny but only to be consistent with conclusions as to evolutionary affinities.

GEORGE GAYLORD SIMPSON

Lucy may be considered a late Ramapithecus.

RICHARD LEAKEY

To consider Lucy a Ramapithecus *is laughable.*

C. LORING BRACE

During the fall of 1977, I had a chance to become more comfortable with some of the new ideas about hominid relationships that had been forced on Tim and me by the analysis we had made that summer. We had not yet actually placed our hominid on the family tree, nor had we named it, but I was relieved that the hard part of the job – the fieldwork, the collecting, the sorting out, the measuring, the comparing, and finally the conclusion – was over. Paradoxically, if we were going to get into trouble, it would be over the easiest part: finding a name. Descriptions and suggested affinities never raise blood pressures the way names can.

If we were to limit ourselves to writing a conscientious and accurate paper that described the fossils fully, explaining how they differed from existing ones, and drew only inferences from those differences, leaving it to others to draw the real conclusions, I was sure that we would be complimented on a job well done and that we could retire from the field without having created any controversy.

But if we did only that, we would have ducked a responsibility. We knew – or were in a position to decide – where the homonids should be placed. Therefore we should do the placing ourselves, and not leave it to someone who might not understand the fossils as well as we did, and who certainly would not have studied them as carefully as we had, and who might give them a misleading name that would plague paleoanthropology for years.

There was no question in my mind that we should place them. But to do that, we would have to name them, and that might be what would burn the barn down. No new hominids had been named since *Homo habilis* had received his label in 1964. Often during that autumn, thinking over *habilis'* clouded career as an illegitimate sprig on the family tree (his legitimacy still questioned by some), I wondered what sort of reception a species named by me would get.

I worried about that. Having run field expeditions for several years and having had my share of minor skirmishes on small issues, I was well aware of how skirmishes can escalate into all-out war on a large issue, how convictions can polarize people, how easily enemies are made, how timid foundations are and how their funding can dry up suddenly in the face of protracted arguments. I wanted none of that. All I wanted was to be able to complete my laboratory studies, get back to Ethiopia as soon as I could, find more fossils and *then*, perhaps, have something to say about them. But, as in a pregnancy, birth comes whether one is ready for it or not. Tim and I had produced a baby. I wondered if I could be a proper mother to it, give it a decent christening and introduce it to polite anthropological society.

That fall I suffered from what a new mother might have called postpartum blues. I had moments of thinking our analysis might have been inflated by too much air. I would have been relieved if someone had come along to let some of that air out with a pin and give us an excuse to back down. Nobody did, of course, and I had to gear myself for what I knew would be the devotion of another chunk of my time and energy to writing a paper that was absolutely airtight, then more time and energy defending it, writing more articles, arguing, arguing, arguing.

And what if we were wrong?

I met Tim again in December. He also had had time for some sober reflection. Like me, he felt a bit daunted at the prospect of shaking the family tree and adding a new species to it. It was par-

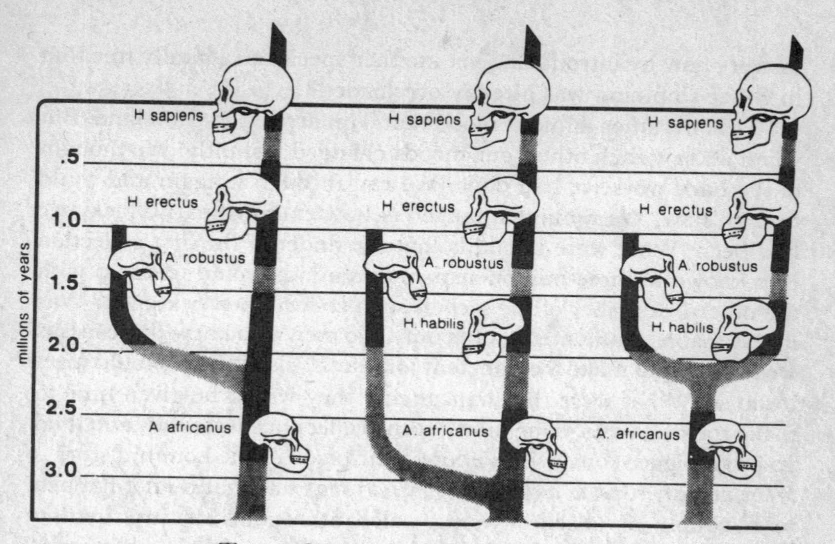

THREE WELL-KNOWN FAMILY TREES

The one-species theory (TREE AT LEFT), espoused by Brace and others, recognises only one branch in the hominid tree, and the existence of only one kind of hominid at a time. Recently, however, Brace has permitted the branch ing off of *A. robustus*, because it is too unlike the *Homo* types that follow for it to fit on the main line. The virtue of this tree is its simplicity. Brace, a 'lumper', does not accept *H. habilis* as a valid species. He lumps it with *A. africanus*.

THE CENTER tree, reflecting the views of John Robinson, attempts to resolve the awkward contradiction which results from the observation that robust types have more 'primitive' back teeth than gracile types, although, on the evidence of South African caves, they are younger. Robinson's solution was to place the robust type correctly in time (2.0 to 1.0 million years ago) and then assume a common ancestor with the gracile *africanus*.

The most widely accepted tree in the 1960s and 1970s (RIGHT) reflects a growing consensus that *A. africanus* was ancestral to both *A. robustus* and *H. habilis*. It acknowledges that increased molarisation of the teeth through time was an australopithecine trait, not a human one, but assumes that this tendency, already evident in *africanus*, was not enough to dislodge it as a human ancestor.

ticularly awkward for him because he had done his postgraduate work under C. Loring Brace of the University of Michigan. Brace, an early proponent of the one-species theory, has held for years that the evolutionary line leads straight from *Australopithecus* to *Homo erectus*, and rejects *Homo habilis* out of hand. Tim had defended a doctoral thesis before Brace. How could he, a Brace student, confuse matters now by introducing yet another species to a family tree that in Brace's opinion was already overloaded?

It was in rather dubious mood that Tim arrived in Cleveland. But when we saw each other, our moods changed instantly. We thought of the hard work we had done. We ran through it again and could find no flaw. We reminded ourselves how extraordinarily lucky we had been. What were the odds that the finder of the first collection ever known of three-million-year-old hominids would team up with the official describer of the second such collection ever known? Was it reasonable to assume that the only two men who knew those fossils well enough to make a competent joint analysis of them should meet to do so? What were the chances that they would be given time to make their analysis, conclude that the collections were one, and then go on to bigger conclusions about their place in the hominid story – *before anybody else had a chance to do it?* Things like that do not happen to a couple of obscure young paleoanthropologists just getting started professionally. But it had happened to us. We had been presented with an exceptional opportunity to do something for paleoanthropology, and we felt ourselves uniquely qualified.

As a first step in nailing down our second question – how did our fossils relate to the other? – we decided to place all the African hominids on a diagram according to their age, their type and their location.

To place the South African fossils, we used the most recent estimates of Elizabeth Vrba, a South African biostratigrapher, for the ages of both the robust and the gracile types. For Omo we had good dates but poor specimens, particularly of the type that Clark Howell had tentatively labeled gracile (*africanus*), so we put a question mark next to it. For Lake Turkana we took Curtis' potassium-argon dates and the fossil pig evidence rather than the Fitch–Miller dates, and accepted Richard Leakey's identification of the fossils there. At Olduvai, Hadar and Laetoli we took the published potassium-argon dates.

The next step was to simplify by ignoring locality and pulling the fossils together by type. The diagram that resulted (page 288) represented, in our opinion, the simplest – or, to use a word favored by scientists, the most parsimonious – way of arranging all the existing fossils and at the same time being respectful of their differences. A still more parsimonious arrangement would be achieved if one used that of Loring Brace (diagram, page 285) and extended the

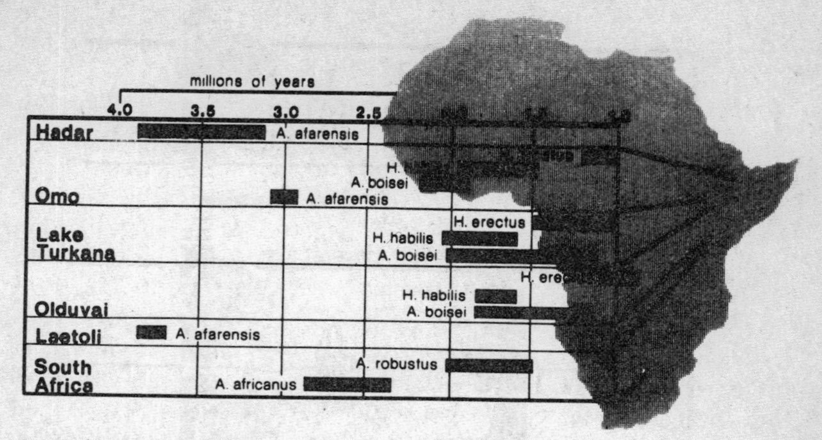

Realizing that none of the family trees on page 285 would accommodate the new evidence produced by their studies of the Hadar and Laetoli fossils, Johanson and White decided to draw a new tree of their own. As a first step they plotted all the known African hominids by type, by location and by age. That gave them the pattern shown above, which they then proceeded to simplify by consolidating the types. For the result, see page 288.

Australopithecus bar beyond three million to include Laetoli-Hadar. That would acknowledge a widely variable and long-lasting species of *Australopithecus* which at its older end would take in the Hadar fossils. As far as parsimony goes, it would be a marvel of simplicity. But it would ask one to swallow something that neither Tim nor I can. At its extremes the gracile section would not work: Lucy, who would be at one end, is simply not a human being; Richard Leakey's 1470, who would be at the other end, is. The two are quite different, and cannot sensibly be regarded as members of the same species. Furthermore, it would be awkward to encounter between them – almost like a barrier – an *africanus* that has large molars and small incisors in a pattern which does not resemble what went before it or what came after. Can molars start small, swell, and then shrink again? They can, if three dietary adaptations to explain them can be postulated, but it is extremely unlikely. That is parsimony in reverse. More digestible – and more parsimonious – would be two diets and two kinds of teeth: a generalised set that favored generalised eating, and an increasingly specialised set that favored increasingly specialised eating. That is the thinking which produces a diagram like the one on page 288.

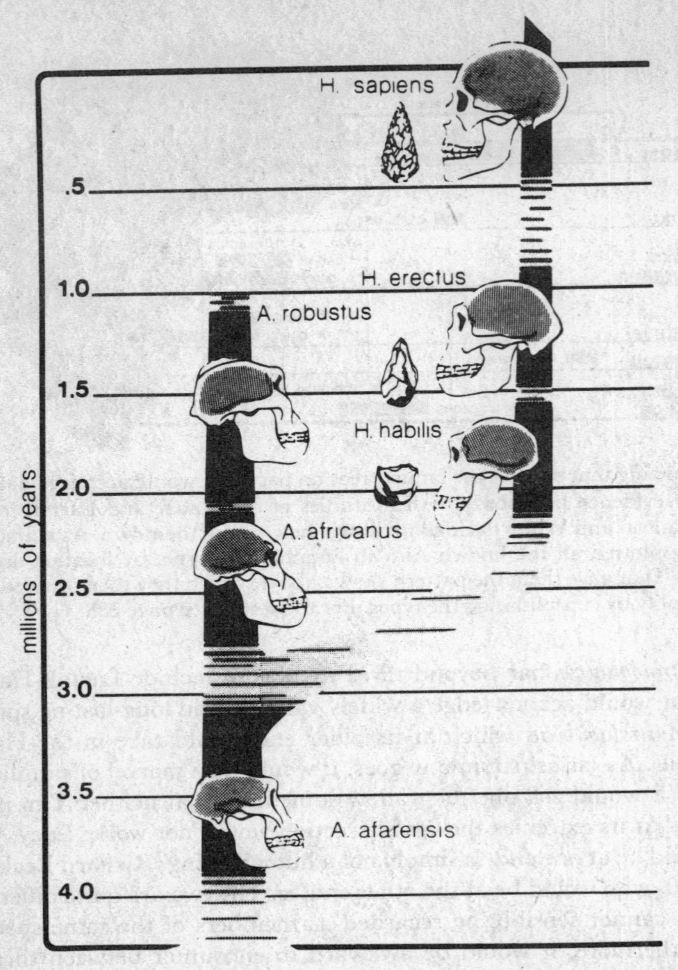

This is the result of consolidating the information in the chart of page 287. All the *afarensis* fossils can be lumped between four and three million, all the *africanus* ones between 2.7 and 2.2 million, all the *robustus* ones between 2.1 and 1.0, and so on. Johanson and White are convinced that *afarensis*, the oldest and most primitive hominid known, was ancestral to all the others. They reason that increased molarisation was a late australopithecine phenomenon, and they have located the types that display it accordingly, with *robustus*, the most heavily molarised, at the end of that line. That leaves the *Homo* types, with back teeth essentially unchanged from those of their ancestor *afarensis*, on a line of their own, with the increasingingly advanced *erectus* and *sapiens* evolving out of *H. habilis*. Tools, as the chart indicates, are a *Homo*, not an australopithecine, invention.

That diagram makes our position clear: that the Laetoli-Hadar specimens represent a common ancestor to the later australopithecines and to *Homo*; that the divergence between the latter two types probably began around three million; and that *africanus* represents an intermediate stage on the way to *robustus*. We do not believe it was ancestral to humans.

We do believe that the emergence of human beings began sometime after three million years ago. By two million it had been accomplished. By then creatures recognizable as *Homo* walked on earth. So did their cousins, the robust australopithecines. For about a million years they appear to have walked side by side. By one million there were no australopithecines left. They had all become extinct.

With these conclusions, after more than two years of work and thought, Tim and I had arrived at what we considered a satisfactory answer to our second question. There remained question three: what to name the new species.

The first task was to decide what genus to put it in: *Homo*, *Australopithecus*, or *Something else*. We quickly eliminated *Something else*, which would have required finding major differences between our fossils and all the others. Those did not exist; all were bipedal hominids, closely related. *Homo* was also eliminated – but not quite so easily. We were still vaguely considering that possibility when Owen Lovejoy walked into the lab late one afternoon.

'With the family tree you've just built, you can't call it *Homo*,' he announced.

'Why not?'

'Because you've made it ancestral to the other australopithecines. They would have to become *Homo* too. Try that, and everybody will begin beating on you. They'll turn you into a couple of knuckle walkers overnight. You'll never be able to straighten up again.' His bray of laughter was like a dash of cold water. The logical choice was *Australopithecus*. The fossils proclaimed it. We accepted it. Farewell Old *Homo*, vanished at last at about two million.

For a species name we cast about among several. I suggested *Australopithecus laetolensis*, saying that it would please Mary Leakey.

'I don't think that's such a hit idea,' said Tim 'You have all the best fossils. Give them *your* name.'

'*Johansonensis*? Come on.'

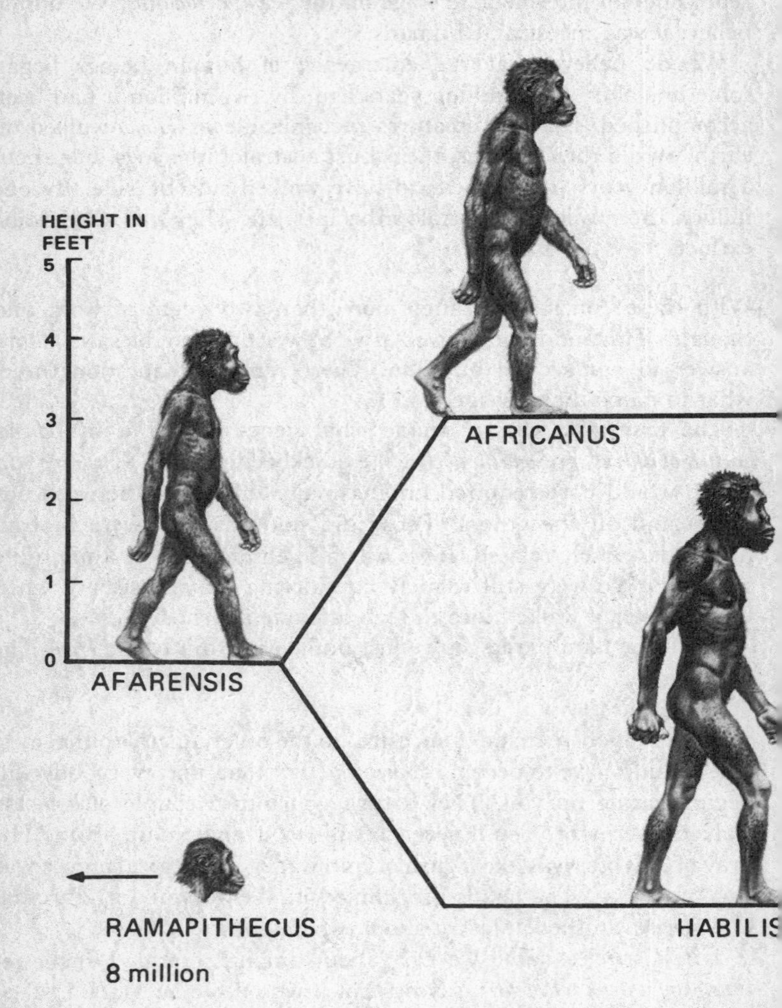

If the family tree on page 288 is turned on its side, and the fossils turned into fleshed-out individuals, they would look like this – maybe. No one can be sure just what any extinct hominid looked like with its skin and hair on.

Sizes here are to scale, with *afarensis* about two feet shorter than the average modern human being. *Ramapithecus*, so far known only from teeth, jaws and bits of skull, was smaller yet, perhaps not much more than three feet tall.

'No, no. I mean the name of your place. *Hadarensis.*'

That did not satisfy me, because I knew that huge yet-untapped areas around Hadar were fossil-bearing and in future years might be even more important than Hadar. I felt we should acknowledge that possibility and use the name of the entire region: the Afar triangle.

'Okay, *afarensis*,' said White.

Australopithecus afarensis was agreed on.

Next there had to be agreement on the type specimen – the specific fossil that would be used to describe the species. Several of the people in my lab urged me to use Lucy, on the ground that her magnificent condition would make possible an exceptionally detailed picture of the species. I refused. I pointed out that Lucy was not typical of the collection, being extremely small. Furthermore, a complete scientific description of her – a necessary prelude to naming – had not yet been published. I recommended LH-4, the best of the Laetoli specimens. It had been fully described by Tim, illustrated and published.

That would draw Laetoli in, and that was important. Having convinced ourselves that the two collections represented the same type, we felt we should formalise that. The best way to tie them together was to award the species name to one place and the type specimen to the other.

Also, Laetoli was then believed to be older. We felt that the time-span represented by the two populations should be acknowledged, since it seemed to say something important about hominid evolution: stasis for about a million years, then a sudden burst of evolutionary activity during the next million. If that was not acknowledged, and if future fieldwork produced a whole avalanche of material from the Afar, people might lose sight of the fact that there was this other population in another part of Africa. Recoveries from it would probably never be large or in very good condition; it might be overlooked. If it was not accepted formally by us as being the same as the Afar material – particularly if somebody later put a different name on it – then things would be in a real mess. We could forestall that by naming the Laetoli material ourselves.

I also thought it would please Mary Leakey. As it turned out, I couldn't have been more wrong.

With a family tree and a name agreed on, Tim and I turned to the task of writing up our findings in a paper that would review all the early African hominids, describe the new fossils in detail, show why they deserved to be considered one type, then show why that type deserved assignment to a new species and what that assignment should be. Each of us realised that the paper would be a watershed in his career. We also realised that if its analysis were to be accepted by other scientists, it could not fail to be an important paper; it would change the way man looked at his origins. Thus we felt a heavy obligation not only to ourselves but also to science.

Our first draft was a bloated, rambling tract. We had the magazine *Science* in mind as a target, and quickly realised that what we were writing was far too long to be accepted by that publication; the descriptions of the new fossils alone took up pages. Should we leave them out? No, that was impossible. Publish them separately? That was better. Those descriptions had to be in print so that they could be cited, but they did not have to be published together with the analysis. Our solution was to publish a purely descriptive paper of the entire Laetoli-Hadar collection in *Kirtlandia*, the scientific house journal of the Cleveland Museum of Natural History, and save the analysis for a second paper which we would submit to *Science*.

I thought of Mary Leakey again. Since she had found some of the fossils, would it not be a scientific courtesy to ask her if she would like her name added to the *Kirtlandia* paper? I wrote her a letter. Soon she was in the United States, and when she stopped at Berkeley, Tim brought up the matter again. She seemed pleased by the idea, but only if the paper was a descriptive one, not an interpretive one. Later she came to Cleveland, and told me that she would be pleased to have her name used as a contributing author as long as the paper did not say that any australopithecine was ancestral tc any *Homo*. I assured her that the *Kirtlandia* paper would be purely descriptive. Then Tim and I settled down to write it. We finished it in the late spring of 1978 and sent it to the printer.

We also completed our interpretive paper, having taken the precaution beforehand to check the name *Australopithecus afarensis* with Ernst Mayr of Harvard. Mayr is a distinguished zoologist who is also the last court of appeal on the scientific naming of things. In May 1978, with Mayr's approval of the name, we submitted our paper to *Science*, aiming straight at the top, at the most prestigious scientific publication in the United States.

That done, I went to Sweden, having been invited to talk about the Hadar fossils at a Nobel Symposium convened by the Royal Swedish Academy of Sciences to commemorate the death of Carolus

Linnaeus, who pioneered the now universally used binomial system (use of two Latin names) for the classification of all living things. This event had considerable emotional impact on me. Being of Swedish descent, and being concerned with the classification of some important fossils of my own, I felt myself to be a link in the chain of scientific inquiry that led straight back to that other Swede, now dead two hundred years. It was a thrill to stand up and utter the name *Australopithecus afarensis* for this first time in public. I described Lucy and the other Hadar fossils, and went from them to Laetoli. I was not aware that Mary Leakey, sitting in the audience, was flushed red with anger to hear me talking about the Laetoli specimens and the Laetoli footprints. They belonged to her, and I should not have been discussing them.

Tim found that out when he reported to Laetoli a few weeks later to continue his work there.

'You should not associate with that man,' Mary Leakey told him. 'You should not have coauthored that paper with him. I shall have to have my name removed from it.'

'If you're serious about that,' said White, 'you'd better send a cable. I think the paper is already on press.'

It was. When I got the news, I had to instruct *Kirtlandia* to print the paper over again. Publication was delayed until the fall.

That episode was as puzzling as it was distressing. Since the Laetoli material had already been fully described in published scientific papers, I considered it to be in the public domain, on the ground that the purpose of such publication is to acquaint other scientists with the material and to encourage discussion. That my discussion of 'her' fossils and 'her' footprints at a symposium should have upset Mary Leakey came as a complete surprise. When Tim arrived back in the United States we had a long talk about the matter, and concluded that the trouble had been with the nature of my presentation. I had used Mary Leakey's material to advance interpretive ideas she did not support. No one, looking at the new family tree, could escape its two major messages: first, that man was descended from a form of australopithecine, a proposition that the Leakeys had always

Their arguments about Lucy over, Johanson (left) and White are ready to announce a new hominid species to the world. The fossils on the table are mostly Hadar specimens except for ape and human jaws (extreme right) and the human and ape skulls in front of Johanson.

rejected; second, that man was not nearly as old as the Leakeys preferred to have him.

A journalist friend of mine was in the laboratory at the time my discussion with Tim took place. 'It seems to me that you and the Leakeys are headed for a big shoot-out,' he said.

'I guess we are,' I said.

'What if they're right?'

'We'll have to change our minds. But the only way to prove they are right is with more and better fossils. After all, trees are just reflections of the best current knowledge. They are made to be redesigned. We don't know ours is right; we just think so.'

'Amen to that,' said Tim.

'I mean, it's got that big gap in the middle between three and two million. There are no good fossils in there. We assume two branches coming out of a Laetoli-Hadar ancestor, but we have nothing along the way to prove it. All we have is a strong case. I think that when the gap is filled our case will be stronger.'

Tim was asked if, being so skeptical by nature of anything he could not nail down with hard evidence, he was happy with a tree that had a gap in the middle of it.

'On the whole, yes,' he said. 'The paper we wrote isn't just any little Mickey Mouse paper. We worked for months over it. We think it's airtight. We know it's logical. All the arguing we did served a useful purpose. Instead of having somebody else pick on us in print after publication, we decided to pick on each other before publication. We've done our homework. Any criticism of this paper will have to be backed up with fossils that justify the criticism – fossils whose existence we were not aware of when we wrote the paper. I don't think there are any. If any should be found later, then we'll just pull in our horns and redraw the tree.'

'A tree is not an end in itself, I suppose,' the journalist said.

'No, it's not, but it's useful,' Tim said. 'It's a way of clarifying, of simplifying. Once it's made, it looks logical, and it seems so easy. I want to tell you it's not easy. I never would have been able to do this by myself. All I had was data on a small collection of very old, very peculiar hominids. What was I to do with them? I'm not good at theorising. I'm not good at mathematics. I'm not even very good at laboratory work. But I'm conscientious in the field; I want to increase the data base with some cold, hard facts that reflect well-done fieldwork.

'When I met Don, I was suspicious of him. I'm suspicious of any anthropologist who wears Gucci loafers and Yves Saint Laurent pants. He had to prove to me that he had the same attitude toward careful fieldwork that I had. Well, he did prove it. His fossils are very well described. Their dating is exemplary, and they have been carefully cleaned. Look at some others here and there, and you'll appreciate how important careful cleaning is. A lot of specimens that you encounter are broken. They have corners knocked off, or tooth-wear facets have been obliterated by careless matrix removal. Don's originals are beautiful. Even his casts tell you a lot more than a lot of other people's originals do.

'All that caught my attention. When I saw that Don's fossils were like the Laetoli fossils, I realised that this *was* an expanded data base that gave us a chance to do a useful study. If we didn't do it, the descriptions would just have dribbled out, little by little, here and there, over a period of years. Nobody would have known what to make of them.'

I pointed out that having two people work on a difficult problem instead of just one was important, that if you have somebody to stimulate your brain all the time, you get further than you do by yourself.

'That's right,' said Tim. 'In our case Don supplied the imagination and the energy that I lacked – the daring.'

'Tim kept me in line,' I said. 'He also worked on my bias. He saw earlier than I did that we had one species, not two. It wasn't easy to wean me away from my original thinking. I'd already published that there were two species. Was I now going to publish again and say there was only one?'

'That's what you had to do.'

'But what a wrench.'

Our visitor left. He told me later that that short exchange had brought home to him how important our partnership had been. He said that it reminded him of James Watson and Francis Crick, the two men who had worked out the double-helix pattern of the DNA molecule, in the sense that neither of them could have done it alone. In fact, there was a geneticist, Rosalind Franklin, working at the same time they were, who all on her own nearly broke the DNA problem several times. She came *so close*, but she had nobody to talk to. She needed that last push – another mind. She never got it. Tim and I, in having each other, had been extraordinarily lucky.

15 The Reaction

Every man has the right to utter what he thinks truth, and every other man has the right to knock him down for it.

SAMUEL JOHNSON

We expect rough treatment from our colleagues whenever we produce something shoddy . . . The essential factor which keeps the scientific enterprise healthy is a shared respect for quality.

FREEMAN DYSON

Dyson makes a good point, but he should have extended it to say that criticism shouldn't be shoddy either. If it is, it should be treated just as roughly.

TIMOTHY WHITE

In December 1978 we heard from *Science*; it would accept the interpretive paper. Indeed, the editors thought it important enough to make it their lead article in the issue of January 26, 1979. A drawing of one of the Hadar jaws appeared on the cover. That was *Australopithecus afarensis*' official coming-out party. It had been described fully and named in *Kirtlandia*. Now it had been evaluated as well. Any paleoanthropologist in the world could weigh our arguments that the Hadar-Laetoli finds deserved identification as a new species. Neither of us was prepared for the explosion of interest that followed the formal disclosure of *afarensis* in print. *The New York Times* ran an article on its front page with a picture of a reconstruction of an *afarensis* skull. In succeeding days other articles followed in *Times*, *Newsweek* and elsewhere. I was invited to appear on a number of television talk shows. But the headline in *The New York Times* said it all: NEW-FOUND SPECIES CHALLENGES VIEWS ON EVOLUTION OF HUMANS – and went on to give a succinct review of the paper's salient points:

A previously unknown human ancestor that lived in Africa three to four million years ago and had an unexpected combination of a small-brained apelike head and a fully erect body has been discovered by two American anthropologists.

The discovery, the first species of human ancestor to be named in fifteen years, deals a major blow to the old but still widely held belief that erect posture, which would theoretically free the hands for tool-making, evolved in tandem with an enlarged brain . . .

The next view is that the bones are not only too apelike in the jaws, teeth and skull to be considered *Homo*, but also that they are even more primitive than the previously known remains of another human-like lineage called *Australopithecus*.

Most of the press followed the line of *The New York Times*, treating the announcement as a piece of groundbreaking scientific news, accepting it as reported, and making no effort to criticise it. An exception was *Time* magazine, which belittled the *Science* paper as a rehash of old fossils that had been described before and did not really say anything new.

My first reaction was that the *Time* writer probably had not read the paper. If he had, he would have known that a great many of the fossils in it had never been described before, and that the paper's conclusion was *all* new.

The *Time* article went on to quote a 'distinguished anthropologist' who said that that was the kind of thing to expect from Johanson, who always had been a publicity hound.

'Why would they say that?' I asked a journalist friend who once worked at Time Inc.

'Maybe it's because of Richard Leakey. They ran a picture of him on the cover of their magazine not long ago, with a long article on his theories about Old *Homo*. Your announcement of *afarensis* blows Old *Homo* right out of the water. It destroys Richard Leakey's central idea. Don't you realise that when you publish about *afarensis* you're on a collision course with Leakey? He's *Time*'s pet anthropologist. When they want to say something about hominid evolution, they go to him.'

'Okay, but why that crack by the unidentified 'distinguished anthropologist'? Who the hell is *he*?'

My friend said he would try to find out – and did. He learned that it was Elwyn Simons of Duke University. Simons is indeed distinguished. He and David Pilbeam are the world's leading authorities on Miocene apes. But when questioned, Simons denied the quote. When asked later if *afarensis* began to make sense, he said, 'Yes'.

Others thought *afarensis* made sense too. I had been on tenterhooks waiting for the reaction from Clark Howell; of all opinions, Clark's would count the most with me. When it came – 'a significant advance in interpreting the pattern of human evolution' – I was vastly relieved. Pilbeam also offered enthusiastic comments, as did Bernard Campbell, a leading British authority.

One who did not was C. Loring Brace, the single-species adherent. He arrived in Cleveland one winter day just as the paper was about to be published, in response to my invitation to inspect the entire collection. Tim was there, waiting with mixed feelings to defend the new species against the attack of his old teacher, who certainly would not be made happy by the introduction of yet another name to a family tree that contained too many of them already.

Now in his early fifties, Brace has become one of the widely respected senior men in the field. He is one of the few paleoanthropologists alive today who can be considered 'educated', in the sense that he is at home in all corners of English literature and the classics. He is a lover of serious music. He has heard and committed to memory nearly every ranchy limerick ever written, and has composed a few hundred of his own. He is a handsome man with weary blue eyes, white hair, and a full beard. He wears his hair long, pulled into a ponytail at the back and secured by a rubber band. He walked into the lab wearing a flowing woollen cloak with a pointed hood, and looking like a persecuted but forgiving and worldly-wise monk who had somehow escaped from the twelfth century.

It would have been nice, I thought at the time, if he had done what Robert Broom did for the Taung Baby – thrown himself down on his knee in adoration of our ancestor. But Brace is not one for dramatic gestures. He is soft-voiced and patient. He calmly inspected all the fossils as Tim laid them out, shaking his head over their extraordinary quality.

'She certainly is primitive,' he said after a long examination of Lucy. 'But I don't see why you have to invent a new name for her. What's wrong with *africanus*?'

'She's not like *africanus*,' said Tim. 'Her teeth are entirely different. By the time you get to *africanus*, you've already moved in the direction of specialised back teeth. Lucy hasn't even started that. Also, she has a primitive first premolar. You don't get that in any *africanus*. They have bicuspid premolars like *Homo*. You certainly can't call Lucy *Homo*.'

Brace pushed back his chair and looked at White with a tired smile. He had been through this before. 'That depends on what you mean by *Homo*,' he said.

'I mean *Homo habilis*. Richard Leakey's big-brained, two-million-year-old skull 1470.'

'I don't consider *Homo habilis* a valid taxon. I accept only *Homo erectus*.'

'That puts 1470 in with the australopithecines.'

Brace sighed. 'It's a bit awkward, isn't it?'

'It puts 1470 and Louis Leakey's Zinj in the same species. How can you do that?'

'I would take that one out,' said Brace.

'*Homo habilis*?'

'No, Zinj. All the robust ones. I'd leave your fossils in as early primitive examples of *africanus*. It's mostly a matter of names, after all.'

'I think it's a matter of morphology,' said Tim – 'of differences in teeth.'

'They are not all that different. I think that when we know *afarensis* and *africanus* better we will find that the fit isn't too awkward.'

Tim persisted: 'That still leaves Lucy and 1470 in the same species. How do you deal with that?'

'I'll have to think about it.'

After Brace had gone, Tim said, 'I think we shook him a little.'

'Wait until he reads the full paper,' I replied. 'I think we'll shake him more.'

'I'm not so sure. The fossils are their own best argument. It's hard to look at those jaws very long, or at Lucy's tiny primitive head, and think of them as human beings – although you managed it for quite a while, didn't you?'

'All right, White, all right. We've kicked that one around enough.'

Later Brace wrote to thank me for the visit and to compliment us on our paper. But he was careful not to concede the validity of the taxon *afarensis*. The single-species theory had received a rude shock, but in Brace's view it was still standing. Tim and I decided we would have to score that skirmish as a draw.

*

The controversy heightened in February. Richard Leakey was in the United States on a lecture tour. In Boston he was asked about the new species. He replied flatly that he disagreed with us. He insisted that *afarensis* was not ancestral to *Homo*. He said that there were three separate species, not one, from that general time frame – and that he had the fossils to prove it.

Boyce Rensberger, a reporter for *The New York Times*, asked him what those fossils were. Leakey replied that he did not wish to discuss them. He would make his own scientific announcement in *Nature* in two months. Rensberger telephoned me and asked, 'What is he talking about?'

'I think he's talking about a small collection of isolated teeth from Kubi Algi, a fossil site just south of Koobi Fora on Lake Turkana.'

Rensberger went back to Leakey. 'Are you talking about your Kubi Algi teeth?' Leakey admitted that he was.

That seemed hardly consistent to me. I remembered a day at the Bishop Conference a few years before when Clark Howell was giving a talk on the hominid fossils he had found at Omo. Many of them were single, isolated teeth. He could not deduce much from them, he admitted, but he was proceeding to give tentative assignments to them when Richard Leakey stepped in and with great force said that it was a mistake to try to make species determinations on the basis of single teeth. Leakey apparently felt strongly about that. He repeated it in his *Scientific American* article, published in 1978, arguing 'for caution in the making of taxonomic judgments . . . when the evidence at hand is a few isolated teeth'.

'I guess he forgot he said that,' said Tim. 'I agree with him about isolated teeth; you can't make taxonomic judgments on them. But here he is, making them.'

'It's a small sample, too,' I pointed out. 'Only eight teeth.'

'Maybe he's got something else. Maybe he's pulling a Louis Leakey – holding out on a superfossil that he's going to zap us with.'

I doubted that. I told Tim that that was something we shouldn't be worrying about. We knew the fossils – by this time better than anybody. We knew the case they made. I said that we would just have to wait for Richard Leakey's paper in *Nature* to see what sort of evidence he could produce. (A year later that paper had not yet been published.) Meanwhile, we would in all probability be hearing from him in a written criticism of our own paper. It is customary, when papers are printed in scientific journals, for the editors to throw their pages open to discussion by critics, and to rebuttal of that criticism by the authors of the paper. I felt quite sure that Richard, having committed himself in opposition to us in a newspaper interview, would follow it up.

We did not have to wait long for the followup. Richard and I met at a symposium in Pittsburgh on February 17, 1979. He reminded me of my earlier paper that had suggested *Homo* affiliations for the Hadar jaws and a different species for Lucy.

'I think Don was right the first time,' he said at the symposium.

I explained that I hadn't known the fossils as well then, and that a longer examination of them had forced me to change my mind.

He said again that I should have stuck to my first view, and that Lucy's difference in size and jaw was too great to allow her to be put in one taxon with the others; clearly there were two types at Hadar.

It was funny to find myself on the other side of an argument I had waged so fiercely with Tim earlier. But I was certainly glad we had had it. Now I felt very confident, and replied that the evidence had forced me to change my mind – particularly the 333-site First Family fossils. I had not yet discovered them at the writing of the first paper. But now, with that collection in front of me, I could see clearly that there was a range of variation among them that accommodated Lucy very comfortably.

I told Leakey that the fossils were of one type and offered to review them with him feature by feature. He declined to do this, and referred again to new fossils found by associates of his in Kenya. I had a momentary thought that there might be a superfossil held out for just a moment like this, as Tim had suggested. But when Leakey said, 'The material I've got is very insignificant,' I knew that my original hunch about the Kubi Algi teeth had been right. There are only eight of them, and they prove very little. But apparently Leakey was prepared to hang his case on them because he continued: 'There's enough to challenge Don with. It gives me the right to offer my opinion.'

Of course he had that right. But I would have been happier, and I think everyone else would have been too, if we had been given a chance to settle the dispute by letting the fossils themselves speak. To be told that there were fossils that would refute me, but to be told also that this was not the time to discuss them, was disappointing.

The first formal assault on our paper came on March 7, 1980. *Science* sent us a criticism signed by Richard Leakey and by Alan Walker, who had been Leakey's collaborator on his *Scientific American* article and who is now an anatomist at Johns Hopkins. It also sent one signed by Mary Leakey and by two anatomists from St Thomas' Hospital, London: Michael Day and Todd Olsen. I had hoped that this time there would be an argument based on the characteristics of the fossils. Again I was disappointed.

Richard Leakey and Alan Walker disputed us on two grounds. The first was an important but subtle interpretation of how evolution takes place – whether cladogenetically (by the splitting of populations and their subsequent evolution along different lines) or anagenetically (by a gradual evolution of one type into another along a single line). This is certainly a vital matter, and it is something I would like to discuss in depth with Leakey if I ever get the chance, particularly since his written criticism attacked us in one paragraph for being anagenetic, and then in the next paragraph defended his own view of human evolution with an anagenetic argument. That inconsistency in his own thinking we found hard to accept.

The second Leakey–Walker argument was equally elusive. It depended on their identification of certain small fossil individuals found at Lake Turkana, and on whether they did or did not resemble *Homo erectus*. One specimen in particular, skull 1813, is a small version of the famous skull 1470. Indeed, it is almost identical with 1470 in every respect except size. For that reason most authorities assign both to the same species: *Homo habilis*. Richard assigns it to *Australopithecus africanus*, arguing that it is morphologically different from *Homo erectus*. We never claimed that it was *Homo erectus*. To drag us into a peripheral argument by citing fossils of his own that we don't mention, and by failing to discuss our fossils at all, simply dodges the main point: is *afarensis* a valid species or isn't it?

Our overall objection to the Leakey–Walker review was that it criticised our arrangement of the hominids but failed to offer any alternative. If there had been something tangible, any discussion at all of our fossils that we could have responded to substantively, I think something profitable might have come out of the exchange. As it was, it just left us feeling frustrated.

Mary Leakey's review was even more frustrating. She had previously stated in an interview that Laetoli and Hadar could not be

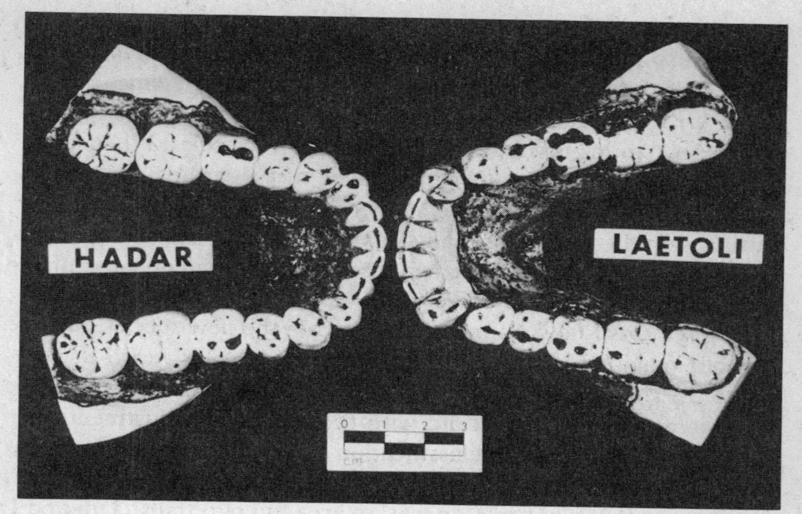

If teeth and jaws are to be the criteria by which hominids are identified, then the Hadar and Laetoli populations must be assigned to the same species. Here are AL-400 (left) the mandible that the little Afar boy found, and LH-4 (right), the best of the Laetoli specimens. The two are virtually identical.

linked sensibly because of their geographic separation and because of their separation in time by about three-quarters of a million years. I was primed to meet her on both these objections: to point out that *Homo erectus* had survived almost unchanged for even longer, and in parts of the world much farther apart than Laetoli and Hadar. I was also ready with 1470, which on Richard's first announcement was a million years older than its counterpart in Olduvai. If he could span a time gap like that without a complaint from his mother, why couldn't I span a smaller one? Finally, I was ready with Richard's assignment to *africanus* of his 1813 skull. It was much more distant geographically from its namesake in South Africa than Hadar was from Laetoli; again, if Richard could have three thousand miles, why couldn't I have one thousand?

Those were not the purest of arguments; they addressed themselves to inconsistencies in Mary Leakey's own position. I would have preferred to talk about the fossils themselves, but again was not given the chance. In fact, I never got a chance to use the time and geography arguments. I have to assume that Mary Leakey had decided not to press time or geography in print – although she had

had a good deal to say about them in interviews – because they simply do not hold up. In any event, she did not repeat them. She hit us instead with a hairsplitting obfuscation about nomenclature, about errors she claimed we had made in naming our new species.

Species naming is subject to the most rigorous rules. When a new one is announced, the namer must make sure that the name chosen satisfies a number of requirements, one being that it in no way infringes on the territory already staked out by a previous namer of a somewhat similar fossil. Further than that, the namer must make sure, if there seems to have been such infringement, that a previously discovered fossil now claimed to fall within the new species does so without making an awkward or contradictory splash with its new name. In short, it should lie more comfortably and more logically with its new name than it did with the old. Such adjustments are constantly being made in classification as new information suggests new arrangements.

Knowing that classification was an area full of pitfalls, Tim and I had done our homework there with great care. We had reviewed all the existing African hominids with respect to the implications of the various names they had been given in the past. We had had to consider *Australopithecus*, *Praeanthropus*, *Paranthropus*, *Pithecanthropus*, *Meganthropus*, *Plesianthropus* and *Homo*. We had reported all our findings to Ernst Mayr in our request to him for an evaluation of the species named *afarensis*. We had even gone so far as to present in our paper nine different possible ways of naming the various species we wished to link together in a tree, and to explain why the one we had chosen was, in our view, the most appropriate. Mary Leakey took exception to four that we discarded, claiming that they should not even have been considered because they did not comply with the rules of nomenclature.

That argument had no bearing on what *afarensis* was or was not. We disposed of it quickly in our rebuttal by citing Ernst Mayr, George Gaylord Simpson, Phillip Tobias and others, all of whom have made statements that support the route we followed.

Although both Tim and I were disappointed with the obliqueness of the Leakey comments, we could console ourselves that since there had been no attack on substance, it might be because the substance was sound. In matters of this sort the scientific journals play an important role. They not only disseminate new information through the steady publication of research papers, but give fellow scientists a chance to attack papers that are felt to be badly prepared, scientifically flawed, or wholly wrongheaded. They also act as screening processes, with the result that very few wrongheaded papers get as far as publication. The journals are edited by experienced people who are dependably shrewd in sifting the startling from the truly nutty. By publishing the startling, they expose it to the criticism that any paper containing original ideas must endure if its ideas are to survive.

Our announcement of *Australopithecus afarensis* fell into the category of the startling. It discovered and defined a new hominid, and it redesigned the human family tree. In so doing, it put a rather constraining lid on some competing ideas. That, I think, is why the Leakeys reacted as they did to our paper: it challenged their long-held view of the specialness of *Homo*.

Louis Leakey had held it. He had ruthlessly pared every competing fossil from the *Homo* line. All the australopithecines were shoved aside, even *Homo erectus*, even Neanderthal Man. While the Leakeys as a family ended by going their separate scientific ways, they did hold to a 'family' view in their attitude toward human origins. Like his father, Richard Leakey preferred to look back and to eliminate the competition. He too felt of all the hominids: if it wasn't actually *Homo*, it could not be a *Homo* ancestor.

That point of view was first made evident to me in 1974, when Richard and his mother visited me at Hadar, got their first glimpse of Alemayehu's jaws and pronounced them *Homo*. I have discussed at length in previous chapters why this was an entirely reasonable conclusion, given the state of knowledge of hominid fossils that prevailed at the time, and given the dramatic inferences about early *Homo* that were made possible by Richard's announcement that 1470 was nearly three million years old. In Richard's shoes, and given such an early date for that extraordinary fossil, I am quite sure I would have done what Richard did. I would have decided that the human line did go way back and that the prehuman or protohuman stage would have to be sought in the still more distant past. With no certified australopithecine fossil from anywhere known to be as old as 1470, I too would have had to eliminate australopithecines as ancestors.

Richard Leakey was made world-famous by 1470. His ideas about human evolution were solidified by it. But when its great age began to erode, he found that very hard to accept. Instead of trying to reorganise his thinking in the face of growing evidence that the date was wrong, he defended the date. Finally he just drew back from it. By not publicly endorsing the new evidence, he has kept the Old *Homo* option open. If he can find older fossils of the 1470 type, or manage somehow to re-establish earlier dates for the ones he has, his view of human evolution can be preserved.

Afarensis slams that door. If Leakey is to accept a nonhuman, extremely apelike creature as the ancestor of *Homo* at three million, then what becomes of Old *Homo* at 2.9 million? There is just no room for him, a certifiable human, in that narrow time slot.

Only if the Laetoli and Hadar fossils can somehow be reinterpreted as *Homo* can the time be expanded again. Do that, and there Old *Homo* is, popping up at 3.0 and 3.7 million. I believe that is why Richard and Mary Leakey fasten so hard on what they see as *Homo* traits in the Laetoli and Hadar collections, and ignore the more primitive ones. I did the same. But when confronted with the entire collection, and with no ideological axe to grind, I was free to change my mind. In truth, I had no choice. I had to.

Following the Leakey viewpoint a step further, if the larger Laetoli and Hadar individuals, best represented by teeth and jaws, are to be called *Homo*, then it becomes necessary to separate Lucy, because she is so embarrassingly un-*Homo*-like that including her would destroy the logic. Nobody has suggested calling her *Homo*. Therefore, Leakey would argue, there must be two kinds of hominid at Hadar, a human and Lucy.

Again, my thought originally; again, a thought I had had to change. I greatly regret that the Leakeys so far have not addressed themselves directly to what is emerging as the most interesting paleoanthropological issue of the 1980s. Perhaps they will in due course. I hope so.

An unexpected thing happened in the summer of 1979. Tim and I had used the three-million-year-date assigned to the Hadar basalt by James Aronson as one of our assumptions in our analysis of the Hadar fossils. Aronson notified us in August that he was changing the date.

It will be recalled that he had emphasised from the beginning that a date of 3.0 million was the minimum that could be accepted, but he had always refrained from saying anything more definite than that. He insisted that he would have to wait for Basil Cooke's final assessment of the Hadar pig fossils to decide whether the basalt fell into the Mammoth period of magnetic reversal of the earth's polarity, or whether it would have to be shoved back to an older period of reversal, the Gilbert. Late in 1978, Cooke released his findings. The pig analysis said that 3.0 million years for the basalt was too young. Stimulated by that evidence, Aronson turned to some exceptionally pure samples of the basalt that had been collected by Bob Walter in 1976–1977 and ran sixteen separate tests on them. All returned an astonishing age of 3.75 million years, plus or minus 100,000.

Radical shifts in dating are usually a shock. This one was a great simplifier. Suddenly the Hadar fossils, in addition to being nearly identical morphologically with the Laetoli ones, became nearly identical with them in age. What a fantastic, blindingly beautiful match! It made the Hadar jaws that were below the basalt close to four million years old! It made Lucy and the 333-site family, which were above the basalt, about 3.0–3.5 million years old.

It was wonderful. But nothing comes free; it raised another problem. It left an even larger hole than before in the middle of the family tree we had drawn. Now there were no securely known fossils on the *Homo* line from anywhere for approximately one million years – between about 3.0 million and about 2.0 million. What was in there?

If our construction was right, any fossils subsequently found from that long period would show evolutionary progress from *afarensis* in either of two directions: either toward *Homo* or toward the later australopithecines. But if something quite different showed up, it would be back to the drawing board.

I did not think that would happen. The fit was too good. It meant that *all* the Laetoli-Hadar specimens were as old as their primitive condition suggested that they should be. That was doubly satisfying. It underscored the similarity of the two collections at the same time that it eliminated any argument that they should be different because they were not the same age.

It also provided some much-needed time for the parent species to evolve in whatever directions its various populations would find it ecologically expedient to take. There would be more time for one line to go from *afarensis*, a near-ape, to *Homo habilis*, a certifiable human. There would be more time for another line to begin the inexorable development of increasingly massive cheek teeth, proceeding through *africanus* to *robustus*. And if Richard Leakey wanted more time to push Old *Homo* back into the past beyond two million again, there was even a spot in there for him. I will accept *Homo* as far back as the fossil record will take him – but on present evidence, only as a descendant of *afarensis*.

Things have a way of straightening themselves out if you wait long enough and work conscientiously enough. The years 1978 and 1979 were tumultuous ones for Tim and me. They had nothing of the excitement of discovery that made the earlier field seasons so thrilling. Instead, they were full of grinding lab work, endless talk, thinking, politicking and worrying. We were just a couple of fresh youngsters taking on the whole paleoanthropological establishment. What right had we to be so bold, to tackle anything as formidable as that?

Whatever right we had, we had done it. On the whole, we were well satisfied with the result. The only thing we had not discussed up to this point was the biggest enigma of all: how did it all get started? What pushed those ancestral apes up on their hind legs and gave them – some of them – an opportunity to evolve into humans?

That question is basic to the entire story of hominid evolution. By supplying some fossils and interpreting them, we had added our bit to the When and Where of erect walking. In future, with more and more older fossils, we might be able to say even more about When, and perhaps even something about How – by catching earlier stages of change in the bones of the pelvis, leg and foot, maybe at just the critical time when an ape that did not yet walk erect was on the verge of becoming one that did.

But the Why of it was something else. Why, of all the mammals that have ever walked the earth, did only a group choose to walk erect? That tremendous enigma stumped us. Paleoanthropology alone could not solve it. It would need the wisdom of disciples that had nothing to do directly with fossils, insights such as a specialist in locomotion like Owen Lovejoy might contribute, because bipedalism has an inherently perplexing component: it is not really the best way of getting around in a hostile world. And yet it is the way our ancestors chose in order to become humans. Why?

PART FOUR

Why Did Lucy Walk Erect?

16 Is It a Matter of Sex?

You don't gradually go from being a quadruped to being a biped. What would the intermediate stage be – a triped? I've never seen one of those.

TIMOTHY WHITE

Man, to put it briefly, is continuously sexed; animals are discontinuously sexed. Man is prepared to mate at any time: animals are not . . . If we try to imagine what a human society would be like in which the sexes were interested in each other only during the summer, as in songbirds, or, as in female dogs, experienced sexual desire only once every few months, or even only once in a lifetime, as in ants, we can realise what this peculiarity has meant.

JULIAN HUXLEY

You might not think that erect walking has anything to do with sex, but it has, it has.

C. OWEN LOVEJOY

'For any quadruped to get up on its hind legs in order to run is an insane thing to do,' said Owen Lovejoy, the locomotion expert. 'From the standpoint of pure efficiency, bipedalism is a preposterous way of running.' Lovejoy had dropped in at my laboratory early in 1980 to discuss locomotion with some of my students. I thought it would be a good idea for me to listen in.

'I mean, it's just plain ridiculous,' he said. 'Even the arguments for it are ridiculous:

' "Man moved out on the savanna and learned to stand up so that he could see over tall grass." Poppycock. It may have helped him see over tall grass when he got there. But if he had to learn to do it *after* he got there, forget it. He never would have made it.

' "Man was a tool user. He had to stand up in order to have his hands free to carry tools and weapons." Ultimately, yes. But originally, rubbish. That idea never did make sense. Now it is exploded by the Laetoli and Hadar fossils. Those animals were bipedal, but that had nothing to do with tools. They were walking that way maybe a million years before their descendants began using tools.

'"Man proceeded through a knuckle-walking stage like a gorilla or a chimpanzee, and gradually worked himself up on his hind legs." Balderdash. The idea that a chimp represents some sort of halfway stage leading to erect walking is idiocy. Knuckle walking is a specialised adaptation to a particular mode of living. It leads nowhere.'

The way to think about locomotion, Lovejoy went on, was not in single cause-and-effect terms, but as part of an overall survival strategy. The paleontologist or the anatomist has a tendency to pick out one feature in an animal, and then identify that animal as a purveyor of that feature: it is a brachiator; it is a biped. That is simplistic. In reality, the animal's locomotor system or its reproductive system turns out to be part of a complex adaptation to a specific ecology. There must be a constellation of attributes working together. In short, one must look at a chimpanzee's total survival strategy before its locomotor system makes sense. Even before that is done, there must be some understanding of what arms and legs are like and how they function. To Lovejoy, the amount of ignorance about how legs work, even among people who teach locomotion, is staggering. He would go back to basics.

Looked at simply, a body is a package of flesh lying on the ground or in the water. If it needs to move to get food, it must have fins or a tail to propel it through the water; on land (snakes excepted) it needs legs. If the only function of the legs is to shove the body ahead, then there will be considerable scraping, friction, and large energy loss every time it moves. That is why an animal that needs to walk or run must have legs that can give its body not only forward motion but upward force to hold it off the ground.

The legs of the earliest walking vertebrates – amphibians and reptiles, and even of newts and alligators that survive today – performed both these functions, but not very well. Their legs stuck out from the sides of their bodies – a poor design for running, as a front view of an alligator makes clear. The sheer effort of supporting a body on legs that stick out to the sides is considerable. It is comparable to what a man would experience if he lay on the floor and tried to do push-ups with his arms extended to the sides instead of under him. That is why an alligator can run neither fast nor far. It proceeds by moving the front left limb and the rear right limb together. It can do this slowly if it wants to walk, or faster if it wants to run, but that is all it can do.

For walking and running, a mammal's legs are a great improvement on the reptile model. They stick down instead of out to the sides, which is why a horse can support itself comfortably on its legs all day, whereas the same effort would exhaust an alligator very quickly. In addition to

standing and walking, a horse can trot and gallop. A trotting alligator is an impossibility.

'Why are alligators so inefficient?' Lovejoy was asked.

'They're not.'

'But you just said—'

'I said they were poor runners. That's a different thing. An alligator is very good at slithering unobtrusively off a mudbank. It is a superb underwater hunter. Go back to what I just said about an animal being a constellation of attributes working together. You have to look at *all* the evolutionary strategies of an animal to understand any of them. A horse is very good at its thing, but its thing is not underwater swimming. Its legs reflect that.'

Mammals' legs also reflect the process by which they achieved better running ability. A mammal's limbs not only have moved directly under the body but have also turned. The hind limb is rotated so that the knee faces forward. The forelimb is rotated so that the 'elbow' faces backward. These new alignments make for a more efficient input of energy into running, and a more efficient consumption of it.

Lovejoy went on to explain that all running is a matter of putting in energy and consuming it. The energy is put in when the hind legs propel the animal forward, and consumed by the front legs as the animal lands. If an animal's body were suspended in space, and a force applied to it, it would accelerate until the top speed attainable by that force was reached. Then, if it were in a vacuum, it would coast forward forever. But real animals do not float in vacuums. When they run they must constantly touch the ground, and each time they do they lose much of the forward motion they have achieved. Therefore, to move the body forward another step, momentum must be put in again. But the body will fall to the ground if the forelimb is not thrust forward again and its muscles used to hold the body up – and so on: energy put in to move the body ahead, energy absorbed to hold it up.

To understand how energy is absorbed by the forelimb, one might regard it as a kind of shock absorber acting to cushion the impact of the body's landing on it at the completion of a step. As the limb joint begins to buckle under that descending force, its muscles act to resist it. That muscular resistance, the act of trying to straighten up the front limb as the weight of the body presses it down – negative energy, one might call it – is what absorbs the energy that was put in by the forward thrust of the hind limb.

To act like a proper shock absorber, the forelimb should not be fastened directly to the skeleton, and it is not. It is held in place only by muscles. If it were attached directly, there would be no way for a muscle to begin smoothly resisting the weight of the descending body, and there would be a horrible jolt as all the energy was absorbed in one crashing contact. By having its forelimb separated from its shoulder (rather than nestled in a bony socket, as the leg is), an animal is given an opportunity during each step in running to absorb energy smoothly.

A mammal's hind limb, by contrast, does not need to be a shock absorber. Its function is to put in energy, not absorb it. Consequently, the top of the thighbone can fit directly into the socket of the pelvis, and is held there by tendons and ligaments. Direct fits are stronger, and would be preferable for the forelimb too, were it not more important that it be an energy absorber. Football players know about the disadvantages of an indirect coupling. They are constantly suffering from shoulder separations when the muscles that hold their arms in place are torn. But the disadvantages of a direct coupling for a shock absorber are even greater. The reader need only jump stiff-legged from a one-foot step to discover that the impact will rattle every bone in the body. With the leg attached directly to the hip and the knee locked, there can be no absorption of energy; there is nowhere for it to go.

Given the preceding model of a typical quadrupedal mammal – limbs sticking straight down, rear ones rotated forward and front ones rotated backward; rear ones attached directly to the skeleton, front ones attached only by muscles – it becomes possible to understand why, when the model is converted to bipedality, it becomes less efficient. That is best shown by an examination of how the energy used in running is applied. If a great deal of it is put into forward motion and only a little absorbed by holding the body up, the result will be a high rate of speed. But if most of the energy put in is consumed by holding the body up, then high speed is impossible. That is what is wrong with bipedality: too much energy devoted to holding the body up.

That is demonstrated by a man walking. In the middle of a step, with one foot on the ground and the other swinging forward, there is some leftover momentum from the previous step that carries the body forward also. That momentum is added to (energy input) when the man uses his leg muscles to straighten out his leg at the knee and ankle, making it longer and thus pushing his body ahead. The body is now out of balance, and the other leg, swinging forward, must be planted to support it. That action eats up most of the forward energy. These components of forward motion and upward motion can be expressed as a vector – a combination of the two forces that shows the man's true forward motion, or speed.

If he wants to increase his speed, the man can tilt his body forward, giving a greater component forward and a smaller one upward. That also achieves the beginning of falling, because he has reduced the component that is holding his body up. Therefore, to keep from falling, the man must apply the upward force more frequently – work his legs more rapidly. That is what running is. When a man starts running he can lean way forward (like a sprinter in starting blocks), but as soon as his legs are going as fast as he can churn them, he must assume a nearly vertical position. He can lean forward again to gain more speed, but if his legs will not go any faster, he will simply fall down. Sprinters lunging at a tape do that. At the very end of a race they hurl themselves forward, and they often do fall after crossing the finish line.

A quadrupedal animal does not have that problem. Its body is already leaning so far forward that it is horizontal to the ground. Most of the energy supplied by the hind limbs is invested in forward motion, and very little of it consumed by the forelimbs to hold the body up. They are already out in front, ready to be planted as a kind of pivot on which the animal rests momentarily as its body comes forward and its hind limbs are gathered for another spring. The most highly developed quadrupedal running machine is the cheetah. It has long legs and a long, very supple backbone. When it starts a

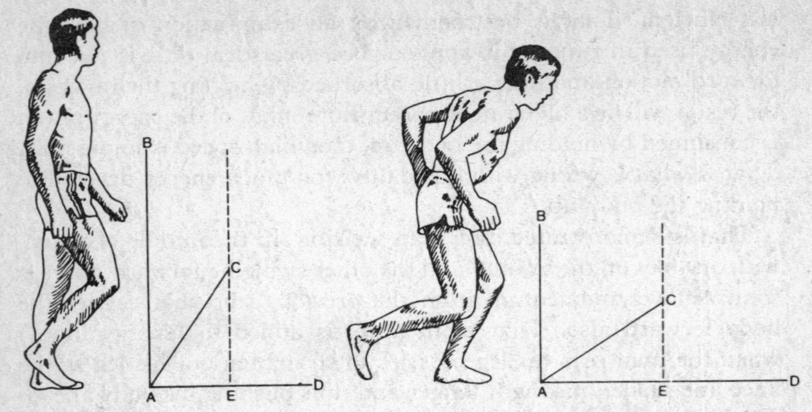

A man walking expends more energy holding his body up (AB) than he does propelling it forward (AD). The combination of those two forces produces a vector (slanted line AC). The vector may be used to measure the man's true speed (AE). When he starts to run, the man leans his body forward and begins to work his legs faster. Less energy is put into holding the body up and more into propelling it forward. The resulting vector slants more in the direction of running, and the distance AE is greater – that is, more speed.

leap forward, its backbone is coiled. As it straightens out its hind legs with a tremendous push, it also straightens out its backbone, and its body is shot forward twenty feet or more in one bound, its forelimbs using only a small part of that energy as they touch down.

'What has all that got to do with the origins of bipedalism?' Lovejoy was asked. 'You've made it clear why quadrupeds can run faster than bipeds. Why did any of them change?'

'To answer that question,' said Lovejoy, 'you must remember that our ancestors evolved in the trees, and then ask yourself what they were doing up there.'

The earliest ancestors of the primates, he went on, were not tree dwellers. They were insect-eating quadrupeds about the size and shape of squirrels that lived and hunted on the ground. But as huge tracts of hardwood tropical forest developed, providing an excellent home for all sorts of arboreal insects, not to mention a great variety of slightly larger potential prey – small frogs, lizards and snakes – some of those early little ground predators were encouraged to climb into the trees after them. It was only a matter of time before the forest canopy was populated by large numbers of small clawed carnivores that spent all their time aloft.

The cheetah's vector shows how the investment of only a little energy into upward motion and a great deal into forward motion can translate into speed, in its case seventy miles an hour. A Galápagos tortoise, by contrast, has to invest nearly all its energy in holding up its enormous shell. Its top speed is less than one mile an hour.

It is one thing to hunt small prey on the ground. It is quite another to do it in a tree. Although some beetles and all caterpillars are so slow-moving that they may be plucked and eaten almost like berries, lizards are speedy and extremely elusive. One that is not caught at first pounce will dart to the underside of the branch and get away. The pouncer must be able to jump on it from a distance, grab it, hold it and – now being in a three-dimensional world – also be able to hang on itself. In short, a small hunter that expects to do its hunting in trees must become a leaper, a grabber and a clinger. The earliest primate ancestors, which are believed to have resembled the insectivorous tree shrews of Asia, apparently began to develop longer and more prehensile digits on their forepaws as they began to aspire to larger prey like tree frogs and arboreal lizards. Their claws evolved into flat nails on the tops of their toes, and the toes themselves began to resemble fingers. The primate 'hand' gradually appeared.

Its owner, instead of being primarily a runner, was now more of a clinger and jumper. In exchange for better catching ability it gave up some of its nimbleness in running about in the branches. No tarsier or bush baby – to mention a couple of small primates that still jump and cling in tropical forests – can scramble about in a tree nearly as fast as a squirrel can. A clawed animal is always better at that than a 'handy' one. A squirrel can also run up and down the trunk of any tree, which is something that a primate cannot do. The clinging little ones stay aloft, coming to the ground only on the rare occasions when they have to change trees. Big ones, like chimpanzees, that do come to the ground regularly can shinny up medium-sized trees, but large trees defeat them; the trunks are too fat for them to get their arms around. To climb a really big tree, an animal must have claws.

That explains why squirrels never developed hands. Climbing and running have always been more important to them than catching. They eat nuts, pinecones and seeds, stationary things they can pick and nibble at their leisure. A primate, by contrast, is an animal that has managed to develop a way of being arboreal and at the same time preserve a way of being predatory.

Other specialisations have gone with handedness. If one is going to jump and snatch, one had better be able to judge distances accurately. If not, one will come up empty-handed at best; at worst, one will miss the branch entirely and fall. The way to precise distance judgment is via binocular vision: focusing two eyes on an object to provide depth perception. That requires that the eyes be set in the front of the skull and facing forward, not on the sides of the head, as a squirrel's eyes are. Primate ancestors developed such vision. Their skulls became rounded to accommodate the new position of the eyes, and with that change in shape came an enlargement of the skull capacity and the opportunity to have a larger brain. At the same time, the jaw became smaller. With hands, an animal does not have to do all its foraging and hunting with its teeth. It can afford a shorter jaw and fewer teeth. Modern apes and monkeys – and humans – have sixteen teeth in each jaw. Their ancestors had as many as twenty-two.

Safe from ground predators, the emerging primates were extremely successful in the trees. Many of them became bigger-bodied and increasingly specialised as they found different ways of gaining a living aloft. Some of them turned more to fruits and berries for their diet. Some ended up eating nothing but leaves. They developed different ways of getting about in the branches. Some of the little ones, like tarsiers and bush babies, continued to jump and cling; they had rather short forelimbs and long hind limbs, and gave up running entirely. One large group, although it had a primate hand, and even a primate foot with a flat sole and prehensile digits, remained quadrupedal. These were the monkeys. They walked on branches in preference to clinging to them or swinging from them. Their backbones were rather long. They were preserving a way of running because many of them were returning to the ground, and they needed a good deal of speed to enable them to regain the safety of a tree in the face of danger. In retaining quadrupedalism, monkeys are distinct from apes.

Apes are designed more for swinging. Instead of walking along branches, they hang from them, sit on them, stand up in them, go up or down hand over hand. They have shorter spines than monkeys – three or four lumbar vertebrae instead of seven – and as a result they are not nearly as good runners. Some of them, in fact, can scarcely run at all.

There are five kinds of apes in the world today. Two of them – gibbons and siamangs – are true swingers. They move by brachiation, by swinging like pendulums from branch to branch. Their arms are exceedingly long, their hands and fingers elongated and specialised. Their bodies are short and light, their legs shrunken. With that design – a minimum of weight at the bottom of the pendulum – they can move remarkably rapidly through the trees, swinging from branch to branch with a success and smoothness that must be seen to be appreciated, often negotiating gaps of ten feet or more. When they come to the ground, which is almost never, gibbons stand erect, waddling along on their short, weak legs, holding out their long arms to either side for balance. A strolling gibbon reminds one of a tightrope walker.

The orangutan is just as arboreal as the gibbon but in an entirely different way. Is hip joint is close to being a universal joint. It can extend its legs down, backward, forward, straight out to the side, and almost straight up. All of its extremities are good at grasping. It is, in effect, an animal with four arms and four hands. Like a huge orange-brown spider, it spread-eagles itself in trees, holding on with any handy foot or hand and reaching out to grab food – mostly fruit – with any other. On the rare occasions when it comes to the ground, it goes on all fours, slowly and deliberately like an old man walking with a couple of canes (as Sarel Eimerl once said), proceeding on the soles of its feet and the knuckles of its hands, although it quite often balls up its hands into fists and walks on them. As a result, the orang might be called a partial knuckle walker.

The gorilla is a true knuckle walker. It is the one ape that has returned to the ground almost completely. All its food – coarse vegetable matter, roots, bamboo shoots, berries – is found there. It has given up running in favour of large size and enormous strength. Gorillas depend for survival on being big and powerful – up to four hundred pounds for an adult male – and by looking fierce. Young gorillas hang and play in trees, but adults are too large and lethargic for that. Sometimes they will squat in groups on very low, very broad branches, but they really prefer the ground. They are extremely sedentary, stay in the same places for long periods of time, and almost never run at all.

The chimpanzee, also a knuckle walker, is the most adaptable of the apes. It is a sort of all-purpose model that reflects its all-purpose approach to living. It eats a great deal of fruit, particularly figs, and is adept at climbing after them in the upper canopy. Its arms are long enough to enable it to shinny up and down all but the largest trees. Its legs, though short, are long enough to allow it to run surprisingly fast. A man chasing a chimpanzee would have little chance of catching it. So endowed, the chimpanzee spends a large part of its time on the ground, living on grubs, termites, berries, insects, buds and roots when the figs are not ripe. It is also, in a small way, a cooperative hunter. Occasionally a group of male chimpanzees will corner a young baboon or other monkey in an isolated tree, catch it and eat it.

Despite their radically different life-styles and markedly disparate bodily adaptations, all apes are potentially erect animals. They are all forest dwellers that make their living in ways that do not require running. By the beginning of the Miocene, about twenty million years ago, during a period when there was an immense band of forest stretching right around the earth in the tropical and subtropical regions, apes were numerous and successful. Monkeys, by

Primates exhibit a variety of modes of locomotion. The tarsier (foreground), a primitive, nocturnal, exclusively arboreal type, proceeds by leaping and clinging. The gibbon (top left), also is arboreal, is a brachiator, swinging from limb to limb like a pendulum. The orangutan (mid-left) is 'four-handed', being able to grip equally well with hands or feet. The gorilla (lower left) is a true knuckle walker. It is a semierect animal that can stand upright easily and often does. The monkey (top right) is a true quadruped. It does not have the potential for erect posture that the apes have in varying degrees. Instead it has preserved the option for fast running, which it achieves with long legs, a long backbone and a hand that goes flat on the ground at each step. Only hominids have ventured to become completely bipedal.

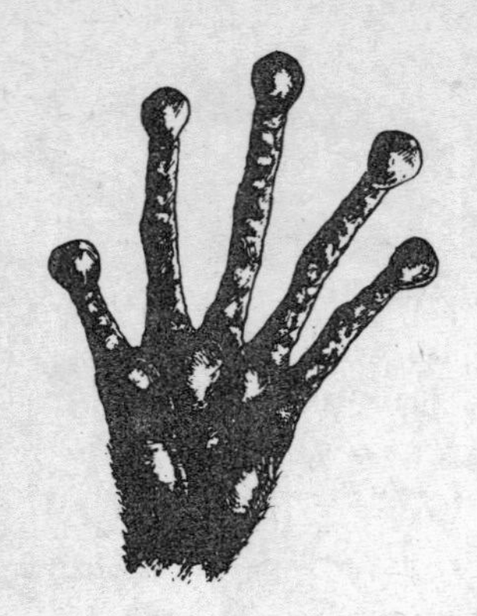

Tarsier

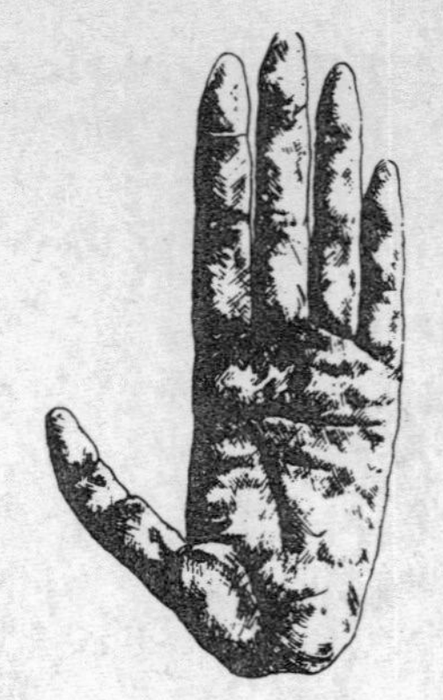

Gibbon

All primates are dexterous, having flexible fingers with flat nails, but the different shapes of their hands reflect the different locomotion and survival strategies of their owners. The tarsier, a leaper and clinger, has extra-large finger pads for gripping branches. The gibbon's hand is almost all long, strong fingers, which it uses as hooks while brachiating. The chimpanzee, a partly arboreal, partly terrestrial animal, has excellent manual dexterity and can even manipulate crude implements. It has a fairly well developed

contrast, were far less so. Today just the reverse is true. Though the decline of tropical forests at the end of the Miocene as a result of changes in climate certainly had something to do with the decline of apes, there were other, subtler forces at work. Lovejoy has attempted to analyse these, and believes he has found an explanation for the origins of bipedalism in the interplay of those forces.

'Thank goodness we're getting back to bipedalism,' muttered somebody. 'I was getting bored with those apes.'

'Bored or not, you have to understand their origins: predators that went up into the trees and managed to find a way of being both predatory and arboreal by becoming semierect swingers with hands, short spines, binocular vision and large brains. Nothing like them had ever been seen before on earth. When some of the bigger ones, like gorillas and chimpanzees, came back to the ground, they were able to carve out new ecological niches for themselves because of their radically new equipment.'

'But they didn't become bipedal.'

'Some did. The hominids – our own ancestors.'

'We know that, but why?'

'Would you like to talk about sex?'

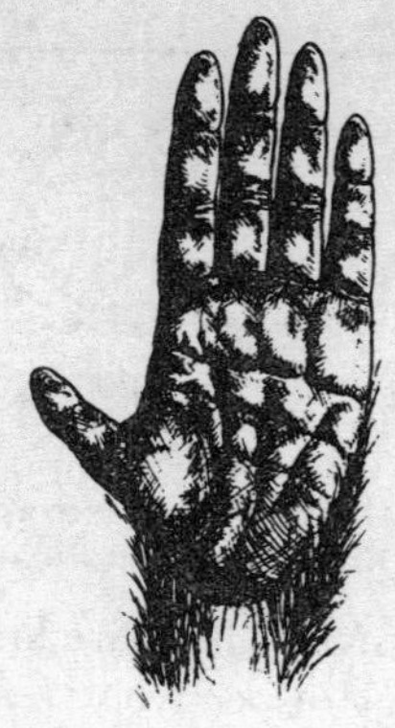

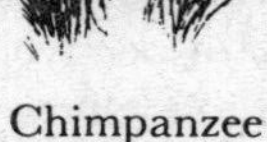

Chimpanzee

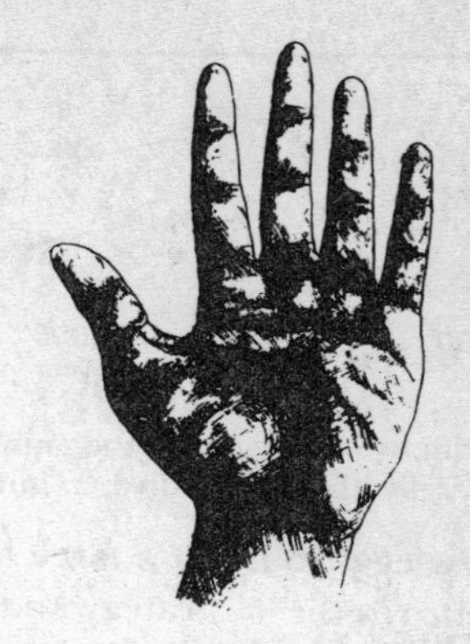

Hominid

opposable thumb, but it is stubby and meets the forefinger along its side, not at its tip. In the hominid hand, the thumb is much larger and is twisted so that it faces the forefinger. This is a logical concomitant to bipedalism and produces a great increase in dexterity. All hominids seem to have had this kind of hand – even *afarensis*, the oldest one now known. Its hand is scarcely distinguishable from a modern man's.

'I'd rather talk about bipedalism.'
'Okay, we'll talk about sex.'

Talking about sex is getting to basics with a vengeance. It recalls the shrewd remark made nearly a century ago and attributed to Herbert Spencer: 'A hen is an egg's way of getting another egg.' Which is to say: look at the survival of the species not from the point of view of the individual but from the lowly all-important view of the gene. Consider that what counts in the long run is the preservation of genes, the getting of another individual like oneself.

Lovejoy, in this matter, thinks like Spencer. Every living individual of every species is nothing more than a protective envelope containing the seeds of propagation, skillfully packaged to maximise the likelihood that it will live long enough to produce more of its kind. To do that, it must eat. To eat, if it is a mammal, it must move around. Therefore an animal's locomotor adaptation cannot be thoroughly understood unless its secual strategy is also understood.

There are two fundamentally different ways in which an animal can function sexually. It can produce a great many eggs, with an investment of very little energy in any one egg. Or it can produce

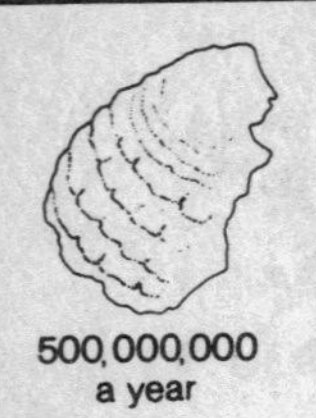

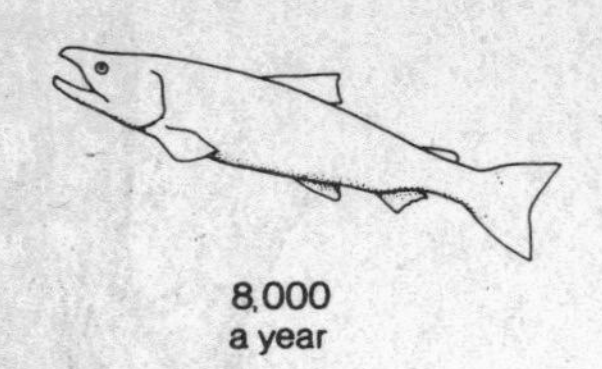

Reproductive strategies in the animal world go all the way from extreme 'r', the strategy that relies on maximum egg output and no parental care, to

very few eggs, but put a large investment in each. These are known to science as the 'r' strategy and the 'K' strategy respectively.

An extreme example of 'r' would be an oyster, which may produce as many as five hundred million eggs a year. The most extreme examples of 'K' are the great apes – the gorilla, chimpanzee and orangutan – which produce only one infant every five or six years. In between is a bewildering mix of 'r' and 'K' among animals that have tried every blend of it. Both 'r' and 'K' work as long as they are not pushed to their limits. Conceivably, an oyster might be able to double its egg output to a billion, but if the energy it had to invest in eggs was fixed, the available energy per egg would be cut in half. That might shortchange all the eggs to a point at which, although there were more of them, even fewer of the very few that now survive would reach maturity. It is clear that pure 'r' is an impossibility, and that it is now being exploited to its limit by such lowly creatures as oysters. Indeed, extreme 'r' can be considered an inefficient way of reproduction. What is the value of producing half a billion eggs if only half a dozen will reach maturity?

The value to the oyster is that it has no other option. It has rudimentary sense organs, it has no brain, it cannot move. It is incapable of taking care of its eggs; it cannot even put them in a safe place. All it can do is pour them out – so it maximises its survival chances as a species by pouring out as many as it *reasonably* can.

Although 'r' works for oysters (they are still with us), it is obviously extravagant. As more highly developed animals evolved, other more frugal options began to be available. When creatures with backbones and brains came on the scene, the beginnings of crude parental care began to be possible. Certain fish make nests for their eggs. Others hold the eggs in their mouths until they hatch and can swim on their own. Some reptiles do even better. An alligator

extreme 'K' in which nearly all emphasis is on care and the birthrate is reduced to a minimum.

makes a large nest mound for its eggs, guards them during incubation, and when the babies hatch and begin squeaking, it hears them and helps them out by tearing the nest apart. But not many reptiles are that attentive, and none protect or teach their young as they grow. That would be left for the even more intelligent and even more 'K'-oriented mammals that succeeded the reptiles.

What dinosaurs did as parents, no one knows, although some of them on the evidence of clutches of fossil eggs found in the Gobi Desert, were certainly stingy egg producers; some 'nests' contain only a dozen or two. Thus, if dinosaurs were no better parents than most other reptiles, they would have been losers in two ways: 'r'-oriented in their probable inability to give their young intensive care, 'K'-oriented in their tendency to lay small numbers of eggs. That could explain their extinction in a world that was beginning to fill up with 'K'-oriented mammals that were better parents.

You don't have to look for sunspots, climatic upheavals or any other weird explanation to account for the disappearance of the dinosaurs,' said Lovejoy. 'They did fine as long as they had the world to themselves, as long as there was no better reproductive strategy around. They lasted more than a hundred million years; humans should as well. But once a breakthrough adaptation was made, once dinosaurs were confronted by animals that could reproduce *successfully* three or four times as fast as they could, they were through.'

'K' is obviously far more efficient than 'r', but it too has its limits. Accidents, predation, seasonal food failure, illness – all take their toll of animals. Losing an infant to one of those hazards after an investment of five or six years is hideously costly compared with the loss of an egg by an oyster. Two or three such accidents in a row could mean extinction for a particular assortment of genes possessed by a particular ape mother, because her fertile years might then be over. For a whole population of apes – an entire gene pool – a slow reproduction rate is as dangerous as it is for an individual. A series of disasters could wipe it out forever.

Indeed, the level of 'K' demonstrated by modern apes is already dangerously high. All of them exist today in perilously small numbers and only in the most favorable environments. Their extreme vulnerability to extinction has been obscured by the role man has played in accelerating their disappearance in recent times. We are constantly being told that we are exterminating apes. That is true, but it is only the final flourish in what apes have been doing to themselves for some millions of years. Left alone, they probably would become extinct anyway. Monkeys, on the other hand, have prospered and multiplied. That difference in success is so dramatic that it is worth comparing the two to see if there is any basic difference between them that might explain it.

It turns out that there are two such differences. Monkeys, as has been noted, are quadrupeds. Apes are potential bipeds. Monkeys also are less 'K'-oriented than apes. They have infants every two years instead of every five or six. They are not quite as intelligent as apes, nor are they quite as good parents, but they are good enough to more than make up for that deficiency with their improved reproductive rate. The modern success of the more 'r'-oriented monkey strongly suggests that the more 'K'-oriented ape has pushed that strategy too far.

Why do something that is bad? This is the second time that question has come up. First it was Why become bipedal if it is inefficient? Now it is Why push 'K' too far if it is dangerous? The two are interrelated, and must be considered together.

First, it should be clear from the preceding discussion of primate evolution that apes, living in forests and spending their time as climbers and swingers in trees or sitting and walking about under the trees, are not inefficient. They have no real need to be good runners. The short spines and semierect postures that they developed earlier in the trees proved to be no handicap to some of them later on the ground, as long as forests were extensive and productive of the kinds of food they ate.

The lesson to be learned from this is that quadrupedalism, although it is apparently advantageous, can be given up – indeed, *will* be given up – if other, more useful adaptations are available. To repeat what Lovejoy said at the start of this chapter: good running ability seems so obviously efficient that to abandon it seems capriciously stupid. But when the total survival strategy of the creature is studied, that advantage disappears. That is what makes possible the appearance of potentially erect apes, on the ground, in Miocene forests. To understand why certain ones went all the way, we must study other needs and other strategies.

The next thing to consider is that all through the Cenozoic, from about seventy million years ago right up into the Miocene, there was among primates a growing trend toward 'K' in their reproductive strategy. It continued to intensify because it worked. As ape mothers became more highly evolved and more capable of solicitous parental care, the survival rate of their infants vis-à-vis the infants of less solicitous mothers was bound to be better.

At this point, the mechanics of a complex feedback loop – in which several elements interact for mutual reinforcement – must be examined. If parental care is a good thing, it will be selected for by the likelihood that the better mothers will be more apt to bring up children, and thus intensify any genetic tendency that exists in the population toward being better mothers. But increased parental care requires other things along with it. It requires a great IQ on the part of the mother; she cannot increase parental care if she is not intellectually up to it. That means brain development – not only for the mother, but for the infant daughter too, for someday she will become a mother. Bringing a large-brained child to term requires a great deal of oxygen and the passage of a considerable load of energy through the placenta – a large investment by the mother. That is because there can be no development outside the uterus of a large-brained child. There are various reasons for this, one being that the neurological system has to be pretty much adult at birth. One does not learn unless one has a good brain to start with; if one is born with too undeveloped a brain, there can be no catching up later on. Therefore the mother has to be responsible for much of that development inside her own body. Since her ability to transfer energy to a fetus is always limited (just as the oyster's ability to invest in eggs is limited), the result of having a larger brain has to be fewer offspring.

To express this in reverse: if one is going to have fewer offspring, one had *better* have a larger brain to take better care of them. That is the feedback principle in action. Each tendency works with, depends on and reinforces the other.

In the case of primate evolution, the feedback is not just a simple A–B stimulus forward and backward between two poles. It is multi-poled and circular, with many features to it instead of only two – all of them mutually reinforcing. For example, if an infant is to have a large brain, it must be given time to learn to use that brain before it has to face the world on its own. That means a long childhood. The best way to learn during childhood is to play. That means play-mates, which, in turn, means a group social system that provides them. But if one is to function in such a group, one must learn acceptable social behaviour. One can learn that properly only if one is intelligent. Therefore social behaviour ends up being linked with IQ (a loop back), with extended childhood (another loop), and finally with the energy investment and the parental care system which provide a brain capable of that IQ, and the entire feedback loop is complete.

All parts of the feedback system are cross-connected. For example: if one is living in a group, the time spent finding food, being aware of predators and finding a mate can all be reduced by the very fact that one is in a group. As a consequence, more time can be spent on parental care (one loop), on play (another) and result ultimately in fewer offspring (still another). The complete loop shows all poles connected to all others (diagram opposite).

'Look at it another way,' said Lovejoy. 'Imagine the feedback loop without one of its essentials, and see what you get. Imagine a baby chimp brought up in a laboratory incubator without any opportunity to learn by playing with his friends. Give him a few critical years in there and then let him out as an adolescent to join regular chimp society. He won't be able to join; he won't know what to do, how to behave. He won't be able to jump around and climb as well as the others. He won't know how to fight or how to avoid fights, how to be respectful to his elders. A young gorilla brought up in isolation doesn't even know how to mate. He has to be shown movies.'

So, Lovejoy continued, there were a number of very good reasons to explain how 'K' could get rolling. It went hand in hand with the improvement and increasing specialisation of vertebrates through time. Frogs are smarter than oysters. Alligators are smarter than frogs. Rabbits are smarter than alligators. Monkeys are smarter than rabbits. Apes are smarter than monkeys. More brains, fewer eggs, more 'K'.

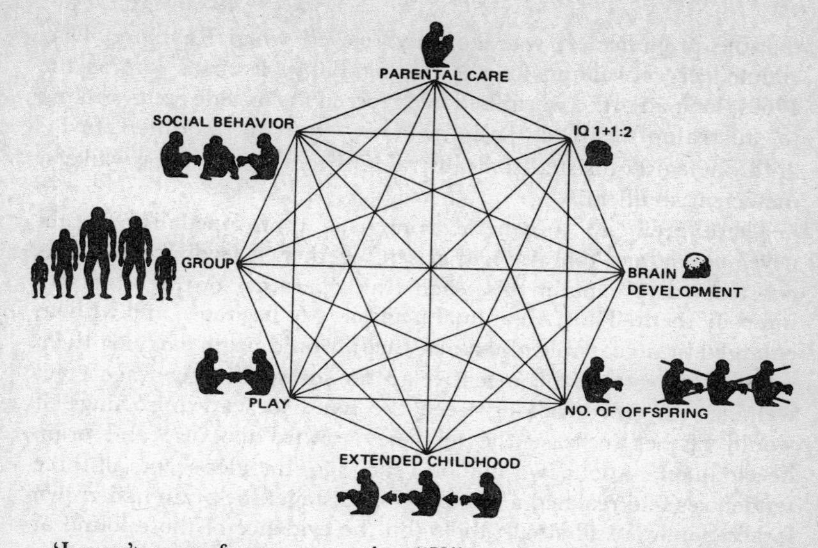

'I won't go so far as to say that "K" *produces* intelligence, but the two are certainly related. Since intelligence is obviously adaptive, it is not hard to see that under the right circumstances, it will be intensified. Those circumstances apparently were right for certain primates. By the Miocene, although we have only the indirect evidence of fossils, we can speculate that intelligence already had gone pretty far among apes. Those fossils give us a picture of animals that were not unlike modern chimpanzees. They were semierect, manually dexterous, and already had larger brains than monkeys. We can also speculate that they were strongly "K"-oriented. It has taken us a while to get here, but we are now in a position to examine what it was in one line of apes that led it to become truly bipedal. In other words, what did "K" have to do with erect walking?'

Bipedalism has plagued anthropologists for about a century. Not long after Dubois discovered the Java ape-man in 1891, he came up with a leg bone indicating that it walked erect. So little did people believe that such a primitive skull could be associated with erect walking that the general reaction was to deny the connection between the two fossil parts: the leg belonged to a later manlike creature and had gotten buried near the skull by accident; nothing as primitive as the Java ape-man could possibly have walked erect.

Accidental association or not, the reaction reflects a long and durable prejudice. It was stormily revived when Raymond Dart claimed erect walking for the Taung Baby. Even as late as the 1960s, long after the world had been forced to concede erect walking to australopithecines, it hung on – but in modified form: anthropologists insisted that australopithecines were poor walkers, inefficient shufflers.

There grew up a general consensus that bipedalism, brain development and tool use had arisen together. A feedback loop was constructed to explain this, each trait operating on the others to intensify them. Thus, an animal beginning to use tools, and with an enlarged brain capable of making them, would begin carrying them around more, and the incentive to do so would encourage erect walking. As erect walking freed the hand for carrying things, it would further increase the tendency toward tool use and brain development. About two million years ago, the idea went, all three tendencies had reached a semideveloped stage: the brain had grown larger but was still small; tools (on the evidence of those found at Olduvai) did exist, but were about as primitive as they could be while still deserving to be called tools; erect walking (if the pattern of hand-in-hand development with the other two emerging traits was to be followed) was also probably just emerging, and was not yet perfected.

In the early 1960s, Sherwood Washburn of the University of California, in the midst of a long and eminent career as a student of primate evolution and behaviour, hypothesised that a clue to erect walking could be found in the knuckle walking of chimpanzees and gorillas. He suggested that this could be an intermediate step on the road to true bipedalism, and pointed out that traces of it could still be found in human behavior: in the knuckle-down three-point stance of a football lineman, or in the way a man leans over a desk, supporting himself on his knuckles and his thumbs. The role of tools in the gradual elevation of knuckle walkers to true bipedalism, Washburn thought, was significant. 'Locomotion and tool use were both cause and effect of each other,' he wrote in 1974.

Other advantages were also hypothesised. When better knowledge of geology and climate suggested that australopithecines were probably open-country dwellers, it seemed sensible to derive another benefit from bipedalism: an erect posture that would make individuals taller and thus able to spot predators better in tall grass. To Lovejoy, not long out of graduate school and already becoming interested in all aspects of locomotion, there was something unconvincing about all those ideas. He could not shake himself free of the conviction that no hominid could ever have ventured out on the savanna as a stumbling, imperfect walker and learned to do it better there. If it had been unfitted for erect strolling on the savanna, it would not have gone. If it had gone, it would not have survived the trip.

The conclusion he arrived at was that hominids *already* were erect walkers when they moved onto the savanna, and that they had perfected that odd gait for reasons that had nothing to do with savanna dwelling – and perhaps not with tool use either. How could tools have been a factor if, as he suspected, true bipedalism had been perfected before tools began to show up in the geological record?

Lovejoy's problem was that he could find no fossil evidence to support his ideas. He argued them on logical grounds with his peers, but without much success. He threw himself into an analysis of the mechanics of locomotion, and came away with a better understanding of it than most anatomists possess. But the 'smoking gun', as he called it – hard fossils that would prove his contention – eluded him. The oldest well-dated australopithecine was only about two million years old, and it could already walk well. Or so he thought; others did not agree. It was utterly frustrating to Lovejoy that australopithecine leg and foot bones were not well enough represented in collections to nail that point down. So he was quite bowled over when I walked into his office early in 1974 with what I said was a three-million-year-old knee joint from Hadar.

Suddenly, erect walking was a million years older than it had been the day before.

'*Maybe* it was,' Lovejoy said to the group in my lab. 'There was just that one little knee joint. I told Don he would have to go back and find me a whole animal. He obliged by finding Lucy. I said, "Okay, get me some variety," and the next year he found the First Family. When I had a chance to study those bones, it became clear that they were excellent erect walkers. Now I had enough confidence to go ahead and try to figure out why. It couldn't be tools. Don's fossils were older than any known tools. I turned to reproductive strategy. That got me thinking about "K".'

Then he turned to one of the students. 'All right,' he said, 'you're a species with too much "K"; what are you going to do to try to improve that?'

'Well, reduce "K" somehow. Monkeys are more successful; be more like them.'

'You mean, go backwards?' said Lovejoy. 'Return to quadrupedalism? Get a smaller brain?'

'Something like that.'

'You can't do that. Not with that feedback loop you're in. Everything in it is pushing the other way.'

'What, then?'

'How about aiming at the most obvious point in the loop: low birth rate? How about having more babies??

'But,' the student protested, 'extreme "K" prevents that. You said so yourself. You gave a very convincing feedback-loop argument.'

'Okay, suppose we change one thing in a way that might let you have more babies.'

'Like what?'

'Like becoming bipedal.'

That quantum leap, that apparent non sequitur, so confounded the group that Lovejoy had to pause again for a review. He referred back to the Miocene, when apes were abundant and poised to explore various modes of living and locomotion. Out of those Miocene forebears came some that were brachiators, some that were 'four-handed', some that were knuckle walkers, and some that were bipeds. All were well into 'K', and all but the biped must be regarded today as failed evolutionary experiments.

If the biped succeeded, what did it do that was different? The answer: it managed to avoid a trap that the 'K'-oriented others fell into: namely, primate birth spacing, the strategy that sentences all apes to a very low birth rate. Their habit is to bring up one infant at a time. However long that takes, however long the mother must devote herself to carrying the infant around, feeding it, looking out for it, that will be the time spent before she can have another. A chimpanzee mother will not become sexually receptive until her baby is about five years old, the practical reason being that she has her hands full with the first one and cannot cope with a second. The biological reason is that nursing and infant care actually inhibit the onset of estrus.

'If you could find a way,' said Lovejoy, 'to speed up that rate so that the babies could overlap more, that would be a way out. How would you do that?'

'She'd have to toss out the first baby sooner.'

'You don't get a toss-out,' he said. 'You get a mother taking care of two together. She divides her energy first among two, then three and four.'

'How does she do that?'

'She moves around less. That way she uses up less energy.' Also, he explained, the more infants are moved around, the more dangerous it is. There is more chance of their getting lost, of their falling out of trees and of their being caught by predators. If the mother is carrying an infant, either it is holding on to her or she is holding on to it. If she goes up into a tree after figs, and reaches for one, she may drop the infant. If she stays in one place on the ground, there is less chance that the infant will fall and less chance that it will be found by a predator, for both mother and infant have the protection of the group. They also have a better alarm system – thirty pairs of eyes instead of only one. They have other animals scouting around for food. When they find it, the mother can go directly to it with a minimum waste of time or exposure to predators. She spends her life in a small territory that she knows intimately. She knows where the climbable trees are. She knows the shortest distance to each in time of danger.

There was a simple rule for that, Lovejoy went on: 'Less mobility is an adaptation. If you can achieve it, you may be able to have infants more often. There is always the tendency to have them more often anyway. The only thing that holds it back is the mother's inability to take proper care of them. She has to make sure she brings up one infant to the point where she knows it is going to make it on its own before she has a second. Staying in one place improves that chance and makes it possible for her to have a second a little sooner.'

There was a flaw in Lovejoy's logic, and my group picked it up immediately: 'Closer spacing raises a problem. The food requirement becomes greater – the mother has more mouths to feed – just when less mobility makes food harder to get.'

'I was waiting for that,' said Lovejoy. 'What you're saying is that the mother must have more help. You're right. Somebody will have to bring food to her. People will tell you that can't happen. They point to baboons and say, "They live on the savanna like hominids; they don't share food." Or they point to chimps and say, "They're like us, and they don't share food either." Well, the proper way to look at that is to examine the reproductive strategies of all primates. They vary, and you also find a varying amount of sharing and

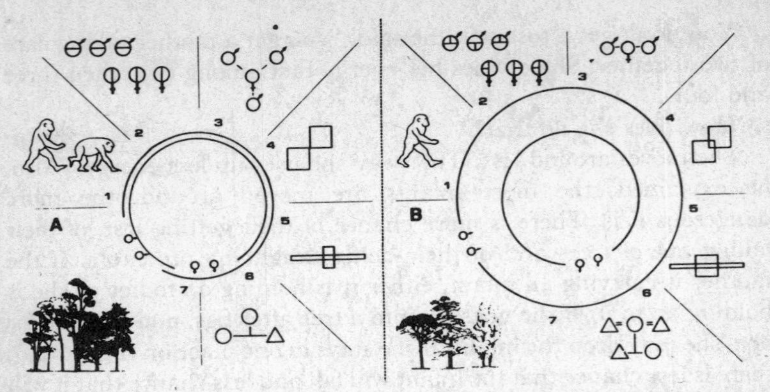

These four diagrams, prepared by Owen Lovejoy, illustrate some of the conditions he believes acted in concert to produce a fully erect hominid. Section 1 of each diagram traces the probable progress of erectness itself. Section 2 refers to epigamic differentiation, the symbols for males and females become increasingingly individualistic as time goes on, and leading to greater selectivity in sexual pairing. This is reflected in the mating patterns shown in Section 3, where males at first mate indiscriminately with females but later settle down as mated partners of specific females. Section 4 deals with the spacing of infants. In Diagram A there is no overlap, in Diagram B there is little, and finally, in Diagram D, there are families of three and four children.

These changes are accompanied by changes in sexual activity (Section 5). The female menstrual cycle is represented by the long bar, with a portion in the middle representing her actual period of sexual receptivity. In Diagram A, virtually all sexual activity takes place during estrus. In the subsequent diagrams the period of receptivity increasingly lengthens, eventuating, in Diagram D, in a condition of continuous sexual receptivity on the part of the female, entirely independent of her estrus cycle. Meanwhile, kinship relationships are changing (Section 6). In Diagram A, the only kinship bonds are between a mother and her children. In Diagram B there is a

caring. Among marmosets, a kind of New World monkey, the male is the primary parental caretaker. It's an ecological adaptation. Marmosets are small, active animals that have worked out a way of beating the primate birth-spacing problem by having twins. Because of their small size and their activity, they must eat a lot and eat often. With one infant, the mother might make it, but not with twins. So the father has to carry them around. The mother forages for herself and comes back to nurse them. It's the same with the owl monkeys and some others. Primates are exceptionally plastic in their behaviour. You find that it runs the gamut from the father being the

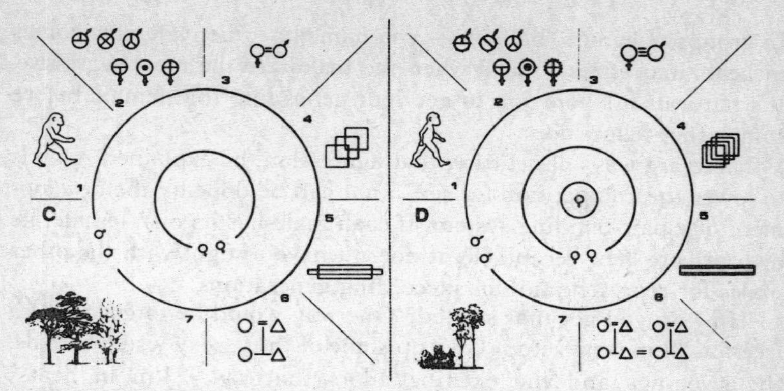

beginning of a male–female relationship, but no true bonding. In Diagram C the male is established as a parent, and a family group is formed. In Diagram D there are many such groups.

All these activities take place within (and probably under the influence of) a changing environment (Section 7), where the tropical forest in which quadrupedal apes casually mated and formed no families gradually changes to open woodland and savanna. The circles in the centers of the four diagrams have to do with foraging and movement. The inner circle marks the core area of the group, where females and infants spend their time. The larger circle represents the limit to which males forage. In Diagram A the two circles are nearly the same size, because the female is not being given any food by the male and must do all her own foraging. In Diagram B the male ranges more widely to remove himself from the core area and leave more food for the female – which she must have because she is beginning to have more infants to feed. In Diagram C, this behavior intensifies. The male is now bringing food back to his mate. To do this he ranges still farther, and she can afford to move around less, thus exposing herself to less danger. Diagram D shows the establishment of a true home base, where the mother can leave her infants in the care of aunts or older siblings. Now she is free to range more widely herself.

principal caretaker to the mother being the exclusive caretaker.

'So you inject some male behaviour into the infant-rearing cycle,' he continued. 'But as soon as you do that, you get other problems. If the female has to have help, then there must be a good level of group cooperation. That's obvious. If a bunch of apes are going to live together as an intelligent, mutually supportive social unit with all the advantages we have already spoken of, then those animals are going to have to get along. There can't be a lot of fighting within the group. Unfortunately, when you mix up males and females – and you spell that S-E-X – that's when the fighting is most apt to start.

In groups of lemurs, baboons – you name it – when a female comes in heat (into estrus), that's when the males are the most aggressive. It's natural. It's your job to get your genes into that female before some other fellow does.'

There are ways of defusing that aggression, he explained. One is to lower the competition for sex. That can be done by the development of a pair-bonding system. If each male has its own female, its own private gene receptacle, it doesn't have to fight with the other males for representation in succeeding generations.

'How do you get that started? One way would be to get rid of a free-for-all mating setup. Give up some of those sexy visual signals, the swellings and the exciting odors that say, "I'm in heat", because that's what drives all the males crazy. But if only half the males are excited by a certain female, then you reduce the potential for fighting by half right away. You can do that by making the sexual symbols more individualistic. Concentrate more and more in the individualisation of sexual responses, and after a while you may find that a particular female is sexually exciting to only a few males, maybe only one. She's just not attractive to others; they ignore her. The development of stimulating systems that are specific to individuals is called epigamic differentiation. I call it being in love.

'Baboons don't fall in love, they fall for sex – sex with any female displaying the signs of estrus. Among baboons all the girls, whenever they come in heat, appeal to all the boys. Baboon societies deal with that explosive situation by working out male-dominance hierarchies. The top, or alpha, male has gotten to the top by fighting with other males, and he stays there by the threat of further fighting. He's the strongest and most intimidating in the crowd, and the one who gets the first crack at the females without any argument. The other males defer to him in a clear order of descending rank. The trouble with that system is that the alpha male's authority is enforced only by his presence. If he goes down to the river for a drink, he loses it. Some other watchful fellow is always hanging around. By the time the alpha male gets back, his chance for having any offspring may be gone.'

It was becoming clear to the group that the level of social cooperation among animals whose sexual aggressiveness is high and whose sexual discrimination is low has to be limited. Introduce pair bonding into such a society, and social harmony can grow. Males can leave the group for short periods of time without forfeiting their chance for sexual representation in the next generation. Male parental care and food sharing become possible. As a result, the females can afford to become less mobile.

'You're now in another feedback loop,' said Lovejoy. 'We can look at it by starting with mobility. If you become less mobile, you can become more bipedal. Why? Because if you don't have to run much, you can afford to be less efficient at it in order to do other things that now begin to have more survival value – like holding and carrying the extra food you need as you increase the number of children you're nurturing. We've just spoken about pair bonding. If your mate is now walking upright, he's better equipped to carry food, and more likely to bring some to you. Meanwhile, you're better off too. As a quadruped you had only one free hand. You walked on one hand and carried with the other. Now you have two hands – one to provision yourself, the other to hold your baby. And it's getting more important that you do hold him because he's losing the ability to hold on to you. His foot is becoming a bipedal mechanism and losing the ability to grasp. A hominid baby can't hold on to his mother with his feet at all; they are no longer shaped for holding. In fact, he can scarcely hold on with his hands; he's too helpless as a small infant. A baby chimp is much better at that, but even it has to have a lot of holding by its mother. An Old World monkey, by contrast, can hold tight with its hands and feet. It has to; its mother doesn't hold it at all. If a baby monkey in a tree ever lets go, that's the end of it. Therefore, when I saw bipedality in Don's Hadar fossils I knew something about the social and reproductive strategies of those creatures. I knew the babies couldn't hold on. I knew the parents had to give their infants considerable care.'

'That's all very well,' said a member of the group, 'but there's something not quite right about these feedback systems. If everything depends on everything else, what triggers it off?'

'There's no trigger,' said Lovejoy. 'Just a very gentle flow. You have a lot of time – maybe hundreds of thousands of years. You have slightly different environments here and there, with slightly different emphases being put on different behaviors. You also have pretty smart animals. That's important. They can observe and imitate one another. They can try things. Not in the sense that they say to themselves, "Wow, we're bipedal, we've got a free hand, let's take some lunch home to the little lady." It's nothing like that. No animal is ever remotely conscious of the evolutionary processes it is undergoing. Those take place through tiny increments. Any slightest bulge anywhere in the loop, if it is adaptive, will have an effect on that part of the loop and, spreading out from there, to all the other parts.'

There was still some dissatisfaction with this analysis. 'Okay, but you've left out something important. Why *should* the male bring food to the female? If so much hinges on that, and if there is no sign that it is practiced by any other ape, how can you just assume it for this bipedal one?'

'I said we'd talk about sex.'

'Right.'

'Let's continue to talk about sex. This whole thing is a reproductive strategy. We're beginning to understand that you don't talk about locomotion just as a way of getting around; it's part of your entire species-survival mechanism, which involves mating and socialising just as much as it involves running and eating and child play. We have seen what pair bonding can do in decreasing aggression and improving the chances for male cooperation. What you're asking – and you betray that with your previous question – is: how does the male *get started* bringing food? Right?'

'Exactly.'

'Maybe he likes the female.'

'Now, wait a minute.'

'I mean it. We spoke a minute ago about epigamic differentiation, about being in love. The only purpose of all the sexual signals that a female puts out when she is in estrus is to attract males, to generate a maximum amount of sexual activity so that the egg she is producing at that very moment will get fertilised. If she doesn't wave those flags pretty energetically, she may miss. If she misses, she will have to go until her next estrus before she gets another chance. All mammals are efficient in advertising their sexual availability. That is what estrus is for: *to guarantee sexual activity*. If there is none, a female will ovulate in vain. For an ape that is perhaps too far into "K" already, that would be a terrible extra strain on its already low rate of reproduction. The ape that comes into heat has got to be impregnated immediately for the good of all apes.'

'But,' said a student, 'if you increase sexual discrimination, you decrease the number of males that are interested in that female, and you lower her chance of being impregnated.'

'Not necessarily,' said Lovejoy. 'You may increase a particular male's interest in her. Even among chimps, although they regularly mate indiscriminately, you find an occasional honeymoon couple that goes off by itself for several days when the female is in heat. That's a little epigamic differentiation right there.'

'Even so –'

'What if you extend the period of a female's sexual attractiveness to a given male? Suppose she flashes the sexual signals for a little longer. Suppose she gradually begins to switch from signals like odors and swellings that are part of the estrus cycle and begins to rely on some permanent features of her body – her hair, her skin, her shape. That's more like what humans do. Think about that for a moment. Human males and human females are sexually attractive to each other constantly, regardless of time or season. That has nothing to do with the estrus cycle. No other mammal is like that. Why are humans?'

No one had an answer.

'Come on. What have we just been talking about? A pair-bonding system was arising among those early hominids and prehominids as a way of keeping a male attracted to a female and ensuring that she be impregnated by him through the strategy of fairly continuous mating instead of a frenzy of it at the peak of her ovulatory cycle. You can't have a tremendous amount of fighting and indiscriminate copulating, *and* have pair bonding and food sharing. They just don't go together. So, what better substitute is there than a system that would bring the male back? And if he brought food, the selective value of that behavior would begin to show up pretty rapidly in the presence of a greater number of his genes in the gene pool.'

'So there's no trigger,' I said.

'None at all,' said Lovejoy. 'I think this is a gradual thing. We can speculate that it took a long time. It starts when a female continues to look sexy for a week or so at the end of her estrus cycle. Slowly that stretches out. Finally the estrus flags don't count; she has permanent ones that keep her man – her hominid – interested in her all the time.

'And she'd better keep him interested, because she's fertile for only about three days. If she copulates once every two weeks, her chance of getting pregnant is pretty low. She can't afford that. It's her job to get pregnant quickly, as soon as she can handle the next infant.'

'What you're saying,' said a student, 'is that bipedalism causes pair bonding and that causes food sharing, and that causes more babies.'

'I'm not talking cause,' said Lovejoy. 'That's the wrong word. You don't get cause in a feedback loop. You get reciprocal reinforcement. I'm saying that all these things happened and that each had an effect on all the others. I'm not saying which came first. I don't know which did. I doubt if anybody ever will know. I suspect there wasn't a first. But they did happen. We know that because you and I are sitting here talking about them. We are the results. And if I list the principal behavior differences between apes and hominids, and ask you to find a better way to explain how we got where we are, I'd like to hear it.' He started listing items on a piece of paper. 'Look at these. They all happened more than two million years ago. They've got to be interrelated:

Hominid	*Pongid* (*Ape*)
Exclusively ground-dwelling.	Some predominantly in trees. Some predominantly on the ground. None exclusively terrestrial.
Bipedal.	Not bipedal.
Pair-bonded, leading to establishment of nuclear families.	Not pair-bonded. No nuclear families except in gibbons.
Increasing immobility of females and young. Possibility of a home base.	Females move to secure food and take infants with them No home base.
Food sharing.	No food sharing.
Beginnings of tool use and tool making.	Tool use absent or inconsequential.
Brain continues to enlarge.	Brain does not enlarge.
Continuous sexuality.	Sexuality only during estrus.
Multiple infant care.	Single infant care.

'What I'm trying to say,' Lovejoy concluded, 'is that you don't go off and adopt what seems to be a stupid way of walking for no reason. If you want to know why Lucy stood upright, you have to consider a lot more than just how she moved around. If you put it very crudely, you might say she stood up so that she could have babies more often. But that would be a direct cause, and we want to get away from that. If you want to put it more accurately, you'd say that in an animal with a complex of specialisations that lead it too far into "K", bipedalism is a way out, a way of reducing "K" again.'

'Or,' I said, 'in a highly intelligent social animal that requires an extended childhood and long maternal care, an animal that already has the potential for upright posture, an animal that is so "K"-

oriented that it can live only in the most favoured environments, an animal that shares its food, that–'

'Yes, yes, all of that,' said Lovejoy. 'It's what I've been saying.'

'I was just summarising,' I said, 'because I think you left out one point.'

'What's that?'

'You made the point that apes, because of their "K" problems, are practically extinct. But you left out that hominids, because they are more "r"-oriented, are everywhere now.'

'I'll accept that, but it's another ball game. It's a late thing. It gets us into the importance of tools and culture as encouragers of survival and population spread. Without them we'd still be stuck on the tropical savanna, somewhat better off than apes, but not all that much better. Our topic today was bipedal locomotion, something that happened to man *before* he became a man. Tools and culture have to do with him *after* he became a man.

'Let me summarise. I think that hominids learned to locomote bipedally in the forest, not out on the savanna, even though that's where you find them living later. They went there as bipeds. They did it – and we learn this from your fossils, Don – close to four million years ago, maybe much earlier. They did it without the benefit of tools. I think the reason was primarily sexual and social. Once they had perfected upright walking, they were free to walk wherever they chose, because upright walking, in itself, is not inefficient. I can walk all day as well as a dog can. It's only when I try to run that the dog has it over me. So, if I don't have to run on the savanna, I can go there. And I probably go there because it's beginning to get crowded in the forest. I'm breeding pretty well now, and it's getting congested in there with those other apes. You know I did move out, because by one or two million years ago I'm spread all over the place.

'If I spread only twenty feet a year – that isn't much – in two million years I've spread four thousand miles. Tools probably helped that spread. Did I use tools to begin with? Probably no more than a chimpanzee does. But now I'm bipedal. Once you get upright, you can expect something different. I've got a much better ability to hold and carry and throw things. There was a period somewhere between four million and, say, one and one-half million when tool using went from being wholly insignificant to being critical in hominid development. I didn't get my start by being a tool-using ape, as people have been telling you for so long. I was a socially and sexually innovative ape who became a biped and, as a result, managed to propagate my kind better than other apes. It was sheer luck that my ability to stand up and use my hands led to a later development of tools and culture, and a still larger brain, and ultimately to four billion others like me. But all those things came as a by-product of what it was really all about: a better reproductive strategy.'

Lovejoy put on his coat. 'You could say his "r" innovations made man possible. Then it would be appropriate to say that tools refined him. Tools were responsible for *Homo erectus*, not for *Homo* himself.' After Lovejoy had gone, I was asked if I believed everything he had said.

'In general, I do,' I said, 'although Owen himself would not ask you to "believe" it. That word is just as inappropriate as "cause". How can you believe something you can't prove? You accept it as probable if it satisfies the data and if it seems logical.'

'But you believe in evolution.'

'Yes, I do. That's because the theory has been kicked around for so long. It has withstood all that violent treatment, and yet it is *still* logical. In fact, the longer you look at it – the more you learn about it – the more logical it becomes. Give Owen's theory that same test of time; if it stands up, then I'll believe it. Right now I'll accept it as the best explanation of bipedalism yet formulated.

Tim White was asked his opinion.

'You know me. I don't believe anything I can't measure, and sexual behaviour leaves no fossils. But in general I agree with Don; it's the best game in town. I think Owen's entirely right when he says the bipedalism developed back in the forest. You don't get it lurching around on the savanna. Stand up to run away when you're just learning how to run that way? That's absurd.'

'What did bring hominids out on the savanna?' I asked.

'As Owen said,' said Tim. 'It was getting crowded back there in the forest. Not only because there were more hominids, but because the forest was shrinking. By the Pliocene – by the time when all this may have been happening – that huge Miocene tropical forest had shrunk way down. What I don't entirely buy is that theory about bringing food back to females and young. You don't need it. I think that carrying objects, carrying children, carrying food was enough of an incentive for bipedalism.'

'You'd leave out the estrus argument?' I said.

'Yes. I've never seen an estrus fossil.'

PART FIVE
Unfinished Business

17 Electron Microscopes, Black Holes, and a Return to Hadar

The excitement of the chase is properly our quarry; we are not to be pardoned if we carry it on badly or foolishly. To fail to seize the prey is a different matter. We are born to search after the truth; to possess it belongs to a greater power.

MONTAIGNE

Although the haul of hominid fossils and the body of knowledge about them have become almost a river in the last twenty years compared with the trickle of the previous century, there is still a huge amount of work to be done. Anyone who thinks the golden years are over is misinformed. I could work for the rest of my life in the Afar and barely scratch it. It is true that there may never be anything to jolt the mind quite the way the Java ape-man did, the first really old human fossil from anywhere in the world; or the way the Taung Baby did, the first two-legged relative that was not a human being. I hear young people say that it has all been done, that all the great finds have been made. That is utter nonsense. Paleoanthropology has never been healthier. It has never commanded a wider public interest. It has never had as generous funding. There has never been so much to do as there is right now. At Hadar alone we have two enormous problems. First, we must enlarge our knowledge of what we have by going back to find more and better material and by completing our interpretation of it. We have done our work on the jaws and teeth, but there is a long way to go on the rest of the skeleton.

Second, by placing a new species at roughly 3.5 million years, we have emphasised the existence of two deep black holes, one on either side of it. The big job of the 1980s is going to be opening up those holes and shining some light into them. Those holes ask:

What happened between 3.5 and two million?

What happened before 3.5 million?

Any way you look at them, those are questions of the first order.

They can be answered only by a return to the field. Meanwhile, we still have the Hadar fossils to finish up.

The work on the Hadar collection has not gone evenly. There has simply not been enough time or enough qualified personnel to do it. Tim and I have done the jaws thoroughly, but that is because teeth and jaws are our speciality and because our identification of a new species was based on teeth and jaws. But the rest of *afarensis* lags. Still hidden in the various parts of its skeleton are many answers to the how, the why and the when of ape-to-hominid transformation. Detailed studies of those other parts are now under way, and some very large questions are still unanswered. For example, Bruce Latimer, one of my graduate students, is, as I mentioned earlier, working on the *afarensis* foot, the world's oldest clue to the origin of erect walking.

The foot in question is unique – there is nothing like it in any other fossil collection anywhere. It consists of thirteen bones from a single individual, extracted by Latimer and White from one chunk of matrix. Getting those pieces out and cleaning them took several months. Getting acquainted with them has taken even longer. They consist mostly of phalanges (the long bones at the base of the toes) and metatarsals (the bones in the foot to which the phalanges attach). There are five pairs of these, showing exactly how those bones fitted together – again a unique collection.

Looking at those small, knobbly, rough, blackish objects, one wonders what on earth may be inferred from them. But to someone like Latimer they say a great deal. They say, first, that Lucy and her kind were fully erect, with feet that were functionally the same as those of modern humans. In fact, they are so much like modern feet that they leave unanswered the question of how they got that way.

Thinking about that – about the process of change – one must wonder, first, what the ancestral foot was like. There one can only speculate; no ancestral foot exists. The nearest model is probably the chimpanzee foot, there being good grounds for suspecting that it has not evolved much. In many respects the chimp is the least specialised of all the apes. More and more, as we become better acquainted with pongid evolution and life-styles, we become impressed by the chimp's *lack* of oddity. It is the general-purpose ape, the one to which we look with increasing confidence for anatomical characteristics it may have inherited from a hypothetical

Selati, the Afar guide on the 1980 survey, watches for marauders from the top of a Land-Rover.

common ancestor to all apes.

A second reason for looking at the chimpanzee foot is that it is almost identical with the gorilla foot. Since the two animals have diverged considerably in other respects, it is interesting that in their feet they have not. This further suggests that the foot they now share must resemble fairly closely the one they inherited.

All in all, Latimer feels comfortable in accepting the chimpanzee foot as a reasonably faithful prototype from which he must derive a hominid foot. In other words, he must account for the evolution of a partially grasping foot into one that can only walk, that serves as a platform, that gives a good push-off in striding, but that cannot grasp at all.

'What would I have to do to make that change?' he asked recently. 'First I would have to lengthen the big toe. Then I would have to bring it in line with the other toes instead of having it stick out like a thumb. Finally I would have to rotate it so that it faced downward instead of facing sideways toward the fingers, again like a thumb.'

In Lucy, he went on, all those things have happened already. Does that mean that Lucy's foot is like a modern one?

Generally, yes. Functionally they are alike. Both provide a modern stride and a modern footprint. But structurally they are a little different. The *afarensis* phalanges are arched, and proportionally a good deal longer than those in modern feet. They might almost be mistaken for finger bones. The arching was for the same reason that a bridge is arched: to make them stronger and better able to deal with the pull of very large muscles whose presence is betrayed by markings along the sides of the phalanges.

From this evidence Latimer concludes that *afarensis* was an exceptionally strong walker, and that its elongated toes may have been of service to it in moving over rough stony ground, or in mud, where some slight gripping ability would have been useful.

Although similarly curved phalanges and muscle markings are found in the chimpanzee – reflecting the chimp's ability to climb trees – Latimer warns that this does not mean that *afarensis* was a tree climber too. Indeed, examination of the joints of the *afarensis* toes shows that assumption to be wrong. The metatarsals that abut the phalanges are flared at their ends, which severely inhibits the movement of the phalanges. Lucy probably could not wiggle her toes any better than a modern woman, but she could walk at least as well – all day long, probably, without tiring. Whether she could run as fast is debatable. The entire *afarensis* skeleton suggests extreme ruggedness, compactness and strength, rather than speed. By comparison, a modern human skeleton seems attenuated and brittle. We are lighter, slenderer, faster, but far less powerful for our size, and certainly less durable.

On the wide-open question of how some of the *afarensis* foot bones were shaped for bipedalism, there seems, fleetingly, to be a clue. The metatarsal joints of a human are broad on top, narrow on the bottom. In chimps the reverse is true. *Afarensis* shows an in-between condition, and it is tempting to conclude that this is a bone in transition; that the loss of toe dexterity may have come about gradually as a result of this change in the shape of the end of the metatarsal. Again, warns Latimer, that may be wrong. The chimpanzee foot is one thing, designed specifically to help chimps do what they do. The human foot is another, again a specific design for a specific creature. The *afarensis* foot, says Latimer, may be yet another. It is not necessarily something in transition, on the way from one evolutionary plateau to a higher one. It may simply represent the best design for a hominid living the life it did in the place it did. He does not yet know how fast the hominid foot evolved, or when, or exactly how. For us to learn those things, he says, more and older bones will have to be found.

At a desk close to Latimer's in the lab, Bill Kimbel has been working for a year on a juvenile partial skull, one of the 333-site finds. In 1979 he took a cast of it to South Africa to compare it with the Taung Baby, something that had never been done before because there was nothing to compare the Taung Baby with. Of all the fossils found in South and East Africa since 1924, there is still no other skull of an australopithecine child.

Kimbel's comparison was bound to be interesting to Tim and me. We were sensitive to Richard and Mary Leakey's objection to our naming of *afarensis*, also to Loring Brace's contention that it resembled *africanus* too closely for comfort, and to remarks in the same vein by Phillip Tobias, who, as a young man, was Dart's assistant and is now professor of anatomy at the University of Witwatersrand in Johannesburg.

Knowing that juvenile chimpanzees, australopithecines and humans look more alike than their adult forms because of differential rates of growth in different parts of the head, Tim and I reasoned that if the 333-site child turned out to be distinctly different from the Taung Baby, then the case for *afarensis* as a separate species would be strengthened. We were delighted when Kimbel returned to report that there were many morphological differences between the two and that a taxonomist would have little trouble in distinguishing between them.

Meanwhile, at Kent State, Owen Lovejoy continues his work on Lucy's pelvis and leg bones. Lovejoy is an unusually thoughtful and imaginative man who says, 'When you live with fossils twenty-four hours a day, you find yourself asking with increasing force the same two questions: what and why?'

Lucy's pelvis gave him much to think about. He spent about a year reconstructing it. When he was through, he realised he had something whose shape and proportions needed explaining, because previous ideas about pelvic shape, erect walking and infant development did not fit Lucy.

The standard explanation had always been that in its evolution to the hominid condition the pelvis had to be enlarged to permit the passage of large-brained babies through the pelvic girdle at birth. That theory prospered before the discovery of Lucy, who has a humanlike rather than an apelike pelvis, although she is not large-brained herself nor very large in the pelvis either. In fact, her pelvis is very shallow from front to rear and is large only from side to side – a peculiar, extended oval shape.

As always, Lovejoy attacked this problem by asking himself questions about mechanical function: in this case, what was the relationship between the pelvis and the muscles that moved the legs? His first step was to look at the chimpanzee pelvis and note that the ilium (the large flat bone that sticks out to either side of the lower spine and to which the abductor muscles that move the upper leg are attached) is well placed for a knuckle walker but not for a bipedalist. A chimp can walk erect, and often does for short distances, but it fatigues quickly because the abductor muscles leading down to the leg from the pelvis blade are very badly placed for that kind of striding. To make those muscles function well, the ilium, instead of sticking out to the side as it does in the chimpanzee, must be swung forward to make more of a box, or girdle, of the pelvis as a whole. That shortens the abductor muscle and also gives it a better working angle.

Turning to Lucy's pelvis, one immediately sees that the blade of the ilium has turned in just the way one might expect if the mechanical requirement were to be to provide better muscular attachment for erect walking.

But nothing is ever that simple. Change the shape or the position of one part, and other parts are affected. When the blades of the ilium are swung forward, they squeeze the entire lower abdomen. Therefore they must become larger and longer as they swing. That too has happened in Lucy, helped along by a considerable enlargement of the sacrum. The result: efficient walking and an adequately large abdomen – and a peculiar, elongated pelvic girlde which no longer seems peculiar once this explanation is offered.

Lucy, in short, demonstrates the first of two steps that must be taken in the evolution of apes into humans. Her pelvis has evolved sufficiently for her to become a biped. But she has not yet taken the second step: further evolution of the pelvis to permit the birth of large-headed infants.

Lovejoy does not have a good explanation for the condition of Lucy's hand or arm. Compared with a human arm, hers is rather long, so he again asks the question: what did she use that long arm for? Among apes, long, strong arms are essential for climbing. But Lucy's sexual strategy has turned her into a biped. Did she climb at all? If so, how much? Logic suggests that she may have done a good deal to escape predators or to find food. But if she did, how does one

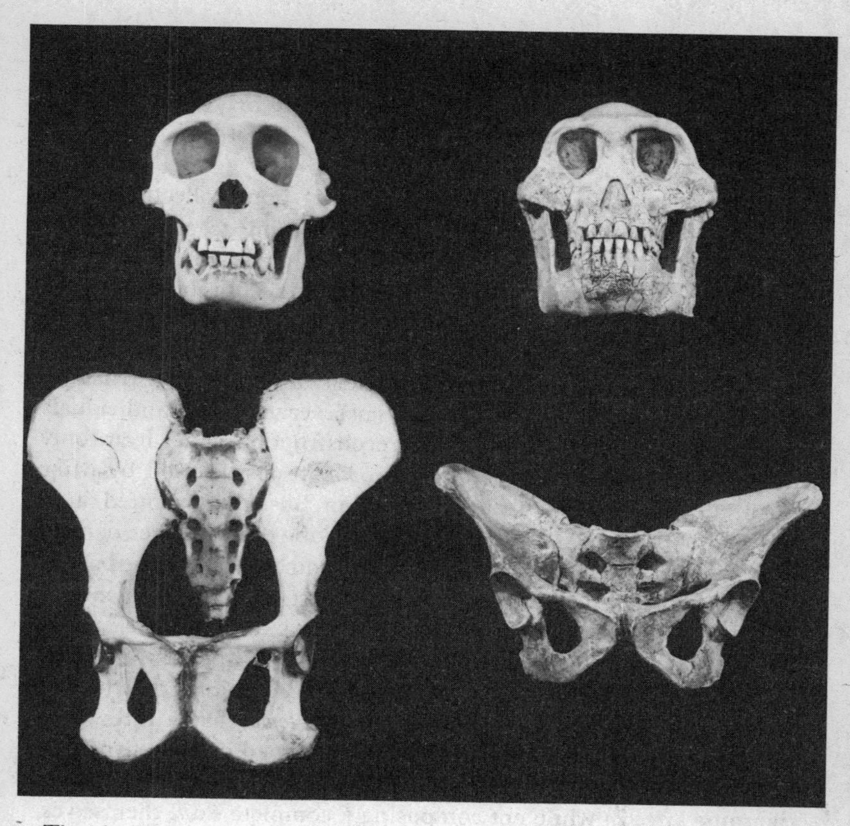

The idea of *afarensis* as a creature with a manlike body and an apelike head is given weight by this comparison. At left is a chimpanzee. Its eyebrow ridges are prominent, its crown low, its brain small, its face prognathous. *Afarensis*, at right, has a skull of very much the same cast but with one compelling difference: it has manlike teeth. Below the neck, however, the difference is overwhelming. The chimpanzee has the pelvis of a quadrupedal ape, whereas *afarensis* has one that is almost the same as a modern man's.

explain her hand? It should have long, apelike fingers, but it does not. They are rather short. Furthermore, although it has the fully opposable thumb of humans, the muscles at the base of the thumb appear to have been small. That means that precision gripping between thumb and finger was probably excellent, but that power gripping which involved the thumb and the entire hand was poor. This is exactly opposite to what scientists hitherto have thought.

Prevailing opinion has been that the power grip came early, and that the precision grip was perhaps the last adaptation of the fully evolved human hand. Lovejoy admits that it may take years to unravel that problem.

One of the most serious gaps in the Hadar bone collection is the absence of a complete *afarensis* skull, something to rival Zinj, or 1470, or Mrs Ples – each of which produced great jolts of astonishment at the sight of something relatively whole, entirely new. In the fall of 1979, Tim White set out to fill that gap. He decided to try to put together an *afarensis* skull from parts in the Hadar collection. This turned out to be a long and hideously difficult reconstruction, and ended up employing several parts of several different individuals – although the great bulk of the reconstruction came from only three. A vital piece was the original 333-site fossil found by Mike Bush. It will be remembered that when Bush first spotted it, it revealed nothing but a couple of teeth sticking out of a piece of rock. The rock was brought back to Cleveland, and the teeth turned out to be attached to a larger piece of bone, part of a face, which was uncovered by a technician in the lab after six months of cleaning. The second important piece Tim used was a well-preserved partial mandible. The third was the crushed 'weird cranium' that he had looked at before and sworn could not be *Homo*. These three were selected by Tim and by Bill Kimbel, who helped in the reconstruction, for two reasons: 1) they came from adult individuals that were the same size; 2) while not composing a complete skull themselves, they provided enough material to convince the two men that the reconstruction could be made.

Tom and Bill's first job was to 'uncrush' the cranium, which had been badly squashed, broken into many pieces and then cemented into its unnatural shape by rock crystal. They took it apart piece by piece – 107 in all – got rid of all the encrustation of rock, and then were able to put the cranium together again, this time in its proper shape. This took four weeks of uninterrupted work, and had to be done over and over again before Tim was satisfied with it. A plaster cast was then made of it. This formed the basis of the reconstruction, to which other parts would be added.

Next, they turned the half-mandible into a full mandible by carving and casting in mirror image the missing teeth and parts of the jawbone from existing ones on the other side of the jaw.

With the jaw complete, it too could be set into the master cast, because the upper hinge of the jaw attaches directly to the skull. Now the cranium and jaw were joined, and thus an accurate position for the lower teeth was obtained. From that they could put Mike Bush's fossil face in place by setting its few remaining upper-jaw teeth in their proper position on top of those in the lower jaw.

Finally – although there was no direct connection between the face and the pieces of cranium above, because the forehead was missing – it was still possible to complete the upper face from inferences about the size, shape and position of the eyes, derived from a section of lower eye socket that luckily had been preserved in the 'weird fossil face.'

When he had everything in place, Tim's original idea about the weird fossil was confirmed. It certainly was no *Homo*; anybody accepting the reconstruction would have to accept that. Rather (forgetting its hominid teeth), it looked very much like a small female gorilla.

Tim's reconstruction, one of the most conscientious and painstaking ever undertaken in paleoanthropology, was shown to several colleagues one afternoon late in December. They were astonished at the way he had managed to put it together, the more so because they had not seen it in its various stages. Tim had removed himself from the laboratory and for the last weeks had been working in seclusion at a makeshift table in my cellar. There he spent his days adding plaster and carefully scraping it away, doing this again and again – red-eyed, his clothes white with scrapings – until he was satisfied. Finally he took it back to the lab and unwrapped it. To his colleagues he identified all the specimens he had used. He explained all the cross-fits he had had to accommodate. It was a moment of immense gratification to him: months of work all coming together in an extraordinary-looking skull, all complete except one bit of cheekbone that remained to be filled out.

He returned to the lab the next morning to do that, turned his back momentarily – and the plaster reconstruction rolled off the table, smashing itself to pieces on the floor.

Tim was so undone by that catastrophe that for a while he could not even bring himself to pick up the pieces, let alone try to fit them together. Finally he telephoned me, and I hurried to the lab. Together we looked at the fragments scattered on the floor, at bits of powdered plaster that were forever beyond restoration.

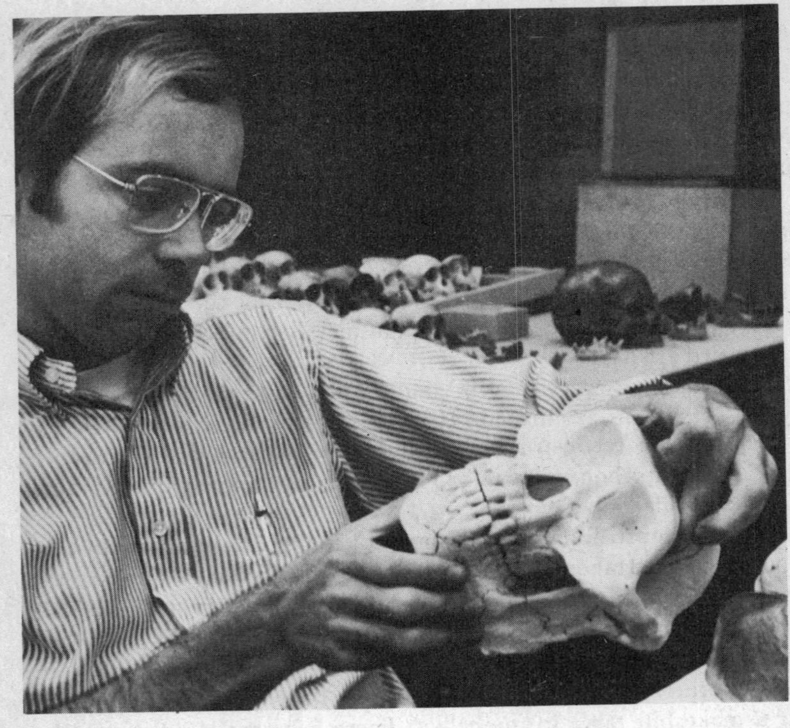

Tim White made this reconstruction of an *afarensis* skull, using parts of several individuals. The section at the back of the skull is the crushed cranium mentioned in the text. It contains 107 separate pieces.

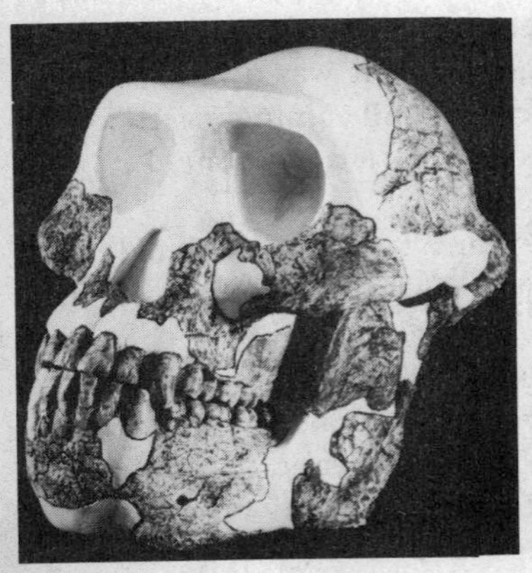

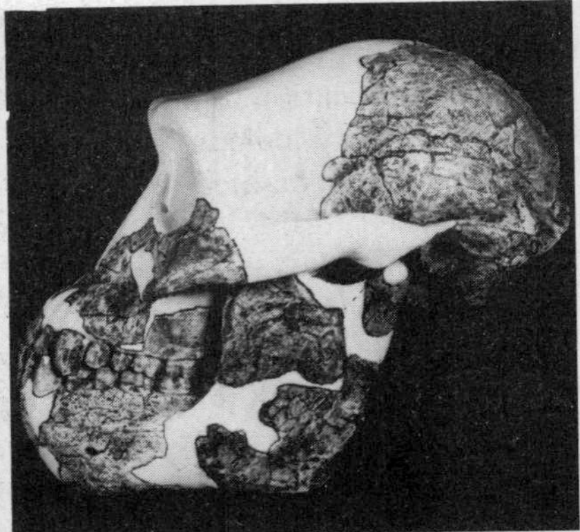

'Leave me alone,' said Tim. 'I can't talk to anybody. I don't know what to do.'

'You'll put it together again,' I said.

'I won't. I can't.'

I reminded him of the day that Thomas Carlyle finished writing his lifework, *The French Revolution*, and of how the maid mistakenly threw the whole manuscript into the fire.

'To hell with Carlyle,' said Tim. He slammed the door to the casting room, leaving me outside.

After a while, he began picking up some of the more salvageable pieces. By the end of the day, he had begun putting them together. Eventually he reconstructed the entire cast.

It was later made the subject of a series of drawings by the anthropological artist Jay Matternes. Matternes is an anatomical expert who relies on his knowledge of anatomy to build up muscles, tissues and features from clues supplied by fossil bones. His work on *afarensis* and on other early hominid species gives a more plausible look to them than any other, although he confesses that there is no way of telling just how a nose was shaped or how hair was distributed over a face, or how large females' breasts were. Whatever the case, Matternes' studies of *afarensis* repeat what the bones say: smallish, spare, exceptionally powerful human bodies topped – as Sherwood Washburn has remarked – with the heads of apes.

Overall, a hominid? Yes, because it walked erect. But a transitional one, with ongoing riddles about many of its features. Another decade of the laboratory studies now under way should provide answers.

'You have to be patient,' says Lovejoy. 'Sometimes you can't hit solutions directly. You have to come at them from an angle, maybe with new techniques that haven't even been thought of yet.'

Techniques will certainly improve, and they will certainly come from surprising directions. Dating by radioactive elements, that epochal and totally unexpected shaft of light that illuminated the 1960s and 1970s, is a spin-off from atomic studies of the preceding thirty years. Now we may be on the verge of another breakthrough that will make such dating not only more reliable but more widely usable. Today, in many fossil localities, there are volcanic samples that could be used for dating if they were not badly altered or badly contaminated. A way of purifying them and making them useful has

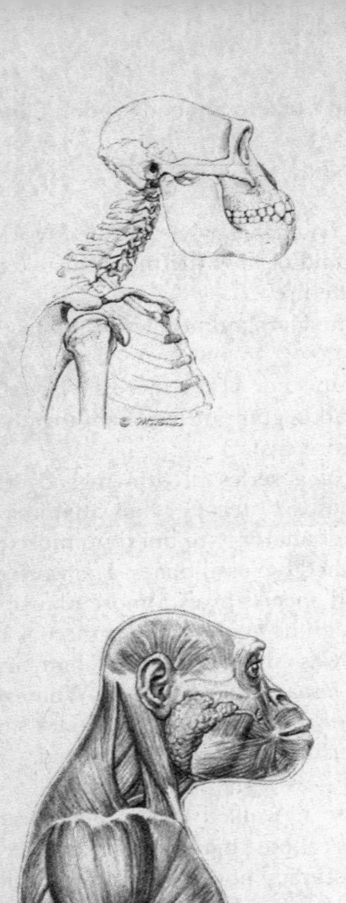

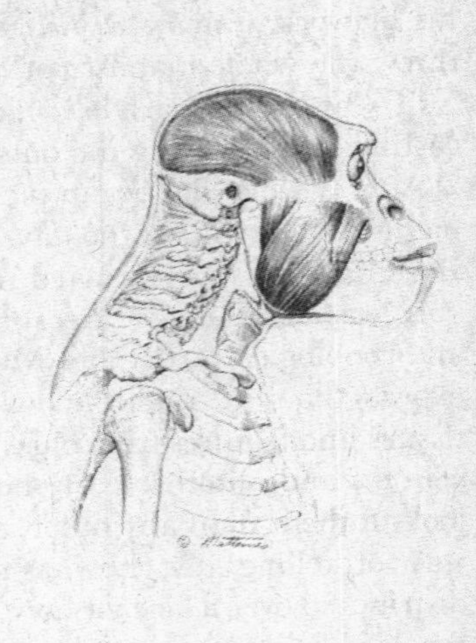

Using White's reconstruction as a model, the scientific illustrator Jay Matternes made these drawings to show what *afarensis* probably looked like. Matternes has specialised in this work for twenty years and brings to it a highly refined knowledge of anatomy. His first step was to determine the attitude of the head, which he obtained by fitting it to a horizontal line running through the bottom of the eye socket and a point at the back known

as the porion. This produces a forward-jutting face which requires elongated spinal vertebrae and a thickly muscled neck to support it. Next, the muscles are laid in, giving the head the beginnings of its eventual shape. A detailed anatomical explanation of the steps taken to obtain this restoration will be found in the Appendix on page 382.

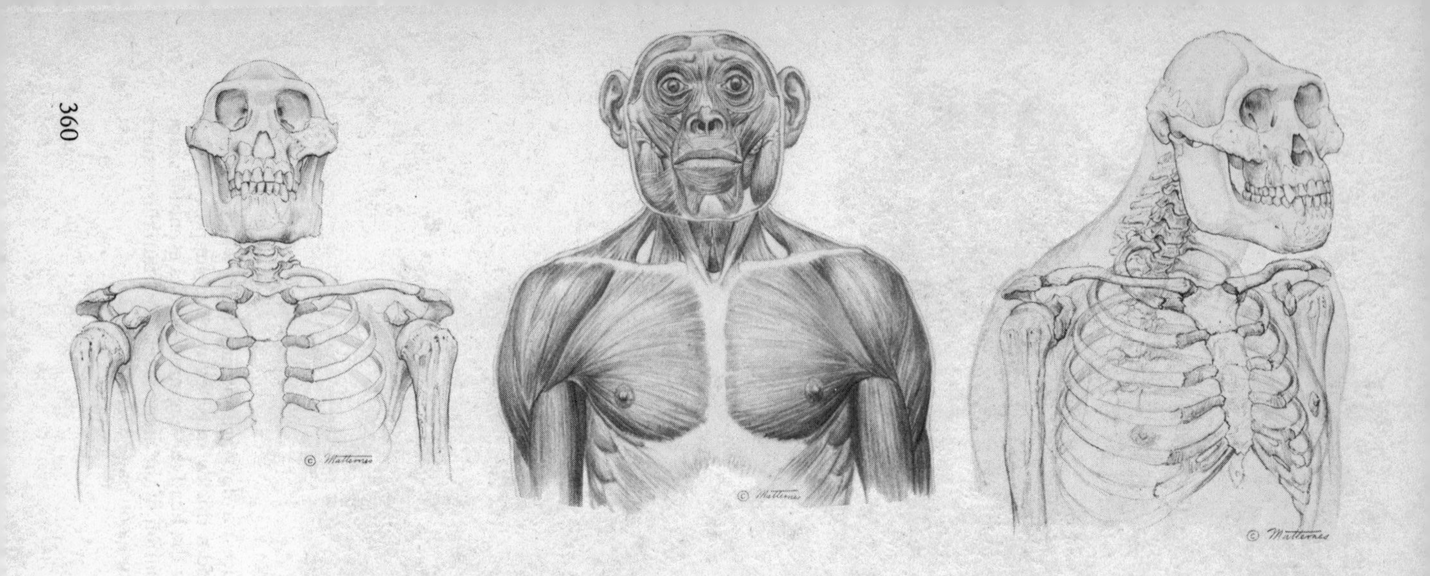

A front view of *afarensis* continues the restoration. The eye is positioned in its socket. The mouth is drawn closed, with the lips meeting just below the line of the upper incisors. Since the corners of the mouths of apes meet just back of the canines, Matternes has done the same for *afarensis*. He has given his specimen a fleshy earlobe – hominidlike rather than apelike. He has concluded that the skin of *afarensis* was dark and rather sparsely covered with hair – on the assumption that an open-savanna animal in hot country would have to have a dark skin for protection against sunlight, but that the dark skin would soak up a great deal of heat. That would require an efficient sweat-gland system, which would preclude the retention of dense body hair. A final three-quarter view of *afarensis* completes the restoration.

been a dream of daters for years. It has preoccupied Bob Walter, James Aronson's fission-tracking associate. Walter has learned that every volcano in the world writes its signature in the exact combination of elements that are fused in its lava at particular temperatures and under particular pressures at the time of eruption – not to mention the nature of the raw materials that are deep in the earth at the particular spot where the eruption takes place. As a result, no two volcanoes are exactly alike; each leaves its chemical fingerprint behind. Walter is working on ways of identifying those fingerprints by the analysis of single grains or crystals of ejecta from volcanoes.

If he can do this easily and accurately, he can then purify a sample by identifying and throwing out all the 'bad' grains. What a boon that would be. At Hadar it could lead to the accurate dating of those badly contaminated tuffs, the BKT-1 and BKT-3. They could add a priceless sequence to the dates now obtained from the BKT-2, the only one that can be used under present methods.

One of the troubles with the famous KBS tuff at Lake Turkana is not only that it is badly contaminated but that it is also not located where the fossils are. Therefore, in addition to doubts about the accuracy of its date, there has been the added problem of figuring out, by means of matching stratigraphic patterns, the age of fossils that have been picked up as much as twenty miles away. In between, in a number of spots at Lake Turkana, are traces of other volcanic events of various ages. If those can be purified to yield dates, the Turkana geology, which is ferociously complex, will become more legible.

Just as the potassium-argon machine and the mass spectrometer have given unexpected insights into dating, another instrument promises to reveal information about what hominids ate and how their jaws moved. This instrument is the scanning electron microscope. Like any microscope, it enlarges things dramatically–as much as several thousand times if desired – but its greatest virtue lies in its ability to provide remarkable depth of field. Usually when one looks through any form of high-powered magnifier, one can see only a razor-thin layer. If the skin of a paramecium is brought into focus, the hairs that stick up from it are totally blurred, and vice versa. A scanning electron microscope, because it does not use light at all, but a beam of electrons, which has a much shorter wavelength than light, provides a view of the paramecium in the round. Its top, its sides and its hairs protruding in various directions are all in sharp focus.

This marvelous instrument has been used lately to provide enlarged views of the surfaces of fossil hominid teeth. Dr Alan Walker of Johns Hopkins has recently concluded that the polishing effect he finds on the teeth of robust australopithecines and modern chimpanzees indicates that australopithecines, like chimps, were fruit eaters.

That news comes as a surprise. Everything we have learned about australopithecines – that they were ground-dwelling, bipedal, savanna-frequenting creatures – suggests that they were omnivores. They may have eaten fruit, and undoubtedly did in quantities when it was in season, but they probably ate a great many berries and seeds and roots and tubers, and a good deal of dirt and sand along with those things. If they were primarily fruit eaters, as Walker's examination of their teeth suggests they were, then our picture of them, and of the evolutionary path they took, is wrong.

More recently, a second and more ambitious scanning-electron-microscope study has been made of hominid teeth by Al Ryan, a doctoral candidate at the University of Michigan. Ryan began his study with a larger control base than Walker had used. He examined the teeth of a large population of Indians dug up from burial mounds in the Midwest. He then examined the teeth of living Eskimos. In both he found evidence of micro-flaking, the chipping off of tiny bits of enamel as a result of the use to which both those peoples put their teeth. They used their jaws like vises, gripping objects in their teeth while weaving baskets, cutting hides, tying knots, holding pieces of wood and bone. Eskimos today use their teeth for things that would make any dentist blanch – like opening gasoline cans and turning bolts. All these practices cause micro-flaking.

Ryan then turned to the teeth of apes. He found that chimpanzees, because of their soft-fruit diet, have polished incisors, but that gorillas do not. Gorillas are ground dwellers that eat coarse vegetable matter, whose even coarser ends they pull out through their teeth, stripping off what they wish to chew. This action, because of the sand adhering to the vegetation, and because of the abrasive action of the silicon in the vegetation cells, causes an easily recognisable pattern of pitting and scratching on the biting surfaces of the front teeth. Even the direction of the scratches – straight across the tooth from inside to outside – is suggestive, becoming fainter as it goes, like the tail of a comet. Gorilla teeth are not polished like chimpanzee teeth. Gorillas do not eat enough fruit, and their diet has too much grit in it.

A third form of wear occurs from the crushing of small hard objects like seeds. This activity gradually exposes the dentine and forms pits or depressions in it because it is softer than the surrounding enamel. A man whose dentine is exposed and badly worn can feel with his tongue the small depression in the biting edge of his incisor.

Having established that there are consistent wear patterns in the teeth of various apes and humans, Ryan then examined *afarensis* teeth under the scanning electron microscope, and found that they exhibit all three kinds of wear: micro-flaking, pitting and scratching. He believes he has demonstrated that these early hominids were not fruit eaters, but savanna omnivores, just as other studies and conjectures about them have indicated.

Ryan would like to further refine his work to see if he can figure out what the ingredients of the australopithecine mixed diet were. He has already perfected a machine which has demonstrated that grit particles of a specific size will produce tooth scratches of a specific size. The machine also demonstrates that different kinds of tooth action – grinding, nipping, stripping, and so on – produce their own kinds of wear. If he can calculate with a high degree of accuracy the sizes of the silica particles that are contained in various plants, then he will have a clue to the sources of the variously sized microscopic scratches he finds in teeth. He will start with apes, whose diets he can observe. He will analyze the silica content of their diet and try to obtain correlations. If he is successful, he can apply them to australopithecines, fitting that information to what fossil-plant experts tell him about the prevalence of different kinds of vegetation during the Plio-Pleistocene. It seems almost incredible that *any* laboratory procedure could reveal what an animal of three million years ago dined on; but Ryan is hopeful. Others are watching his work with great interest.

When one hears about people like Al Ryan and Bob Walter one begins to understand that paleoanthropology is just barely getting under way. I cannot even imagine what some of the youngsters now in grade school will do in this science when they grow up. For all we know, its techniques and its conclusions may be as unrecognisable to us as a lot of the things we now take for granted would have been to

Darwin. There's room for everybody: for lab people like Ryan and Walter, for field people like me. Especially for field people. Those two black holes I spoke of earlier are shouting for attention. I am sure that Hadar can throw light into both of them.

Black holes get their name from celestial phenomena, from certain spots in the heavens where enormous amounts of radiation are pouring out but which are dark to conventional telescopes. As energy sources they put a thousand or a million suns to shame and should light up the sky like torches, but they are invisible. Astronomers, trying to explain that paradox, suggest that a black hole is actually a collapsed mass of matter so densely packed and with such overpowering gravitational force that light rays cannot escape from it – an entire earth crammed into a suitcase. These conclusions must be theoretical, since black holes cannot be observed directly. Hence the metaphor for paleoanthropology: a pall of ignorance blotting out certain parts of the past so effectively that no light comes from them.

One black hole in which fossils are virtually absent is the period between three million and two million. In East Africa there are a few robust fossils from just over the two-million mark, and some extremely questionable bits from Omo and other places that seem to say gracile. In South Africa, of course, there are also gracile forms placed at 2.5 million, but the dating is unconfirmed. Otherwise the hole is black indeed. There is no present evidence of *Homo* in it at all, other than the stone tools found by Roche and Harris. On the proposition that emerging *Homo* made stone tools and australopithecines did not, any hominid fossils found in association with those earliest known tools would have to be ticketed as the bones of the earliest known *Homo*.

Would that creature – right on the edge of humanity – be more like *habilis* or more like *afarensis*? That is a question I would dearly like to answer.

Another question: on the far side of Lucy, back beyond 3.5 million, what is there?

That second question lands the inquirer squarely in the middle of the second black hole. If the first one is deep and dark, this one is three or four times as deep, and positively stygian. It extends backward from Hadar and Laetoli for about six million years, well into the Miocene, and confronts the investigator for the first time in his backtracking on the hominid line with something that is no longer a hominid. Somewhere in that second black hole swim forms that may be too primitive to qualify. Evidence for that appears in the fossils that one begins to encounter on the far side of the black hole, at about nine or ten million. They were apes.

In between, in the hole itself, there is almost nothing. During that immense period of six million years in East Africa and Ethiopia, only three scraps of presumably hominid material have come to light. Just beyond four million is an arm bone from Kanapoi; at 5.5 million from Lothagam is part of a jaw with a single molar in it; at Lukeino, at six million, is another molar. These three fossil bits are so fragmentary, so worn, so lost in the wastes of time that they are unable to say anything about themselves beyond what logic could have said anyway: that some kind of ape-into-hominid was developing in East Africa during that period. Just how and when is as opaque as the fossil bits themselves.

Back at nine or ten million, on the far side of the black hole, the fossil record picks up again, because field scientists have been able to find productive strata of that age in several places. In them they have found apes, several kinds of them – primitive ones that do not exactly resemble any that live today, but that do have characteristics foreshadowing modern ape forms, although not clearly enough to make it possible to guess that this one or that one may have been ancestral to an orangutan, a gorilla or a chimpanzee. Another, *Ramapithecus*, has teeth that begin to whisper 'hominid'. Here are the first traces of that mystifying enlargement of the molars and thickening of the molar enamel that would eventually develop to almost grotesque proportions in later robust forms.

Clearly *Ramapithecus*, and a couple of other kinds with similar tooth characteristics, were doing things that the rest of the Miocene apes did not do. What that behaviour was, how it related to that of apes which preceded them or lived contemporaneously with them – indeed, the sorting out of the entire Miocene ape tangle – has been the lifework of David Pilbeam, an erudite, soft-spoken Englishman who teaches anthropology at Yale.

Pilbeam's work is frustrated by the poor condition of his fossils, by their rarity and by their fuzzy dating. They come from three continents and are sprinkled through about ten million years in time. They have been collected by specialists and amateurs on and off for nearly a century. They have been given conflicting names, stored in the drawers of a dozen museums, and forgotten. It took another Yale anthropologist, Elwyn Simons (now at Duke University), to rescue some of them, and recognise that at least two of them, with different names, actually belonged together under the label *Ramapithecus*. By doing that he was able to assemble reasonable facsimiles of both an upper and a lower jaw, and several teeth to go with each. So reconstructed, *Ramapithecus* emerged not only with those telltale large, heavily enameled molars, but also with small canines and a hominidlike arched palate. Whether or not it walked erect or what its skull was like is unknown. Like many other apes of the Miocene, *Ramapithecus* was represented only by gnathic (tooth and jaw) parts.

If Owen Lovejoy's statement about learning through living with fossils applies to anyone, it applies to David Pilbeam. For twenty years he has steeped himself in a small collection of long-vanished ape ancestors, trying to understand them, modifying his views of them as his understanding has grown. In 1978, at the same Nobel Symposium at which I spoke publicly for the first time about *afarensis*, Pilbeam delivered a long and thoughtful paper about the apes of the Miocene. What he has managed to extract from those ancient fragments is remarkable. He is an extremely cautious man who hedges his conclusions all around with warnings that they may be wrong. But he does draw a picture, and the picture he draws looks something like this:

Miocene apes start at about twenty million. One of the earliest forms known is *Dryopithecus africanus* (also known as *Proconsul* after a captive ape that lived in the London Zoo), found by Louis Leakey at Lake Victoria. Other *Dryopithecus*-like forms turn up in later millennia, sprinkled down through eighteen and seventeen million all the way to about nine million. They existed in several shapes and sizes, and were scattered widely in Africa, in central and southern Europe, and in Asia. Like all apes, they frequented forests, and their distribution follows the range of the great tropical forest belt that girdled the earth at the beginning of the Miocene. Pilbeam has lumped their several forms into a family – the dryopithecids – named after the commonest member of the group.

The overall impression that these creatures give is that they were large and small versions of vaguely chimp-shaped animals. But they were not chimps. In many important features they were not like chimps at all. Pilbeam believes that the early dryopithecids actually resembled monkeys more closely than they did modern apes. And yet the assumption must remain that they were the ancestors of modern apes. Some, indeed, seem to foreshadow orangs; others, gorillas; others, chimps. But this is impossible to prove. Dryopithecid fossils disappear eight or nine million years ago. There are no in-between types known. There are, in fact, *no ape fossils from anywhere* after about eight million. One contributory reason for this may have been the scarcity of apes; as the tropical forest began to shrink during the Miocene, the dryopithecids shrank too, perhaps already beginning to show the long-term dangers of an extreme 'K' reproductive strategy when faced with a less-than-ideal environment. But surely more important in explaining the rarity of all forest fossils, ape or otherwise, is that tropical forests do not preserve them. The soil is too acid. Bones are eaten away by the acid and by bacteria before they can begin to undergo the slow process of fossilisation.

At any rate, modern gorillas, orangs and chimpanzees spring out of nowhere, as it were. They are here today; they have no yesterday, unless one is able to find faint foreshadowings of it in the dryopithecids. Pilbeam assumes that the relationship exists, and has so indicated in a chart he has constructed – although he does leave a huge gap in it, and makes no attempt to link any specific dryopithecid with any living ape. He contents himself with the observation that dryopithecids are primitive apes with certain things in common, things that they do not have in common with a second group of Miocene apes that he has also succeeded in sorting out and lumping together: the ramapithecids, named after the aforementioned *Ramapithecus*.

What is the distinction? It is a simple but overwhelmingly important one. With the exception of their premolars, which are apelike, all of the ramapithecids have peculiar unapelike teeth: big molars, heavy enamel, small canines. They foreshadow hominids. The dryopithecids, with apelike teeth, foreshadow modern apes.

So simple, and yet so difficult. Twenty years of study and juggling have gone into that deceptively plain arrangement. What is significant about it is *where* Pilbeam has placed the different types, ranging them from left to right according to how apelike or hominidlike they are. Thus, something called *Limnopithecus* is placed at the extreme left of the chart because it is the most apelike of the Miocene types, and *Ramapithecus* at the extreme right because it is the most hominidlike, and the best candidate for being a human ancestor.

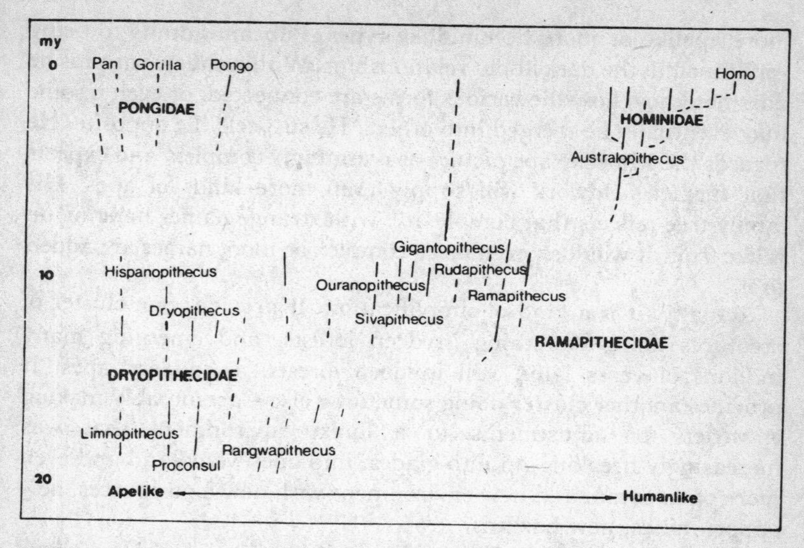

The most recent attempt to sort out the confusing Miocene ape tangle was made by David Pilbeam. He divides these fossils into two major groups: the Dryopithecidae (those which do not have hominid dentition and presumably were ancestral to modern apes) and the Ramapithecidae (those whose dentition does have hominid traits). Of the latter, *Ramapithecus* is considered perhaps the most humanlike; therefore it is farthest to the right on Pilbeam's chart. What degree of confidence has he in this arrangement? Not much. The true hominids are placed even farther to the right, and there is a large gap between them and *Ramapithecus*.

The overall lesson that Pilbeam's Miocene family tree teaches is that well before ten million, there had already been a major separation of apes into two groups. Because of the paucity of fossils, that distinction had not been entirely clear until Pilbeam worked it out; until he had gathered together specimens from other places; until he had himself gone to Pakistan – where some of the most productive ramapithecid fossil beds are located – and enlarged the collection from there; until he had found a few fossils of another ramapithecid, *Sivapithecus*, and recognised that it should be moved out of the more apelike group and into the more hominidlike group because of similarities in its dentition. When that had been done, he was able to say, 'The [fossils] I have clustered in Ramapithecidae exhibit a unique and previously unrecognised constellation of traits.'

Other than that basic division, and the left–right placement of more apelike or more hominidlike types, Pilbeam admits to being very much in the dark about relationships. Within the two groups he does not know how the various forms are connected, or even if some may eventually be merged into others. He suspects the opposite. He regards the Miocene ape picture as dauntingly complex, and expects that further fieldwork will supply even more kinds of apes. His family tree reflects that complexity, with strange names hanging on it like fruit; it will look even more complex as more names are added to it.

Actually, it is a kind of simplification. It provides one cluster of creatures doing one thing in deep forests, and emerging many millions of years later, still in deep forests, as modern apes. It provides another cluster doing something else – presumably making a variety of adjustments to a forest environment that was increasingly breaking up into glades, into open woodland and even more open savanna: a new environment with new food sources, new opportunities, new hazards.

It is known that the Miocene forest did begin to break up about fifteen million years ago. Pilbeam's arrangement of the ramapithecids makes the strong inference that their very existence as a dentally distinct cluster indicates that they had already made a detectable evolutionary adjustment to that forest breakup: they were living on the ground, out in the open, eating different things with differently designed teeth, as much as fourteen million years ago.

Were the ramapithecids really the ancestors of hominids? Another look at Pilbeam's family tree gives a clue to the degree of confidence he has in that: only moderate. He has not put the hominids directly under the ramapithecids, but even farther to the right. The width of the gap between suggests the shakiness of a possible direct connection between the two. It also suggests the possibility that there may be something else, as yet undiscovered, in between.

Even the placing of *Ramapithecus* at the right-hand edge of its group is questionable. Pilbeam's recent work at the Potwar plateau in Pakistan has produced more *Sivapithecus* fossils than any other kind. He has begun to look more closely at that creature, to wonder if its claim is not as good as that of *Ramapithecus*.

Another Miocene ape that intrigues him mightily – although not as an ancestor – is *Gigantopithecus*. That animal was much larger than the others. It was built on the scale of a female gorilla, but its molars were larger yet, far larger. Where did it come from? Where did it go?

Like all those others, it shuffles in the shadows, disconnected, disembodied. The only parts of this hulking enigma so far collected from Pakistan and India are a couple of its cobblestone teeth and part of a femur; they are about nine million years old. But other, younger teeth of the same animal crop up millions of years later, thousands of miles away in China. *Gigantopithecus* appears to have been the ape equivalent to the elephant – the largest savanna vegetarian among the primates, and a very successful animal. It became extinct as recently as half a million years ago in China – presumably eliminated by its smaller but more adaptable and more intelligent cousin, *Homo*. There are a few who think that *Gigantopithecus* lingers on in remote corners of the world, as the Abominable Snowman of the Himalayas; or as Bigfoot, the Sasquatch of the forests of northwestern America. Most anthropoligists dismiss this as fanciful.

Of the others, *Sivapithecus* is now the best represented, with more than sixty fossil pieces recovered to date. As usual, virtually all are jaws and teeth. But Pilbeam has recently begun turning up some leg, foot and toe bones and some bits of cranium that fit either a large type, a medium-sized type, or a small type. The large type, obviously, is *Gigantopithecus*. The medium one is *Sivapithecus*. Pilbeam describes it as chimpanzee-sized, possibly more terrestrial than a chimp.

The small one, *Ramapithecus*, is now also better known than ever before. New jaw finds have reinforced and improved on Elwyn Simons' reconstruction, although informative postcranial remains continue in short supply. Pilbeam summarises this animal as follows:

> *Ramapithecus* was a very small form, no larger than a medium-sized dog, at 30 lbs. or so. As far as we can tell at present it was not a biped but an agile four-footed animal perhaps equally at home in the trees and on the ground . . . I suspect that it climbed easily and frequently in the trees, slept, rested, played, socialised, fled there, even ate there. Yet it also utilised the ground, in woodland and at the forest edge, gathering tough and abrasive vegetable food, perhaps occasionally catching small prey. When on the ground it probably frequently moved, as do the smallest living apes, on two legs, especially when carrying objects . . . I assume [the peculiarity of its teeth] was an adaptation to a new and tougher kind of vegetarian diet. *Ramapithecus* probably used tools no more than does a chimpanzee.

This creature was unlike anything living today, or before, or since. If not the earliest hominid, it probably resembles it. It could be transformed easily into an early australopithecine by becoming completely bipedal.

Our ancestor – maybe.

The great difficulty in working in Pakistan is that the fossils are poor. They are really terribly fragmentary. I carry around with me the constant feeling that there are far better ones in the Afar, in older deposits than any we have worked so far, deposits that have been set down as much as seven million years ago and provide a time link with the Miocene apes being studied by Pilbeam.

Conditions for fossilisation being what they are in the Afar, I cannot help believing that if we should find productive sediments of that age we will find fantastic fossils. I often wonder what a seven-million-year-old knee joint would look like, and how it would jolt Owen Lovejoy's senses if we were to find one. Suppose we stumbled over something comparable to Lucy but four million years older. Would it be a hominid or would it be an ape? Would it be bipedal? Those could turn out to be delicate questions, hinging on how one chose to interpret creatures hovering right on the borderline between apes and hominids. It has frustrated me that for three years political conditions in Ethiopia have kept us from going back to make an effort to answer them. I keep thinking of the fossils that, during those three years, may have been washing out of the slopes in the Awash River valley, carried away by seasonal rains to be trampled by cattle, ground to dust and scattered over a hundred square miles of gravel. After 1976 I thought of that often, although I hated to do so.

I was lifted right out of myself in the summer of 1979 by a letter from Maurice Taieb. Prospects for return to Ethiopia, he said, were getting brighter. He had maintained contact with some of the people in the ministries. It was beginning to look as if we might be able to go back – if only as far as Addis Ababa, to discuss the possibility of a future field season. Was there any particular time that I might join him?

I replied that I would be available at any time. Assuming that such a visit would be followed, in due course, by a field season, I immediately began planning for one, thinking about people I would like to take with me and about what the work priorities would be.

The 333 site would carry a high priority. Digging into the upper section of that hill to establish one way or the other if there were any more fossils buried up there would have to be high on any list.

So would working out the geology at the spot where Jack Harris and Hélène Roche had found stone tools. If those tools were to be definitely established as the world's oldest, a dependable fix on their age would have to be obtained. Indeed, working anywhere in the black hole between 3.5 and 2 million would have a high priority; good hominid fossils found during that time range would certainly go far to either confirm or wreck White's and my entire theoretical organisation of hominids during the late Pliocene and early Pleistocene.

Finally, there was that other black hole – the one stretching backward from Hadar for six million years into the Miocene. Maurice had assured me that there were deposits in remote sections of the Afar that were between four and seven million years old. Choosing among alternatives as glittering as those three would not be easy. Perhaps I should plan on doing something about all of them.

On January 3, 1980, I flew to Paris. With me, carefully packed, was the entire Hadar collection of hominid fossils, scheduled for return to the Ethiopian authorities.

I picked up Maurice, and we flew on to Addis. There we were met at the airport by our old friend and expediter Richard Wilding, and by representatives of the Ethiopian Ministry of Culture. The presence of Ethiopians at the airport was unusual, and seemed to me to be an exceptionally good sign. The next few days we spent conferring with other scientists who had flown in from several countries: Desmond Clark, an English archeologist; Jack Harris, the stone-tool man from New Zealand; Raymonde Bonnefille, a Taieb associate and fossil-pollen expert; Bill Singleton, who had worked at another site in the Afar a few years earlier.

Over a weekend we hammered out a written proposal for future paleoanthropological and archeological work in Ethiopia. It was presented to the authorities at a meeting that began tensely for us, because we realised that the entire future of fieldwork in the Afar might hang on the response to this paper.

Our anxiety quickly subsided. In two years there had been a dramatic improvement in the Ethiopians' willingness to pay serious attention to foreign scientific work within their country. Previously, as fallout from the turmoil and violence that went with the political convulsions, the authorities had been too preoccupied with their own domestic problems to involve themselves with the work of foreign scientists, and the men lower down in the ministries too cowed to do much on their own. Now, with the political situation calm and under control, the Ethiopians made it clear that they welcomed paleoanthropological research. More than that, they encouraged it. They would help with permits, with fuel and other supplies. They hoped that young Ethiopians would be taken along for training, and they stood ready to find suitable young men for this work. They accepted our proposal with only minor alterations. As an unexpected bonus they restored the permit that had been wrenched away from Taieb and me several years earlier by Jon Kalb, which meant that we could work at Bodo, in the middle Awash River valley, where Kalb had actually found a hominid fossil: a *Homo erectus* skull. I was overwhelmed by the courtesy and cooperativeness of the Ethiopians. I left the meeting bursting with excitement.

The next days we spent in gathering supplies, in combing over the camp equipment that had been left in storage in 1977, and in trying to assemble two workable Land-Rovers out of several defunct ones. This was finally accomplished, and late one afternoon a small two-vehicle expedition chugged out of Addis to begin the long eastward descent from the central highlands, a trip that I had often wondered if I would ever take again. With me were Taieb; Jean-Jacques Tiercelin, a French geologist; Raymonde Bonnefille; Bob Walter; and an Ethiopian representative from the Ministry. We stopped for the night at Awash Station and rooted out our old friend Kabete, our cook on previous expeditions. We then went on to Gewani and inquired about a man named Selati who had been to the Middle Awash with Maurice previously. Selati knew the way there. He also knew where the Bodo skull had been found. He signed on as a guide.

As we rattled along in the Land-Rovers, the temperature crept up and up, the air grew drier, the landscape emptier and more hostile. Dropping down into one airless gulch after another, then climbing up again as Selati led the way to Bodo, the expedition ground slowly across an unrelieved expanse of broken rock and dust. I ate that dust for two days and tried to curb my impatience.

Bodo was finally found. Subjected to a quick survey, it proved to be an extraordinarily rich mid-Pleistocene area with an abundance of mammals and Acheulean stone tools in the .5-to-1.5-million-year range. Bodo Man yielded up no more parts of himself, but the area was carefully mapped, and likely sites for intensive collecting in late 1980 agreed on.

At that point, one of the Land-Rovers collapsed. Maurice was forced to drive the other one to a distant plantation in the hope of finding a spare differential. He found one, but it was almost as feeble as the other. When he got back, it was decided that the Land-Rovers were too fragile for us to risk penetration farther into the lunar landscape where the four-to-seven-million-year-old deposits lay. Instead, the caravan limped to Hadar.

No tents this time. Just cots, with mosquito nets over them. I set mine down on the bluff where the old campsite had been, and at dusk went for a short stroll out to the 200 site, where the best of the *afarensis* upper jaws had been found. Here, at last, I was flooded by what I believe to be the most powerful thing the desert can give to a modern urban person: utter silence. I watched a range of mountain peaks, so familiar in their outline against the western sky as to bring a piercing stab of recognition, fade from lavender to black. Then I went back to camp for a dinner of roast goat, and fell asleep to the faint sound of an owl crying down by the river.

The next morning, a file of eight sullen men carrying submachine guns walked into camp. Watching them approach, I felt a lurch of alarm. Selati, who had a gun of his own, began assembling it as casually and rapidly as he could. The men arrived and, to my intense relief, squatted on the ground and began talking. We gave them tea and bread, and during a twenty-minute conversation we learned that they were a patrol looking for bandits from the Issa, a rival tribe. We also learned that Muhammed Goffra, the protector who had been assigned to the expedition in 1976, was still in the neighborhood and still available for protection. His superior, Habib, the iron-faced little Afar chieftain with whom Taieb and I had previously negotiated, was still in control of the region and happy to have us back.

Two weeks of intensive work followed. For its size, and for the time spent in the field, I regard this effort as the most congenial, the most efficient and productive (fossil finds aside) that I have ever been a part of. The group was now shrunk down to Taieb, Walter, Tiercelin and myself, with Kabete and two Afar as the support staff

– tiny, but marvelous. Every man knew exactly what he wanted to do and spent all day doing it. In the evening we gathered for long interdisciplinary discussions. Of my three colleagues, Tiercelin was the only one I did not know well, but during those two weeks I came to have a tremendous respect and affection for him. Tiercelin was a nonstop worker, a brilliant geologist, and an outstandingly congenial campmate who added immensely to the value of the work being done by Taieb and Walter. Out of our evening talks came the following conclusions:

Hadar had changed hardly at all. Evidently there had been very little rain; the outwash of new fossils had been kept to a minimum, and a three-year absence had cost us nothing. In fact, so little had changed that when I was out surveying that desolate landscape, I found prints of my own shoes made in the sand several years before, and saw a bleached cigarette packet that I myself might have thrown away.

We surveyed the 333 site. A good deal of sandstone had crumbled down from the overburden above. It was now scattered in large blocks and smaller chunks over the hillside that had been so carefully screened for fossils two years before. But no new bones were visible as a result of this cascade and turnover of new material. That was a relief. It meant that the previous sifting had been thoroughly done, and that the planned excavation of the fossil-bearing layer near the top could be resumed where the tentative scratchings of 1977 had left off.

The Roche–Harris tool site at nearby Gona was revisited. Here Taieb, Tiercelin and Walter produced a geological plum. When Harris and I had done our quick dig there two years before, we had *thought* that the age of the tools was 2.5 million years, but our conclusions were based on preliminary stratigraphic observations. Not being geologists, we could not confirm them.

When the geologists arrived at the tool site, they were confronted with the bugaboo that geologists so often must wrestle with: no correlations. The strata at Gona had no links with any of the ones in the stratigraphic column at Hadar.

'How are you going to get a match-up?' I asked Walter that night.

'There are a couple of very distinct layers of large pebbles. We'll just have to walk them out and hope that they link up with something.'

Over the next week Walter and Tiercelin did a great deal of walking. They identified an ancient flood plain and established the presence of tools in what had once been a layer of mud and sand there. They then established a relationship between the flood plain and the pebble layers and then began tracing the pebble layers back toward Hadar. After a week of backtracking and false starts they had a linkup. They also had, as a bonus, a collection of volcanic samples from tuff layers that bracketed the pebble layers.

Walter is now quite sure that the tools are two and a half million years old, now definitely the oldest known anywhere in the world – with a possible error of fifty to a hundred thousand years. He will be able to shrink that error down when he runs dating tests on the samples he has collected from the new bracketing tuffs at Gona. One in particular, the AST-1 tuff (artifact site tuff), appears to correlate directly with the BKT-2 tuff at Hadar, which he and Aronson have already dated.

Having that second tuff is valuable in several ways. Not only will it serve as a check against the other one, but it will also unscramble the age of deposits in a much larger area at Hadar. The old BKT-2 tuff may be likened to a yardstick laid against a wallpaper pattern at one end of a large room, and the AST-1 to a yardstick at the other end. By connecting the two – that is, by sighting from one to the other – it becomes possible to lay wallpaper along the walls in such a way that the sheets match at both ends and everywhere in between. That, in effect, was what Walter did at Gona. His wallpaper pattern was a pattern of pebbles; his room, the Gona-Hadar area. I regard his expansion of good dating out into sections where there was none before as the major practical achievement of the field trip.

Not only that: Bob's new samples should be razor-sharp. He has a technique of electronic micro-probing that allows him to focus on a tiny point in a mineral sample and come up with a chemical analysis at that point – literally a single crystal – without damaging the sample. In that way he can recognise impurities and throw them out before they distort his test result.

Walter is also beginning to get an overall picture of the volcanic history of Hadar. He now thinks there may have been as many as ten volcanoes laying down tuffs in the area, some of them erupting several times. Each volcano – each eruption – is different. As he collects samples and analyzes them, he will begin to create a sort of catalogue of volcanism to which he can quickly assign any new sample he gets. This should ultimately permit dating anywhere at Hadar of a precision heretofore not approached.

Of less immediate paleoanthropological value, but as a scientific curiosity that entranced the geologists, came the gradual realization that the Afar plain – that old lake bottom – may once have been as much as 1,000 meters higher than it is today. Raymonde Bonnefille, with her fossil-pollen analysis now under way, has become convinced that the grains she has been studying sample a high altitude vegetation. This is confirmed by Jean-Jacques Jaeger, the French rodent specialist, who has concluded that his collection of rat and mouse fossils says the same thing. What caused the phenomenon Taieb does not know. But since it has occurred in the Afar and undoubtedly is connected in some way with the larger problem of continental-plate movement (his speciality), he thinks there may be a chance of contributing something to overall tectonic-plate theory. But first he will have to prove – geologically – what Bonnefille and Jaeger have been saying. That will require a match-up of deposits in the Afar plain with similar ones much higher up in the distant mountains – a huge and hideously difficult job.

Our group had now been in the field for two weeks. With our work done, we were ready to go, but reluctant to do so – reluctant to break up a team that had worked so harmoniously and productively.

'We've got to get out of here,' I said.

'Just one more day,' someone pleaded.

We never got that one more day. That same afternoon, about five Afar drifted into camp and hung around there nervously. Among them was our protector, Muhammed Goffra. He seemed nervous too. At sundown he spotted someone across the river. He called to him but got no answer. So he fired several shots at him. The man disappeared.

Who was that? we wanted to know.

'He could be an Issa scout. If so, it will be bad for us, because the Issa travel in bands of thirty, sometimes more.'

We held a conference. By that time it was dark, with lightning flashes in the sky. As we talked, it began to rain. That decided us. We could not risk being caught in any of the gulches if it rained hard enough to produce a flash flood. We packed everything as well as we could in the dark, even though the rain had begun to let up, and by eleven o'clock we were gone. We drove for about two hours and slept in safety on the plateau. The next day we were back in Addis Ababa.

On January 29, Taieb and I went to the National Museum. There, at an elaborate ceremony featured by speeches, we handed over the entire Hadar hominid-fossil collection – more than 350 priceless bones – to Comrade Mammo Tessema, Keeper of the Museum.

I was well aware of the symbolic importance of this act, and what it meant for the future of paleoanthropology in Ethiopia. But at the very moment of relinquishment I felt a dreadful sense of loss. Lucy had been mine for five years. The most beautiful, the most nearly complete, the most extraordinary hominid fossil in the world, she had slept in my office safe all that time. I had written papers about her, appeared on television, made speeches. I had shown her proudly to a stream of scientists from all over the world. She had – I knew it – hauled me up from total obscurity into the scientific limelight. Finally, her bones, and all the others I was now giving away, had enabled me to launch a new interpretation of hominid evolution. Standing there in the Museum and listening to the acceptance speech, I felt like a parent signing away a child to an adoption agency. For a few minutes, amid all the handshaking and congratulations, I was quite desolated.

But that did not last long. The door that had been opened up for us by the Ethiopians was too wide and wonderful. The knowledge that we would be back with a full crew in the fall was too exciting. I know exactly what we are going to do and who is going to do it. I will take Tim White and Gerry Eck back to Hadar, with Kabete to cook for us. We will do nothing but survey for hominids – *nothing*. I will try to get some of those Kenyans who have done such magnificent work for Richard and Mary Leakey to join us. (Richard generously made the offer a few years ago, and I will take it up.) Kamoya Kimeu, for example, is eager to go. After seeing Lucy, he once said to me, 'If you found that, think what *I* could find!'

Now that the Middle Awash sites have fallen back into our lap, Desmond Clark and a team of his will excavate there. What makes it potentially valuable is that *Homo erectus*, although spread all over Eurasia and Africa, and although he is a couple of million years younger than the Hadar fossils, is not really as well known anatomically as they are. There are some pretty good *erectus* skulls from here and there, and a lot of teeth, but not much else.

Most of what we know of *erectus* is cultural. His cave sites have been explored thoroughly. Clark Howell did that landmark study at the elephant-slaughter site in Spain. We know what *erectus* ate; we know that he cooked his food and made clothing. We know that he was an excellent hunter of large animals and that he made a variety of stone tools. We know that he showed a considerable amount of variability from place to place. What we don't know is just how or when – and perhaps, even *if* – he emerged out of *Homo habilis*.

We assume he did. We would like to prove it. We would like to find really good *erectus* fossils that sample a long period of time in one place. We would like to find them in association with their habitation sites, to see what they were doing all that time. We would like better dating. The middle Pleistocene is very poorly dated in Africa, but there are tuffs in the Middle Awash that should correct that. Desmond Clark and Jack Harris will be at the Middle Awash while we are at Hadar. I can't wait to see what they come up with, because *erectus* is such a strange and interesting character.

How does one account for *erectus*' sudden assumed jump out of *Homo habilis*? Was it actually all that sudden? Why was it made? Was it a matter of a quick evolutionary spurt taken in tandem with the development of a new and better tool culture? If so, where and why did that new culture start?

Then, an even more interesting question: why did that culture – and the man who made it – stagnate for another million years? *Homo erectus*, it is fairly clear, evolved practically not at all during that immense time. Then, suddenly, humanity took another spurt. About two hundred thousand years ago there occurred a second technological leap, and out of it rose *Homo sapiens*. The Middle Awash beckons because it may hold answers to both those questions.

Meanwhile, the geologists – Taieb, Tiercelin and Walter – will be taking a crack at the second black hole, the four-to-seven-million-year-old deposits in the Afar. These are west of the Awash River, about a hundred miles from Hadar. Tim and I will go there after about six weeks at Hadar. By that time Taieb and company will have a good handle on the geology and should have located some good fossil-bearing deposits. What we find in them could well blow the roof off of everything, because science has not known, and does not know today, just how or when the all-important transition from ape to hominid took place. This is the biggest remaining challenge to paleoanthropology. The gap between ourselves and apes has narrowed in recent years, but it has never been shut. Lucy brings us close. She teaches us the astonishing fact that bipedalism goes back about four million years. But in her we see it already complete, with no clues as to how long it may have existed previously or how fast it may have taken place. The feeling grows that one more step into the past will see its disappearance into a quadruped – into an ape.

Similarly, the Hadar jaws appear to teeter on the very edge, where humanity vanishes forever. If David Pilbeam were to find one of them in Miocene deposits without any associated long bones, he would surely say it was an ape. *Afarensis*, it seems, just barely squeezes through the hominid door. But what will the verdict be on a six-million-year-old foot, or on a seven-million-year-old pelvis? It is those questions that haunt me, and will continue to until they are answered.

The Afar holds the answers, I am sure. Its bones will make Ethiopia the hominid-fossil center of the world, with the entire story told there. We will have something between ramapithecid apes and Lucy at around six million. We will have Lucy herself between four and three. Then we will have later *afarensis* types sliding off toward *Homo* in one direction and *africanus* in the other. Finally, we will have *erectus*.

'Except that you still have to find the fossils,' said the ever-skeptical White.

'Do you doubt that we will? Do you doubt that for a single instant?'

'I doubt everything.'

'I'll bet you that bottle of wine Richard Leakey owes me that they're there. They've got to be. And if they are, we'll find them.'

Appendix

These notes were made by Jay Matternes in connection with his drawings of *Australopithecus afarensis* which appear on pages 358–361.

The reconstructed cranium and mandible on page 356 of an associated (presumed) adult male specimen of *Australopithecus afarensis* by Dr Tim White of the University of California, Berkeley, were the basis of my graphic restoration. Although there are more original fossil cranium and jaw elements present on the left side of the specimen, many of those parts were unclear, warped, or otherwise unconforming to the original long contours, whereas their counterparts on the right side of the skull had been idealised during restoration; the right side, therefore, was better for my purposes.

Page 358 (top left):
The drawings of the skull were made from a projection of a 35mm slide, made from a photograph taken with a long lens to obviate any parallax distortion. The projection and the drawings were oriented in the Frankfurt Horizontal (F.H.), a straight line from the porion (a standard landmark on the skull just above the auditory meatus) to the lowest point on the rim of the orbit (usually the left, but in this case the right). This is thought to be the angle at which the human head is normally carried, but I suspect that with a head as badly balanced as *A. afarensis*, the F.H. would not in fact have been horizontal; that it would have sloped downward more from the vertebral column. But the F.H. is a standard orientation method, and I deferred to it. In any event, the angle of downward slope could not have been very great.

The cervical vertebrae were drawn in with some care to provide an idea of how long the neck might have been, and to approximate the angle of the neck relative to the head. I used the cervicals of an extant human as a general model. Because of the badly balanced head, additional long buttressing would have been necessary. So I extended the neural spines far beyond their normal length in modern man. In similar fashion, other skeletal elements such as the shoulder girdle, humerus, angle of first rib, approximate angle of the sternum, and the like are all patterned after modern man – in whom there is, of course, great variation.

The cartilaginous (sagittal) septum of the nose is here indicated in its farthest forward projection, as a basis for indicating nasal profile.

Page 358 (top right):
The sequential restoration of the head in profile begins with the external muscles of mastication. Note that the insertion of the temporalis muscle on the superior-anterior margin of the coronoid process of the mandible can be seen through the transparentised superficial body of the masseter. (This area of the skull is the only part of the White reconstruction about which I have any question. In most hominid and pongid skulls the space between the anterior margin of the coronoid process and the posterior surface of the zygomatic process of the maxillary is relatively reduced. If that space were to be reduced in this skull, the entire upper face would slope backward more steeply from the rostrum, giving the effect of greater alveolar prognathism and a more apelike profile, and a consequently reduced endocranial length. However, it is evident that Dr White restored the skull at every point as carefully as the available fragments would warrant. Whatever slight distortions of skull proportion there may be are due to crushed and distorted parts of the fossil itself.) A globular adipose body peeps out from the anterior origins of the superficial masseter in the drawing, and fills the hollow between that muscle and the lateral maxillary.

The mouth assumes its extent and shape as follows: The closed lips usually come just below the line of the upper incisors, and the corners of the mouth in most apes extend backward no farther than the third upper premolar or, as here, the rear of the canine.

The eyeball is tentatively located in its fatty cushion; its size and relative recession/projection from the orbital rim must await further overlay drawings.

The powerful guy wire that supports the unbalanced head in the sagittal plane is the nuchal ligament. This is why the neural spines of the seven cervicals had to be extended in the first drawing beyond their usual length in modern man. An even more extreme example of the extended neural spine is to be seen in a male gorilla or orangutan.

The hyoid bone and laryngeal cartilage (and trachea) are indicated to propose the thickness of the throat anterior to the cervical vertebrae. If this distance were too great, several changes of proportion would have to be made: the angle of the cervical column to the skull; the F.H. orientation; the relative forward projection of the first rib – or all three in some degree.

Page 358 (bottom):
This is a further development of the preceding drawing, with the superficial musculature indicated, as well as some of the muscles of facial expression. The latter serve in some degree as a check on the placement and dimensions of the features of the face, even though those muscles, invested in the skin at differing strata of tissue, are extremely variable. The posterior portions of the sheetlike platysma muscle are eliminated here because if shown they would obscure some underlying major features of the neck and the parotid gland (the baroque-shaped body over the superficial masseter). The parotid, in its thickest lower sections, neatly fills in the void between the fleshy mid-part of the sternocleidomastoid muscle of the neck and the (gonial) angle of the mandible.

The ear is assumed to be somewhat like that of a modern man, with a fleshy lobe to indicate its owner's hominid, rather than pongid, affiliation. The entire external ear was located in this drawing around the external auditory meatus, which comes to the surface below the level of the acoustic meatus of the skull, obviously to counteract the effect of running water.

Since the subject is a male, the nipple is located on the sternopectoral muscle between the fourth and fifth ribs.

Page 359:
Even though one has been building toward it in hopeful anticipation, the final overlay drawing in a sequence like this is somewhat of a shock.

Since he was an equatorial animal living in an exposed savanna habitat, my subject almost certainly would have had (or would have been developing) a dark skin. Hairiness is generally regarded as a primitive feature. Since this creature is of such great antiquity, most of my consultants feel that the early African hominids were hairy for that reason alone.

There is one dissenter.

After I had completed some australopithecine drawings for Clark Howell's book *Early Man* in 1964, I chanced to meet at the Smithsonian one day Professor Joseph Weiner, then of Oxford University, a senior physical anthropologist, whose speciality has been human physical adaptation to environment. Weiner took exception to the hairiness of my depictions of the Transvaal australopithecines, and lectured me most persuasively as to why hairlessness would have been a desirable adaptation to the sort of life those creatures were living. I was most impressed by his arguments. Later I looked up some of his general works on human biology. He had some cogent points and much data to back them up.

As I recall, Weiner thought that hairlessness would have been an essential preadaptation to hominisation. An agile, bipedal creature running about in a tropical savanna would need to reduce his hairy coat for efficient heat loss. But as his coat thins out, he also absorbs more ultraviolet radiation, which darkens his skin because more melanin is produced in it. Unfortunately, a dark skin absorbs more heat, which, in turn, stimulates two tendencies: further reduction of body hair, and the development of more efficient sweat glands. I do not recall why this was a uniquely hominid path of development: why, for instance, other competing primate groups such as baboons – of similar habitat, size and weight – did not develop along similar lines.

Assuming (as I do) that Weiner is generally correct, the key question in regard to this reconstruction is *when* hairlessness was acquired by the collateral human ancestor. Since he was so ancient and in so many ways extremely primitive, I think it is safe to assume that he was still imperfectly adapted, that he was a transitional form with respect to hairlessness. That is why I have shown him as dark-skinned and hairy. But the hairs are sparse, to promote heat loss.

I suspect that facial hair in the male, as a sexual indicator, would probably have been minimal in a primitive hairy primate like this. If the males were cooperative social creatures, their intraspecific threat signals would depend more on behavior than on hairiness. Some believe that beards, axillary hair and pubic hair were developed as social signals at a time when general body hairiness was reduced. For these reasons I did not exaggerate the beard. I minimised the upper-lip hair so as not to obscure the shape of the muzzle. Even so, I would expect that *afarensis* females such as Lucy would be somewhat less hairy in the face than males, indicating at least an initial dimorphic tendency in facial hair; an artist would be justified in depicting a female *afarensis* as almost beardless for no other reason than to make her seem more plausibly female to the lay reader.

Cranial hair probably would have been dense as insulation and as protection from solar radiation, a tendency that has persisted to the present. The texture of the cranial hair was inspired generally by that of modern Australoids.

Page 360 (top left):
A front view of the skull. Most of the comments on the first drawing apply here.

Page 360 (top right):
In making frontal drawings of the White reconstruction, I did not feel that the middle stage needed to be done again; what that stage reveals is better seen from the side. This drawing, therefore, combines the second and third drawings from page 358. What does show up better here is the penciled outline of the orbits, the nasal aperture, and the lateral contours of the rostrum relative to the overlying muscles of facial expression.

Page 360 (bottom):
Slight flexure wrinkles have been indicated in the skin over the nose, and deeper ones on the upper lip. The external epicanthic fold might have been even more pronounced and rounded (rather than sagging, an indication of age) to give some natural shading to the eye. Also, the previously mentioned flexure wrinkles on the nose might have been even more marked, but the appearance would have been more apelike (Adolph Schultz speculated that anthropoid apes frequently show deep facial furrows because, unlike man, they have a dearth of fat in the facial skin). Even though the skin of this creature would have been baked by the sun and thoroughly dried out, it is doubtful that any individuals survived long enough to look really aged by modern human standards.

Page 361 (top):
Since the pertinent postcranial skeletal elements were drawn with reasonable care in the profile and frontal views, I felt at liberty to sketch them in more loosely here as a basis for building up the three-

quarter-view bust. Although it looks as though the coracoid processes on both sides are touching the head of the humerus, that is not the intention.

Page 361 (bottom):
It would have been needless duplication to repeat the sequential restoration for this view: the final summation of all that I have learned from the previous drawings.

Acknowledgments

This book could not have been written without the help of many people, starting with those who manned the various field expeditions in Ethiopia: Pam Alderman, James Aronson, Alemayehu Asfaw, Getachew Ayele, Michel Beden, Raymonde Bonnefille, Bobbie Brown, Michael Bush, Françoise Coppens, Gudrun Corvinus, Michel Decobert, Gene Dole, Vera Eisenmann, Tom Gray, Claude Guérin, Claude Guillemot, J. W. K. Harris, J.-J. Jaeger, Jon Kalb, William Kimbel, John Kolar, Jean Maduit, William McIntosh, Nicole Page, Dennis Peak, Germaine Petter, Pierre Planques, Guy Riollet, Hélène Roche, Maurice Sebatier, Tom Schmitt, Becky Sigmon, Emil Tola, Herbert Thomas, J.-J. Tiercelin, Annie Vincens, Robert Walter, Johannes Zekele. And, of course, Maurice Taieb and Yves Coppens, codirectors of the International Afar Research Expedition.

Our sincere thanks to the Provisional Military Government of Socialist Ethiopia for permission and encouragement to work in the Afar; also to the Ministry of Culture and Sports for its encouragement.

For help in funding and mounting expeditions: The National Science Foundation, the L. S. B. Leakey Foundation, the National Geographic Society, Gordon Getty, Hubert Hudson, and Mr and Mrs Willard Brown.

For guidance, lodging and innumerable favors in Addis Ababa: Robert and Martha Caldwell, Richard and Libby Wilding, Gerry and Nellie Decker.

Thanks to Nelson Bryant and Augustus Ben David for supplying facts, and to the following for preparing or helping us obtain illustrations: John Aicher, David Brill, Bobbie Brown, Eric Delson, Gerry Eck, Bruce Frumker, John Gurche, Peter Jones, Anson Laufer, Paul Leser, C. Owen Lovejoy, Jay Matternes, David Pilbeam, John Reader, Larry Rubins, Mary Griswold Smith, Hank Wesselman, Tim White, Rob Wood, and particularly Luba Dmytryk and Steve Misencik for their expert renderings of fossils and charts respectively.

In Cleveland, research assistants Bruce Latimer and Bill Kimbel were bulwarks of support, as was the always flexible and understanding director of the Cleveland Museum, Dr Harold Mahan. Doris Andreoli, administrative assistant extraordinary, caught all the dirty little jobs with unfailing skill and good temper. Sära Kraber provided innumerable good meals.

We are indebted to James Aronson, Bob Walter, and C. Owen Lovejoy for valuable comments on certain portions of the manuscript. Tim White read it in its entirety, and made many important scientific and organisational suggestions.

D. C. J.
M. A. E.

Bibliography

Ardrey, Robert, *African Genesis*. New York: Atheneum, 1961.

Aronson, J. L., T. J. Schmitt, R. C. Walter, M. Taieb, J.-J. Tiercelin, D. C. Johanson, C. W. Naeser and A. E. M. Nairn, 'New Geochronologic and Palaeomagnetic Data for the Hominid-Bearing Hadar Formation in Ethiopia', *Nature*, May, 1977: 323-327.

Bishop, W. W., ed., *Geological Background to Fossil Man*. Edinburgh: Scottish Academic Press, 1978.

——, 'Geochronological Framework for African Plio-Pleistocene Hominidae: as Cerberus Sees It', in *Early Hominids of Africa*, C. Jolly, ed. London: Duckworth, 1978.

Bishop, W. W., and J. D. Clark, eds., *Background to Evolution in Africa*. Chicago: University of Chicago Press, 1967.

Bishop, W. W., and J. A. Miller, eds., *Calibration of Hominoid Evolution*. Edinburgh: Scottish Academic Press, 1972.

Brace, C. Loring, 'Biological Parameters and Pleistocene Hominid Life-Ways', in *Primate Ecology and Human Origins*, I. S. Bernstein and E. O. Smith, eds. New York: Garland Press, 1979.

Brain, C. K., 'The Transvaal Ape-Man Bearing Cave Deposits', Transv. Mus. Mem., No. 11, 1958.

Broom, Robert, *Finding the Missing Link*. London: Watts & Co., 1950.

Broom, Robert, and G. W. H. Schepers, 'The South African Fossil Ape-Men, the Australopithecinae'. Transv. Mus. Mem., No. 2, 1946.

Broom, Robert, J. T. Robinson and G. W. H. Schepers, 'Sterkfontein Ape-Man, Plesianthropus'. Transv. Mus. Mem., No. 2, 1950.

Butzer, K. W., 'Environment, Culture and Human Evolution'. *American Scientists*, September–October 1977, 65: 572–584.

Campbell, B., *Humankind Emerging*. Boston: Little, Brown, 1979.

Clarke, R. J., 'Early Hominid Footprints from Tanzania', *South African Journal of Science*, April, 1979, 75: 148–149.

Cole, Sonia, *Leakey's Luck*. New York: Harcourt Brace Jovanovich, 1975.

Cooke, H. B. S., 'Pliocene-Pleistocene Suidae from Hadar, Ethiopia', *Kirtlandia*, No. 29, 1978.

Coon, Carleton, *The Origin of Races*. New York: Alfred A. Knopf, 1963.

Coppens, Y., F. Clark Howell, G. L. Isaac and R. E. F. Leakey, eds., *Earliest Man and Environments in the Lake Rudolf Basin*. Chicago: University of Chicago Press, 1976.

Curtis, G. H., R. E. Drake, T. E. Cerling and J. Hampel, 'Age of KBS Tuff in Koobi Fora Formation, East Rudolf, Kenya' *Nature*, December, 1975, 258: 395–398.

Dart, Raymond A., '*Australopithecus africanus:* The Man-Ape of South Africa' *Nature*, February, 1925, 115: 195–199.

——, 'The Osteodontokeratic Culture of Australopithecus prometheus'. Transv. Mus. Mem., No. 10, 1957.

——, *Adventures with the Missing Link*. Philadelphia: The Institutes Press, 1967.

Darwin, Charles, *On the Origin of Species*. New York: Atheneum, 1964.

Douglas, John H., 'Ancestors: Shaking Up the Family Tree', *Science News*, June, 1979, 115: 362–365.

Edey, M. A., 'Three-Million-Year-Old Lucy', in 1976 *Nature/Science Annual*. New York: Time-Life Books, 1975.

——, *The Missing Link*, rev. ed. New York: Time-Life Books, 1977.

Findlay, George, *Dr Robert Broom*. Cape Town: Balkema, 1972.

Gleadow, A. J. W., 'Fission Track Age of the KBS Tuff and Associated Hominid Remains in Northern Kenya', *Nature*, March, 1980, 284: 225–230.

Glynn, Isaac, 'The Food-Sharing Behavior of Protohuman Hominids', *Scientific American*, April, 1976: 90–108.

Glynn, Isaac, and Elizabeth R. McCown, eds. *Human Origins*. Menlo Park, California: W. A. Benjamin, 1979.

Halstead, L. B., 'New Light on the Piltdown Hoax', *Nature*, November, 1978, 276: 11–13.

Holloway, R. L., 'The Casts of Fossil Hominid Brains', *Scientific American*, July, 1974: 106–115.

Howell, F. Clark, *Early Man*, rev. ed. New York: Time-Life Books, 1973.

Johanson, D. C., 'Ethiopia Yields First "Family" of Man', *National Geographic*, December, 1976: 790–811.

——, 'Early African Hominid Phylogenesis: A Re-Evaluation'. Nobel Symposium, Royal Swedish Academy of Science, 1978.

——, 'Our Roots Go Deeper', in *Science Year, The World Book Science Annual, 1979*. Chicago: Field Enterprises Educational Corp., 1978, 42–55.

——, 'Odontological Considerations of Mio-Pliocene Hominoids', *Am. Journal Phys. Anthrop.*, February, 1980, 52:242.

Johanson, D. C., and Y. Coppens, 'A Preliminary Anatomical Diagnosis of the First Plio-Pleistocene Hominid Discoveries in the Central Afar, Ethiopia', *Am. Journal Phys. Anthrop.*, September, 1976, 45: 217–234.

Johanson, D. C., C. O. Lovejoy, A. H. Burstein and K. C. Heiple, 'Functional Implications of the Afar Knee Joint', abstract, *Am. Journal Phys. Anthrop.*, January, 1976, 44: 188.

Johanson, D. C., and M. Taieb, 'Plio-Pleistocene Hominid Discoveries in Hadar, Ethiopia', *Nature*, March, 1976, 260: 293–297.

Johanson, D. C., and M. Taieb, 'Plio-Pleistocene Hominid Discoveries in Hadar, Central Afar, Ethiopia', in *Early Hominids of Africa*, C. J. Jolly, ed. London: Duckworth, 1978.

Johanson, D. C., M. Taieb, Y. Coopens and H. Roche, 'Expédition Internationale de l'Afar, Ethiopie (4éme et 5éme Campagne 1975–1977): Nouvelles Découvertes d'Hominidés et Découvertes d'Industries Lithiques Pliocéne á Hadar'. Paris: C. R. Acad. Sci., Série D, 287, 1978: 237–240.

Johanson, D. C., and T. D. White, 'A Systematic Assessment of Early African Hominids', *Science*, January, 1979, 203: 321–330.

Johanson, D. C., and T. D. White, 'On the Status of *Australopithecus afarensis*', *Science*, March, 1980, 207: 1104–1105.

Johanson, D. C., T. D. White and Y. Coppens, 'A New Species of the Genus *Australopithecus* (Primates: Hominidae) from the Pliocene of Eastern Africa', *Kirtlandia*, No. 28 (1978).

Jolly, Alison, *The Evolution of Primate Behavior*. New York: Macmillan, 1972.

Jolly, C., *Early Hominids in Africa*. London: Duckworth, 1978.

Keith, Sir Arthur, *New Discoveries Relating to the Antiquity of Man*. London: Williams and Norgate, 1931.

Kimbel, W. H., and T. D. White, 'A Reconstruction of the Adult Cranium of *Australopithecus afarensis*', *Am. Journal Phys. Anthrop.*, February, 1980, 52: 244.

Latimer, B. M., 'Bonobo or Not Bonobo? The Pygmy Chimpanzee as a Model for Hominid Ancestry', *Am. Journal Phys. Anthrop.*, February, 1980, 52: 246.

Leakey, L. S. B., P. V. Tobias and J. R. Napier, 'A New Species of the Genus *Homo* from Olduvai Gorge', *Nature*, April, 1964, 202: 7–9.

Leakey, M. D., *Olduvai Gorge, Vol. 3: Excavations in Beds I & II, 1960–1963*. Cambridge, England: Cambridge University Press, 1971.

——, 'Footprints in the Ashes of Time', *National Geographic*, April, 1979, 446–457.

Leakey, M. D., R. L. Hay, G. H. Curtis, R. E. Drake, M. K. Jackes and T. D. White, 'Fossil Hominids from the Laetolil Beds', *Nature*, August, 1976, 262: 460–466.

Leakey, R. E., 'Skull 1470', *National Geographic*, June, 1973, 819–829.

——, 'Hominids in Africa', *American Scientist*, March–April, 1976, 64: 174–178.

Leakey, R. E., and Roger Lewin, *Origins*. New York: Dutton, 1977.

Le Gros Clark, W. E., 'Hominid Characters of the Australopithecine Dentition'. *Journal of the Royal Anthropological Institute of Great Britain and Ireland*, 1950, 80: 37–54.
——, *Man-Apes or Ape-Man?* New York: Holt, Rinehart & Winston, 1967.
——, *The Antecedents of Man*. Chicago: Quadrangle Books, 1971.
——, *The Fossil Evidence for Human Evolution*. Chicago: University of Chicago Press, 1978.
Lovejoy, C. O., 'The Gait of Ausatralopithecines', *Yearbook of Physical Anthropology*, 17 (1974): 147–161.
——, 'Hominid Origins: The Role of Bipedalism', *Am. Journal Phys. Anthrop.*, February, 1980, 52: 250.
McDougall, I., R. Maier, P. Sutherland/Hawkes and A. J. W. Gleadow, 'K-Ar Age Estimate for the KBS Tuff, East Turkana, Kenya', *Nature*, March, 1980, 284: 230–234.
Maglio, V. J., and H. B. S. Cooke, eds., *Evolution of African Mammals*. Cambridge, Massachusetts: Harvard University Press, 1978.
Mann, A., *The Paleodemographic Aspects of the South African Australopithecines*. University of Pennsylvania Pub. in Anthrop. 1, 1975.
Mayr, E., *Principles of Systematic Zoology*. New York: McGraw-Hill, 1969.
Millar, Ronald, *The Piltdown Men*. New York: Ballantine, 1974.
Napier, J., and B. Campbell, 'The Evolution of Man', *Discovery*, 1964, 25: 32–38.
Nelson, H., and R. Jurmain, *Introduction to Physical Anthropology*. St Paul, Minnesota: West Publishing Company, 1979.
Oakley, Kenneth P., and J. S. Weiner, 'Piltdown Man', *American Scientist*, October, 1955, 43: 573–583.
Pilbeam, David, 'Major Trends in Human Evolution.' Nobel Symposium, Royal Swedish Academy of Science, 1978.
——, 'Miocene Hominids and Hominid Origins', *Am. Journal Phys. Anthrop.*, February, 1980, 52: 268.
Pilbeam, D., J. Barry, G. F. Meyer, S. M. Ibrahim Shah, M. H. L. Pickford, W. W. Bishop, H. Thomas and L. L. Jacobs, 'Geology and Palaeontology of Neogene Strata of Pakistan', *Nature*, December, 1978, 270: 684–689.
Robinson, J. T., 'The Dentition of the Australopithecinae'. Transv. Mus. Mem., No. 9, 1956.
——, *Early Hominid Posture and Locomotion*. Chicago: University of Chicago Press, 1972.
Roche, H., and J.-J. Tiercelin, 'Découverte d'une Industrie Lithique Ancienne *in Situ* dans la Formation d'Hadar, Afar Central, Ethiopie'. Paris: C. R. Acad. Sci., Série D, 284, 1977: 1871–1874.
Sarich, V. M., 'Molecular Clocks and Hominoid Evolution After 12 Years', *Am. Journal Phys. Anthrop*, February, 1980, 52: 275–276.
Schultz, A., *The Life of Primates*. New York: Universe Books, 1969.

Simons, E. L., 'Ramapithecus', *Scientific American*, May, 1977, 236: 28–35.

Simpson, G. G., *The Major Features of Evolution*. New York: Columbia University Press, 1953.

——, *The Meaning of Evolution*. New Haven: Yale University Press, 1967.

Taieb, M., Y. Coppens, D. C. Johanson and J. Kalb, 'Dépòts Sédimentaires et Faunes du Plio-Pléistocène de la Basse Vallée de l'Awash (Afar Central, Éthiopie)'. Paris: C. R. Acad. Sci., Série D, 275, 1972: 819–822.

Taieb, M., D. C. Johanson, Y. Coppens and J. L. Aronson, 'Geological and Paleontological Background of Hadar Hominid Site, Afar, Ethiopia', *Nature*, March, 1976: 289–293.

Taieb, M., and J.-J. Tiercelin, 'Sédimentation Pliocéne et Paléoenvironments de Rift: Exemple de la Formation á Hominidés d'Hadar (Afar, Éthiopie)', *Bull. Soc. Geol. France*, XXI, 1979: 243–253.

Tattersall, Ian, and Niles Eldredge, 'Fact, Theory and Fantasy in Human Paleontology', *American Scientist*, March–April, 1977, 65: 204–211.

Tazieff, Haroun, 'The Afar Triangle', *Scientific American*, February, 1970: 32–40.

Tobias, P. V., 'Early Man in East Africa', *Science*, July, 1965, 149: 22–23.

——, 'New Developments in Hominid Paleontology in South and East Africa', *Annual Reviews of Anthropology*, 2, 1973: 311–334.

Walker, A., and R. E. Leakey, 'The Hominids of East Turkana', *Scientific American*, February, 1978: 54–66.

Ward, S. C., 'Maxillary Subocclusal Morphology of Pliocene Hominids from the Hadar Formation and Miocene Hominids from the Siwaliks and Potwar Plateau', *Am. Journal Phys. Anthrop.*, February, 1980, 52: 290–291.

Weiner, J. S., and K. P. Oakley, 'The Piltdown Fraud: Available Evidence Reviewed', *Am. Journal Phys. Anthrop., March*, 1954, *12: 1–7.*

Weiner, J. S., K. P. Oakley and W. E. Le Gros Clark, The Solution of the Piltdown Problem. Bull. Brit. Mus. (Nat. Hist.), Geol. 2, 1953, 139–146.

White, T. D., 'Evolutionary Implications of Pliocene Hominid Footprints', *Science*, April, 1980, 208: 175–176.

——, 'Hominoid Mandibles from the Miocene to the Pliocene', *Am. Journal Phys. Anthrop.*, February, 1980, 52: 292.

White, T. D., and J. M. Harris, 'Suid Evolution and Correlation of African Hominid Localities', *Science*, October, 1977, 198: 13–21.

——, 'Classification and Phylogeny of East African Hominids', in *Recent Advances in Primatology*, Vol. 3, D. J. Chivers and K. A. Joysey, eds. New York: Academic Press, 1978, 351–372.

Illustration Credits

American Museum of Natural History: 39, 61 (left), 86; Des Bartlett, © National Geographic Society: 95 (top); David Brill: 16, 23, 190, 203; Bobbie Brown: 206, 207, 319 (left); Bob Campbell, © National Geographic Society: 143; Cleveland Museum of Natural History: 27, 60 (right), 295; Yves Coppens: 122; Luba Dmytryk: 37, 47, 53, 63, 65, 95 (bottom), 105, 161 (top), 161 (bottom), 181, 186, 187, 266-274, 276, 277, 324, 325; Gerry Eck: 119; Gordon W. Gahan, © National Geographic Society: 145; Tom Gray: 231; Melville Bell Grosvenor, © National Geographic Society: 89; John Gurche: 71, 323, 326, 327; F. Herholdt: 61 (right); Donald C. Johanson: 21, 160, 219, 225, 349, 356; Peter Jones: 252, 253; C. Owen Lovejoy: 336, 337; Jay Matternes: 290, 291, 358-361; Steve Misencik: 99, 133, 198, 209, 235, 244, 281, 285, 287, 288; David Pilbeam: 369; Rheinisches Landesmuseum, Bonn: 60 (left); Larry Rubins: 353, University of California Press, *Geology of the Olduvai Gorge*: 93; Hank Wesselman: 123; Tim White: 305; Rob Wood: 315, 318, 319 (right), 331.

Index

READ MORE IN PENGUIN

also published:

Lucy's Child
The Discovery of a Human Ancestor
by Donald Johanson and James Shreeve

In the early 1970s Donald Joanson revolutionized anthropology with his discovery of Lucy, our earliest known ancestor. He presented his findings, with co-author Maitland Edey, in the bestselling *Lucy: The Beginnings of Humankind.* Returning to Africa in 1986, Johanson set up camp in Tanzania and, three days later, stumbled on the fragment of a two-million-year-old elbow. Identified as 'Lucy's Child', it was the first sign of yet another extraordinary hominid specimen and resulted in renewed controversy about our origins.

'Rich in detail and dialogue, conveying a sense of the gritty, sweaty reality . . . A lively review of the latest thinking about human origins and his side of his feud with Louis and Mary Leakey and their son Richard' – *The New York Times Book Review*

'An authoritative and gripping account . . . The past comes alive in this beautifully written book, and so do the people seeking to reconstruct the past – their exciting fossil finds, never-forgiven blunders, personal vendettas and brilliant insights. For an up-to-date, eloquent story of human evolution, this is it' – John Pfeiffer, author of *The Emergence of Humankind*